Erratum

Two figures were inadvertently omitted from "Quasi-exactly-solvable spectral problems and conformal field theory", by Mikhail A. Shifman, pp. 237–262. The figures are reproduced below.

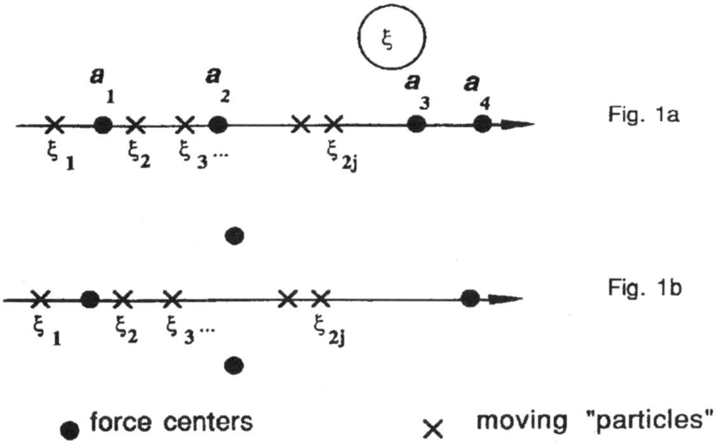

Figure 1

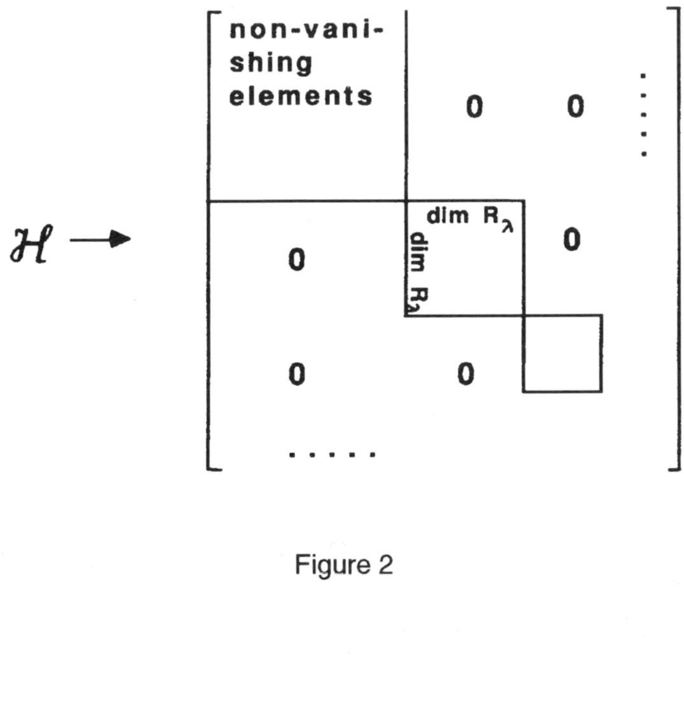

Figure 2

Recent Titles in This Series

160 **Niky Kamran and Peter J. Olver, Editors,** Lie algebras, cohomology, and new applications to quantum mechanics, 1994

159 **William J. Heinzer, Craig L. Huneke, and Judith D. Sally, Editors,** Commutative algebra: Syzygies, multiplicities, and birational algebra, 1994

158 **Eric M. Friedlander and Mark E. Mahowald, Editors,** Topology and representation theory, 1994

157 **Alfio Quarteroni, Jacques Periaux, Yuri A. Kuznetsov, and Olof B. Widlund, Editors,** Domain decomposition methods in science and engineering, 1994

156 **Steven R. Givant,** The structure of relation algebras generated by relativizations, 1994

155 **William B. Jacob, Tsit-Yuen Lam, and Robert O. Robson, Editors,** Recent advances in real algebraic geometry and quadratic forms, 1994

154 **Michael Eastwood, Joseph Wolf, and Roger Zierau, Editors,** The Penrose transform and analytic cohomology in representation theory, 1993

153 **Richard S. Elman, Murray M. Schacher, and V. S. Varadarajan, Editors,** Linear algebraic groups and their representations, 1993

152 **Christopher K. McCord, Editor,** Nielsen theory and dynamical systems, 1993

151 **Matatyahu Rubin,** The reconstruction of trees from their automorphism groups, 1993

150 **Carl-Friedrich Bödigheimer and Richard M. Hain, Editors,** Mapping class groups and moduli spaces of Riemann surfaces, 1993

149 **Harry Cohn, Editor,** Doeblin and modern probability, 1993

148 **Jeffrey Fox and Peter Haskell, Editors,** Index theory and operator algebras, 1993

147 **Neil Robertson and Paul Seymour, Editors,** Graph structure theory, 1993

146 **Martin C. Tangora, Editor,** Algebraic topology, 1993

145 **Jeffrey Adams, Rebecca Herb, Stephen Kudla, Jian-Shu Li, Ron Lipsman, and Jonathan Rosenberg, Editors,** Representation theory of groups and algebras, 1993

144 **Bor-Luh Lin and William B. Johnson, Editors,** Banach spaces, 1993

143 **Marvin Knopp and Mark Sheingorn, Editors,** A tribute to Emil Grosswald: Number theory and related analysis, 1993

142 **Chung-Chun Yang and Sheng Gong, Editors,** Several complex variables in China, 1993

141 **A. Y. Cheer and C. P. van Dam, Editors,** Fluid dynamics in biology, 1993

140 **Eric L. Grinberg, Editor,** Geometric analysis, 1992

139 **Vinay Deodhar, Editor,** Kazhdan-Lusztig theory and related topics, 1992

138 **Donald St. P. Richards, Editor,** Hypergeometric functions on domains of positivity, Jack polynomials, and applications, 1992

137 **Alexander Nagel and Edgar Lee Stout, Editors,** The Madison symposium on complex analysis, 1992

136 **Ron Donagi, Editor,** Curves, Jacobians, and Abelian varieties, 1992

135 **Peter Walters, Editor,** Symbolic dynamics and its applications, 1992

134 **Murray Gerstenhaber and Jim Stasheff, Editors,** Deformation theory and quantum groups with applications to mathematical physics, 1992

133 **Alan Adolphson, Steven Sperber, and Marvin Tretkoff, Editors,** p-Adic methods in number theory and algebraic geometry, 1992

132 **Mark Gotay, Jerrold Marsden, and Vincent Moncrief, Editors,** Mathematical aspects of classical field theory, 1992

131 **L. A. Bokut′, Yu. L. Ershov, and A. I. Kostrikin, Editors,** Proceedings of the International Conference on Algebra Dedicated to the Memory of A. I. Mal′cev, Parts 1, 2, and 3, 1992

130 **L. Fuchs, K. R. Goodearl, J. T. Stafford, and C. Vinsonhaler, Editors,** Abelian groups and noncommutative rings, 1992

(*Continued in the back of this publication*)

CONTEMPORARY MATHEMATICS

160

Lie Algebras, Cohomology, and New Applications to Quantum Mechanics

AMS Special Session
on Lie Algebras, Cohomology, and
New Applications to Quantum Mechanics
March 20–21, 1992
Southwest Missouri State University

Niky Kamran
Peter J. Olver
Editors

American Mathematical Society
Providence, Rhode Island

A Special Session was organized at the AMS meeting at Southwest Missouri State University, March 20–21, 1992, to bring together mathematicians and physicists working in closely related areas.

1991 *Mathematics Subject Classification*. Primary 17B15, 17B56, 81C05, 81C40, 33D55.

Library of Congress Cataloging-in-Publication Data

Lie algebras, cohomology, and new applications to quantum mechanics/Niky Kamran, Peter J. Olver, editors.

p. cm. — (Contemporary mathematics; v. 160)

Includes bibliographical references.

ISBN 0-8218-5169-1

1. Lie algebras. 2. Homology theory. 3. Quantum theory. I. Kamran, Niky, 1959– . II. Olver, Peter J. III. Series: Contemporary mathematics (American Mathematical Society); v. 160.

QA252.3.L55 1994 93-48322

512′.55—dc20 CIP

∞ The paper used in this book is acid-free and falls within the guidelines
established to ensure permanence and durability.
♻ Printed on recycled paper.
This volume was printed from copy prepared by the authors.
Some articles were typeset using $\mathcal{A}\mathcal{M}\mathcal{S}$-TeX or $\mathcal{A}\mathcal{M}\mathcal{S}$-LaTeX,
the American Mathematical Society's TeX macro systems.

10 9 8 7 6 5 4 3 2 1 99 98 97 96 95 94

Contents

Preface

The application of group theoretic methods to quantum mechanics has a long and distinguished history, dating back to the early investigations of Weyl, Wigner and Bargmann. In the "standard" approach, symmetries of the classical physical problem become symmetries (commuting operators) of the associated quantum mechanical Schrödinger operator. In favorable situations, this implies that the bound states of the associated stationary Schrödinger equation decompose into irreducible representations for the symmetry group, and the powerful theoretical methods from representation theory can be effectively used to analyze such problems; vice versa, the applications to quantum mechanics have themselves served as a powerful catalyst and motivation for the enormous amount of mathematical research effort that has gone into the study of representation theory.

The present volume is devoted to a range of important new directions and ideas concerning the applications of Lie groups and Lie algebras to Schrödinger operators and associated quantum mechanical systems. In these applications, the group does not appear as a standard symmetry group, but rather as a "hidden" symmetry group; nevertheless, the representation theory can still be effectively employed to analyze at least part of the spectrum of the operator. Such methods have their origin in the work of Dothan, Gell–Mann, Ne'eman, Barut and Bohm in the mid 1960's. The concept of a spectrum-generating algebra was introduced in nuclear physics, molecular physics, and scattering theory by Iachello, Arima, and Levine, in the 1970's and 80's, leading to the general concept of a Lie algebraic Schrödinger operator. In the mid 1980's, Turbiner, Shifman, Ushveridze, and collaborators introduced the more restrictive notion of a "quasi-exactly solvable" Schrödinger operator, for which the hidden symmetry algebra leads to the determination of a part of the spectrum by purely algebraic means. The synthesis and extensions of these seminal ideas has proved to be of great fruitfulness, not only in physical problems, but in the mathematical theory required to understand and classify the original examples.

In light of the rapidly developing subject, we decided that the time was ripe for bringing together, perhaps for the first time, a group of mathematicians and physicists working in areas closely related to these themes. As you can see from the contributions contained herein, a wide variety of physical applications and mathematical methods come into play, including Lie group methods, Lie algebras and Lie algebra cohomology, representation theory, differential operators, differential geometry, special functions, the theory of orthogonal polynomials, q-series, conformal field theory, quantum groups, scattering theory, classical invariant theory, and so on. The subject is, in our opinion, extraordinarily rich and vibrant in its range of indisciplinary ideas and new applications.

All together, we believe, the papers assembled in this volume will provide the reader with an informative and challenging survey of the broad range of current research in this exciting field.

We would like to thank the speakers for having made this session successful, and for their contributions to the proceedings. We would also like to take this opportunity to thank Donna Harmon at the AMS for her help in the time-consuming task of assembling the volume together in a suitable form.

Contemporary Mathematics
Volume 160, 1994

Hidden Symmetries of Differential Equations

B. ABRAHAM-SHRAUNER
AND
A. GUO

ABSTRACT. Hidden symmetries of differential equations are discussed. These are of two types for ODEs: type I are the symmetries which present in a higher-order ODE are lost when that ODE is reduced to a lower-order ODE and type II are the symmetries not present in the higher-order ODE which appear when that ODE is reduced to a lower-order ODE. The inverse problem for hidden symmetries is described for nonabelian two-parameter subgroups of the projective group for type I hidden symmetries and for energy conserving equations for type II hidden symmetries. For the type I hidden symmetries more general first-order ODEs are reported. Applications to Vlasov-Maxwell equations and the Painlevé-Ince equation are mentioned.

1. Introduction

The hidden symmetry analysis discussed here is part of a general attempt to obtain analytical solutions of nonlinear differential equations (NLDEs) that arise in engineering, physics, biological and chemical systems. The basic approach is to find variable transformations that simplify the NLDEs to a form that has fewer variables for partial differential equations (PDEs) or is of lower order for ordinary differential equations (ODEs). Ideally one reduces the NLDEs to quadratures but simplification may be very helpful also. An alternate path is the reduction of the nonlinear differential equation to a linear differential equation.

The variable transformations are determined by the symmetry analysis of the NLDEs where the symmetry properties are given by Lie groups or the associated Lie algebra. In the classical Lie method the point transformations of particular differential equations are found; this is a direct method. Contact transformations or generalized symmetries are also found by direct methods but are not considered here.

Hidden symmetries are symmetries of PDEs or ODEs not found by the Lie classical method for Lie point groups or by the direct methods for contact or generalized symmetries. This general definition includes previous results where in most cases the term hidden symmetries was not used. The two examples that

1991 Mathematics Subject Classification. Primary 34A43, 22E70.
Supported in part by the Southwestern Bell Foundation Grant.
This paper is in final form and no version of it will be submitted for publication elsewhere.

suggested the two types of hidden symmetries treated in this paper are located on pages 148 and 154 of the book by Olver.[1] The nongeometric symmetries of Maxwell's equations, explained in the book by Fushchich and Nikitin[2] are called hidden symmetries there. The additional symmetries increased the traditional 16-parameter group to a 23-parameter group. Physically the added symmetries correspond to polarization of electromagnetic waves. They also share a common feature of many hidden symmetries in that they involve nonlocal transformations. The potential symmetries incorporate a potential which had already appeared in fluid theory as a stream potential. For example, Kalman[3] used the potential for an early variant of the Lagrangian coordinates for one-dimensional unmagnetized plasma fluid equations. More recently, Bluman, Kumei and Reid[4-6] have applied the idea to a number of nonlinear PDEs. In all cases the equations were put in the conservation form which is characteristic of fluid equations. Vinogradov and Krasil'shchik[7,8] and others have introduced inverse differential operators in a study of nonlocal symmetries. The nonclassical method[1,4] is another approach that introduces additional conditions to the usual direct Lie classical method.

To understand the hidden symmetries a definition of the Lie classical method is needed. This method determines the Lie groups under which a particular set of differential equations is invariant by solving the determining equations found from the invariance condition on the differential equations[1] for the coordinate functions of the group generators. In contrast to the classical Lie method which is a direct method for finding the group under which a differential equation is invariant the inverse problem determines the general form of the differential equation invariant under a group. The particular differential equation is then matched, if possible, to a general form of the differential equation in a table; this method is an indirect method for finding some symmetries of a given differential equation. The inverse method, much used by Lie in fact, can be used for hidden symmetries and is the one suggested here for investigating them. Type I and II hidden symmetries are analyzed here. If an nth-order ODE is reduced in order by a symmetry group two possiblities occur. The reduced lower-order ODE may not retain another symmetry group of the nth-order ODE; this lost symmetry is called a type I hidden symmetry of the lower-order ODE. Alternatively, the lower-order ODE may possess a symmetry group not shared by the nth-order ODE; this is called a type II hidden symmetry of the nth-order ODE. The main topic of this report is a description of the procedure used in the analysis of two types of hidden symmetries that have been identified of nonlinear ODEs.

This paper is organized as follows. In section 2 the motivation for considering hidden symmetries is presented. In section 3 the type I hidden symmetry is defined and its application to first-order ODEs invariant under the nonabelian 2-parameter projective subgroups is examined. In section 4 the type II hidden symmetry is described and the extension of the indirect method for finding hidden symmetries to particular equations is discussed. The conclusions are presented in section 5.

2. Motivation for Examining Hidden Symmetries

The motivation for investigating hidden symmetries of differential equations is the possibility of transforming an ODE with insufficient Lie point symmetries to be solved to another ODE that has enough Lie point symmetries such that it can be solved by integration. In particular the Lie classical method is awkward in determining the symmetries of first-order ODEs. The single determining equation for the coordinate functions for the full symmetry group of a first-order ODE is a linear PDE but nonetheless difficult to solve. The extant list of first-order ODEs with Lie group symmetries has been found either by clever guesses of finite variable transformations or by the inverse method for the most part. Since many higher order ODEs or PDEs reduce to first-order ODEs, the benefit in enlarging the class of first-order ODEs which have hidden symmetries is potentially very important.

One problem that illustrates the need for more symmetries of first-order ODEs is the attempt to find exact solutions of the one-dimensional Vlasov-Maxwell equations for plasmas in a neutralizing background. One method for finding exact solutions of the nonlinear Vlasov-Maxwell equations is to find solutions of the characteristic equations of the Vlasov equations and then require that Maxwell equations hold for the resultant one-particle distribution functions which appear in the integrals for the charge and current densities. The characteristic equation in the one-dimensional case is

$$\frac{d^2x}{dt^2} = \frac{q_j}{m_j} E(x,t) . \tag{1}$$

Previous analysis of the Lie point symmetries[9] reduced this equation to a first-order ODE. The solutions for the reduced ODE were given for two cases: one with an ODE of the energy-conserving form discussed in section 4 and the other of the homogeneous form and invariant under scaling transformations. More solutions were searched for but not found. The need for additional analytical solutions of nonlinear first-order ODEs is extremely common.

3. Type I Hidden Symmetry

The type I hidden symmetry is a 'lost' point symmetry group of an nth-order ODE that is not retained when the ODE is reduced by a separate symmetry group to a lower-order ODE. In the type I hidden symmetries the point symmetry is lost because the ODE is reduced by non-normal subgroup variables.

Hidden symmetries are not found in practice by the usual classical Lie group method. In other approaches a potential has been introduced for nonlocal symmetries and the form of the nonlocal transformation is assumed for the inverse operators. Here, another tack is chosen; the inverse method is employed. Here we assume a Lie group and find the general form of the ODE invariant under that group. Next, we reduce the order of the ODE by one by using non-normal subgroup variables. The reduced ODE no longer has one or more Lie group point symmetries associated with the normal subgroup variables of the original higher-order ODE. If the higher-order ODE is reduced in order by normal subgroup variables, the other symmetries will be retained. These two reduced ODEs are of the same order and can be transformed into each other by a nonlocal variable transformation. This follows because under point transformations the symmetry group of the original ODE retains the dimension of the symmetry group of the original ODE. Since the symmetry groups of the two ODEs have different dimensions (we assume for the moment that no new symmetries arise in either ODE) as one or more symmetry groups are lost for one ODE, the variable transformation between the reduced ODEs cannot be point transformations. In all cases analyzed so far the variable transformation is nonlocal where the nonlocal transformation appears to be characteristic of the type I hidden symmetry. Type I hidden symmetries have occurred in various examples. The main analysis reported here is the search for all type I hidden symmetries that arise for the nonabelian two-parameter projective subgroups. The group generators of the eight-parameter projective groups are:[4]

$$
\begin{aligned}
&U_1 = \partial_z\,,\ U_2 = \partial_u\,,\ U_3 = z\partial_z\,,\ U_4 = u\partial_u \\
(2)\qquad &U_5 = z\partial_u\,,\ U_6 = u\partial_z\,,\ U_7 = zu\partial_z + u^2\partial_u \\
&U_8 = z^2\partial_z + zu\partial_u
\end{aligned}
$$

The nonabelian two-parameter subgroups can be discovered by calculating the commutators of the group generators. These are

$$(3)\qquad [U_a, U_b] = kU_a$$

where a and b take on the values 1 to 8 and k = +1 or -1 for these 8 group generators. Eight different commutators were found which in terms of the four possible normal forms of 2-parameter subgroups[10] reduce to two of those four cases.

For each two-parameter subgroup the general form of the nonlinear second-order ODE has been found, the two reduced ODEs of first-order have been given, the nonlocal variable transformations between the ODEs calculated and general solutions for the first-order ODE with the lost symmetry presented. These results have been reported elsewhere.[11] Here a more general table for the first-order ODEs which lost a symmetry is presented. The procedure by which the two sets of ODEs are determined is also discussed.

We start with one case to illustrate the method. Consider the two-parameter group with the non-normal subgroup generated by U_4 and the normal subgroup generated by U_5. The characteristic equations for the second extension of U_4 are

$$(4)\qquad \frac{dz}{0} = \frac{du}{u} = \frac{du_z}{u_z} = \frac{du_{zz}}{u_{zz}}$$

and the characteristic equations for the second extension of U_5 are

$$(5)\qquad \frac{dz}{0} = \frac{du}{z} = \frac{du_z}{1} = \frac{du_{zz}}{0}\,.$$

These integrate to give the invariants for U_4

$$(6)\qquad z,\ u_z/u,\ u_{zz}/u$$

and the invariants for U_5

$$(7)\qquad z,\ u - zu_z,\ u_{zz}.$$

The general form of the second-order ODE invariant under the groups represented by the two group generators is

(8) $$F(\, u_{zz}/(u - zu_z),\ z) = 0$$

where the invariants for the two-parameter group are those which are common to both groups and the solved form is

(9) $$u_{zz} = (u - zu_z)g(z)$$

and $g(z)$ is an arbitrary function of z. For this case the common invariants can be seen by inspection. A more systematic way is to express the differential equation as a function of the invariants of one of the groups. Then apply the twice-extended group generator of the second group to that differential equation and find the characteristic equations in the invariants of the first group. Integration of the characteristic equations results in the two invariants above in Eq.(8).

Next two first-order ODEs are found by reducing the solved form of Eq.(8) by the invariants of U_4 and U_5. The choice of the new variables is not unique. For the non-normal subgroup variables we choose

(10) $$y = \frac{u_z}{a(z)u}, \qquad x = z$$

where we choose a more general form for y than we had previously, which is allowed, since we can take combinations of the first differential invariant (u_z/u) and the path curve (z). Here $a(z)$ is an arbitrary function of z. For the normal subgroup variables we choose

(11) $$\bar{y} = u - zu_z\ ,\ \bar{x} = z\ .$$

The first-order ODE for the non-normal subgroup variables is

(12) $$\frac{dy}{dx} = c(x) + b(x)y - a(x)y^2$$

with

$$c(x) = -\frac{b(x)}{xa(x)} - \frac{a'(x)}{xa^2(x)}\ ,\ b(x) = -xg(x) - \frac{a'(x)}{a(x)}\ .$$

The other first-order ODE for the normal subgroup variables is

(13) $$\frac{d\bar{y}}{d\bar{x}} = -\bar{x}\bar{y}g(\bar{x})\ .$$

Now as $a(x)$ and $g(x)$ are arbitrary functions of x, we can alternatively view $a(x)$ and $b(x)$ as arbitrary functions of x with $c(x)$ and $g(x)$ defined in terms of them. Even though the Riccati equation, (12), contains two arbitrary functions of x, it is not the most general form of a Riccati equation but is more general than that given in a previous paper where $a(x)$ was set equal to 1. The first-order ODE, Eq.(13), is invariant under the one-parameter group generated by U_4 as expected since it was reduced by the normal subgroup variables from the second-order ODE invariant under a two-parameter group. On the other hand the Riccati equation, (12), has lost a symmetry group or has a hidden symmetry. Hence, Eq.(13) can be integrated by separation of variables but there is no obvious way to integrate Eq.(12). Nevertheless, we can find y(x) here by solving Eq.(13) by separation of variables and then by calculating the nonlocal variable transformation between the normal and non-normal subgroup variables of the first-order ODEs. The variable transformation is found by comparing the two relations for the variables in terms of u, z and u_z. We find

$$y = \frac{1}{\bar{x}a(\bar{x})}\left[1 + \bar{y}\left(C + \bar{x}\int \frac{\bar{y}d\bar{x}}{\bar{x}^2}\right)^{-1}\right]. \tag{14}$$

TABLE 1. The group generators U_α and U_β for the normal and non-normal subgroups and the solved form of the second-order ODEs are listed in columns 2 and 3 respectively where the case is designated in column 1.

Case	Group Generators	2nd Order ODE
I_a	$U_\alpha = \partial_z,\ U_\beta = z\partial_z$	$u_{zz} = u_z^2 g(u)$
I_b	$U_\alpha = \partial_u,\ U_\beta = u\partial_u$	$u_{zz} = u_z g(z)$
II_a	$U_\alpha = u\partial_z,\ U_\beta = z\partial_z$	$u_{zz} = u_z^2\,(u - zu_z)g(u)$
II_b	$U_\alpha = z\partial_u,\ U_\beta = u\partial_u$	$u_{zz} = (u - zu_z)g(z)$
III_a	$U_\alpha = z^2\partial_z + zu\partial_u,$ $U_\beta = z\partial_z$	$u_{zz} = \dfrac{g(u - zu_z)}{z^2 u}$
III_b	$U_\alpha = zu\partial_z + u^2\partial_u$ $U_\beta = u\partial_u$	$u_{zz} = \dfrac{u_z^3}{zu^2}\, g\left(\dfrac{u}{u_z} - z\right)$
IV_a	$U_\alpha = z\partial_u,\ U_\beta = z\partial_z$	$u_{zz} = \dfrac{g(u - zu_z)}{z^2}$
IV_b	$U_\alpha = u\partial_z,\ U_\beta = u\partial_u$	$u_{zz} = \dfrac{u_z^3}{u^2}\, g\left(\dfrac{u}{u_z} - z\right)$

The expression for u in terms of $\bar{x}$ and $\bar{y}$ is found by substituting $\bar{x}$ for z in the relation for y in Eq.(11) and then integrating for u.

The results for the eight cases are collected in Tables 1 to 4. In Table 1 we have listed the two group generators for each two-parameter subgroup with the subscript α and β denoting the normal and non-normal subgroups respectively. The corresponding second-order ODE, which is invariant under this two-parameter group, is also listed where $g(z)$ is an arbitrary function of z The case numbers in column 1 denote one of the eight two-parameter subgroups of the projective group.

TABLE 2. The non-normal subgroup invariants and the first-order ODEs with hidden symmetries are given respectively in columns 2 and 3 where the case is noted in column 1.

Case	Non–Normal Variables	1st Order ODE
I_a	$y = a(u)zu_z\,,\ x = u$	$y' = a(x) + b(x)y$
I_b	$y = \dfrac{u_z}{a(z)u}\,,\ x = z$	$y' = b(x)y - a(x)y^2$
II_a	$y = a(u)zu_z\,,\ x = u$	$y' = a(x) + b(x)y + c(x)y^2$
II_b	$y = \dfrac{u_z}{a(z)u}\,,\ x = z$	$y' = c(x) + b(x)y - a(x)y^2$
III_a	$y = a(u)zu_z\,,\ x = u$	$y' = a(x) + \dfrac{a'(x)}{a(x)}y + \dfrac{a^2(x)}{xy}g\left(x - \dfrac{y}{a(x)}\right)$
III_b	$y = \dfrac{u_z}{a(z)u}\,,\ x = z$	$y' = -\dfrac{a'(x)}{a(x)}y - a(x)y^2 + \dfrac{a^2(x)y^3}{x}g\left(\dfrac{1}{a(x)y} - x\right)$
IV_a	$y = a(u)zu_z\,,\ x = u$	$y' = a(x) + \dfrac{a'(x)}{a(x)}y + \dfrac{a^2(x)}{y}g\left(x - \dfrac{y}{a(x)}\right)$
IV_b	$y = \dfrac{u_z}{a(z)u}\,,\ x = z$	$y' = -\dfrac{a'(x)}{a(x)} - a(x)y^2 + a^2(x)y^3\, g\left(\dfrac{1}{a(x)y} - x\right)$

Table 2 lists the variable transformations in terms of the non-normal subgroup invariants by which the second-order ODEs in Table 1 are reduced to the first-order ODEs. The first-order ODEs have lost the symmetry of the normal subgroup but have the hidden symmetry. The variable transformations for y contain an arbitrary function of z, $a(z)$, as well as the first-differential invariant. For the cases $I_{a,b}$ and $II_{a,b}$ we have rewritten the first-order ODEs in compact form. For case I_a $g(x) = b(x) - a'(x)/a(x)$. For case I_b $g(x) = b(x) + a'(x)/a(x)$. For case II_a $g(x) = -a'(x)/(xa(x)) + b(x)/x$ and $c(x) = -b(x)/(xa(x)) + a'(x)/(xa^2(x))$. For case II_b $g(x) = -a'(x)/(xa(x)) - b(x)/x$ and $c(x) = -b(x)/(xa(x)) - a'(x)/(xa^2(x))$.

The ODEs in Table 2 for the cases with the same Roman numeral can be found from each other by letting y go to $1/y$ to within a sign change. Consequently, there are really only four distinct cases. In fact in terms of normal forms[10] there are only two distinct forms here. The cases $I_{a,b}$ and $II_{a,b}$ are of one normal form and the cases $III_{a,b}$ and $IV_{a,b}$ are of another normal form. The two normal forms for a nonabelian two-parameter group are distinguished by a value of $\delta = 0$ for cases I and II and 1 for cases III and IV. The δ is the determinant of the coordinate functions of the group generators.

TABLE 3. The normal subgroup invariants and the first-order ODEs are given respectively in columns 2 and 3 where the case is noted in column 1.

Case	Normal Variables	1st Order ODE
I_a	$\bar{y} = u_z\,,\ \bar{x} = u$	$\frac{d\bar{y}}{d\bar{x}} = \bar{y}g(\bar{x})$
I_b	$\bar{y} = u_z\,,\ \bar{x} = z$	$\frac{d\bar{y}}{d\bar{x}} = \bar{y}g(\bar{x})$
II_a	$\bar{y} = \frac{u}{u_z} - z\,,\ \bar{x} = u$	$\frac{d\bar{y}}{d\bar{x}} = -\bar{x}\bar{y}g(\bar{x})$
II_b	$\bar{y} = u - zu_z\,,\ \bar{x} = z$	$\frac{d\bar{y}}{d\bar{x}} = -\bar{x}\bar{y}g(\bar{x})$
III_a	$\bar{y} = u - zu_z\,,\ \bar{x} = \frac{u}{z}$	$\frac{d\bar{y}}{d\bar{x}} = \frac{g(\bar{y})}{\bar{x}\bar{y}}$
III_b	$\bar{y} = \frac{u}{u_z} - z\,,\ \bar{x} = \frac{u}{z}$	$\frac{d\bar{y}}{d\bar{x}} = \frac{g(\bar{y})}{\bar{x}\bar{y}}$
IV_a	$\bar{y} = u - zu_z\,,\ \bar{x} = z$	$\frac{d\bar{y}}{d\bar{x}} = -\frac{g(\bar{y})}{\bar{x}}$
IV_b	$\bar{y} = \frac{u}{u_z} - z\,,\ \bar{x} = u$	$\frac{d\bar{y}}{d\bar{x}} = -\frac{g(\bar{y})}{\bar{x}}$

The quantities in Table 3 are the same as in Table 2 but for the normal subgroup variables $\bar{x}$ and $\bar{y}$. All the first-order ODEs in column 3 are invariant under a Lie point group and all can be integrated by separation of variables. The first-order ODEs are the same for the a and b cases for a given Roman numeral.

The solutions of the first-order ODEs in the non-normal subgroup variables are tabulated in Table 4. The solutions for cases I and II are given as $y(x)$ but those for cases III and IV are parametric solutions where x and y are both given in terms of integrals over the parametric variable $\bar{y}$. Sometimes it is possible to solve for $y(x)$ for the latter cases but there is no general relation for $y(x)$.

TABLE 4. The solutions $y(x)$ of the ODEs in Table 2 are given for cases I and II; the parametric solutions $x(\bar{y})$ and $y(\bar{y})$ are given for cases III and IV.

Case	Solutions
I_a	$y = h_1(x)\left[\int \frac{a(x)dx}{h_1(x)} + K\right]$
I_b	$\frac{1}{y} = \frac{1}{h_1(x)}\left[\int a(x)h_1(x)dx + K\right]$
II_a	$\frac{1}{y} = \frac{1}{xa(x)} + \frac{h_2(x)}{x\left[x\int \frac{a(x)h_2(x)dx}{x^2} + K\right]}$
II_b	$y = \frac{1}{xa(x)} + \frac{h_2(x)}{x\left[x\int \frac{a(x)h_2(x)dx}{x^2} + K\right]}$
III_a	$x(\bar{y}) = \frac{h_3(\bar{y})}{\int \frac{h_3(\bar{y})d\bar{y}}{g(\bar{y})} + K}, \quad y(\bar{y}) = a[x(\bar{y})]\,[x(\bar{y}) - \bar{y}]$
III_b	$\frac{1}{x(\bar{y})} = h_3(\bar{y})\left[\int \frac{d\bar{y}}{g(\bar{y})h_3(\bar{y})} + K\right], \quad \frac{1}{y(\bar{y})} = a[x(\bar{y})]\,[x(\bar{y}) + \bar{y}]$
IV_a	$x(\bar{y}) = h_4(\bar{y})\left[\int \frac{\bar{y}d\bar{y}}{g(\bar{y})h_4(\bar{y})} + K\right], \quad y(\bar{y}) = a[x(\bar{y})]\,[x(\bar{y}) - \bar{y}]$
IV_b	$x(\bar{y}) = -\,h_4(\bar{y})\left[\int \frac{\bar{y}d\bar{y}}{g(\bar{y})h_4(\bar{y})} + K\right], \quad \frac{1}{y(\bar{y})} = a[x(\bar{y})]\,[x(\bar{y}) + \bar{y}]$

The h-functions in Table 4 are

$$h_1(x) = \exp[\int b(x)dx], \qquad h_2(x) = \exp[-\int b(x)dx] \tag{15}$$

$$h_3(y) = exp\left[\int \frac{\bar{y}d\bar{y}}{g(\bar{y})}\right], \qquad h_4(y) = exp\left[-\int \frac{d\bar{y}}{g(\bar{y})}\right].$$

The presence of hidden symmetries enables us to integrate the general, linear first-order ODE in case I_a instead of assuming an integrating factor which depends only on x. The solution for y(x) follows naturally from the retained symmetry in the first-order ODE which was reduced by the normal subgroup variables and the nonlocal transformation between the normal and non-normal subgroup variables. In case II_b the approach here retrieves a solution found previously.[12] There if a particular solution $y_p(x)$ was known, then $y(x) = y_p(x) + 1/y_l(x)$ where $y_p(x)$ was a particular solution of the Riccati equation and $y_l(x)$ obeyed a linear ODE. That ad hoc solution emerges here as a consequence of the hidden symmetry.

4. Type II Hidden Symmetry and Particular ODEs

Type II hidden symmetries of ordinary differential equations occur when the order of the ODE is reduced and one or more new Lie group symmetries are found in the reduced ODE. No simple marker as the reduction by non-normal subgroup variables for some type I hidden symmetries occurs here. The existence of the type II hidden symmetry does offer the possibility that even when an ODE is not invariant under a large enough dimensional Lie point group, the ODE may still be reduced to quadratures. The case that has been investigated systematically is that for energy conserving equations.[13] We start with the energy conserving equation

$$\frac{w\,dw}{dv} = f(v) \tag{16}$$

where f is an arbitrary function. This first-order ODE can be integrated by separation of variables and is invariant under the group represented by the group generator

$$U = \frac{1}{w}\frac{\partial}{\partial w}\,. \tag{17}$$

If we next increase the order by using the invariants of one of the groups associated with the projective group generators in Eq.(2), we find second-order ODEs.[13] These ODEs are invariant under the projective subgroup by which we increased the order of Eq.(16). On the other hand the second-order ODEs are not invariant under the group represented by the group generator in Eq.(17). That fact can be verified by evaluating the group symmetries of the second-order ODEs by the classical Lie method or by a procedure described by Bluman and Kumei.[4] For some special forms of f(v) the second-order ODEs were found to be invariant under a two-parameter group but neither subgroup of that two-parameter group could be identified with the group represented by the group generator in Eq.(17). Tables of the second-order ODEs with this type II hidden symmetry are given elsewhere.[13] There v equals the path curve and w equals the first differential invariant.

More general forms of the second-order differential equations are possible than were given in those tables since w can be set equal to an arbitrary function of v times the usual first-differential invariant. For example if we assume that Eq.(16) was reduced from a second-order ODE

$$y'' = F(x,y,y'), \tag{18}$$

then if this ODE is invariant under translations in x, a projective subgroup with the group generator U_1 in Eq.(2) , the invariants are (y, y'). If $v = y$ and $w = y'$, then $y'' = f(y)$ follows from Eq.(16) by the variable transformation. However, more general second-order ODEs than that one are possible. The variable transformation $v = y$, $w = a(y)y'$ for $a(y)$ an arbitrary function of y is an equally good choice and leads to the second-order ODE

$$a(y)^2 y'' + a'(y)a(y)\, y' = f(y). \tag{19}$$

This second-order ODE is of the form of Eq.(6.224) in the Kamke tables.[14] It has a more general form than the ODE found previously but is a natural outcome of the type II hidden symmetry. Increasing the order of Eq.(16) by other projective subgroups gives even more complicated second-order ODEs.

The type I and type II hidden symmetries have so far been discussed in the context of the inverse or indirect method where tables of equations with the hidden symmetry have been prepared. The tables are still not extensive and consequently the question arises as to how one can search for hidden symmetries of particular equations. There is yet no general, direct method which uses nonlocal group generators[1] or some other approach that can consistently find the hidden symmetries. However, hidden symmetries of some differential equations have been evaluated. The key step is to choose the correct groups by which one increases the order of the original ODE. Four examples have yielded useful results so far. These are the Vlasov characteristic equation for plasmas mentioned in Section 2,[9,11] the modified Painlevé-Ince equation which arises from a solution of the modified Emden equation, the Pinney equation which describes trajectories in particle accelerators and plasma beams, and one-dimensional semiconductor transport equations.[15] The order of the ODEs were increased by one by likely groups. The Riccati transformation has been the most useful. One example is the modified Painlevé-Ince equation

$$y'' + \sigma y y' + \beta y^3 = 0. \tag{20}$$

If we increase the order of this ODE by the Riccati transformation

$$y = \frac{\Gamma u_z}{u}\,, \qquad x = z\,, \tag{21}$$

we can then reduce the third-order ODE by the invariants of the translation group in z to a linear ODE. The net transformation is

$$\bar{y} = \left(\frac{\Gamma u_z}{u}\right)^2, \qquad \bar{x} = \Gamma \ln u \tag{22}$$

where the new second-order ODE

$$\frac{d^2\bar{y}}{d\bar{x}^2} + \frac{\sigma d\bar{y}}{d\bar{x}} + 2\beta\bar{y} = 0 \tag{23}$$

where the linear equation is familiar in circuits, etc. and can be solved easily.

5. Conclusions

The hidden symmetries of ordinary differential equations have been discussed. The inverse method by which tables of ODEs with hidden symmetries can be prepared has been presented. Type I hidden symmetries, those Lie point symmetries lost when the order of the ODE is reduced, have been investigated for the eight nonabelian, two-parameter projective subgroups. Second-order ODEs invariant under the the two-parameter subgroups have been reduced by normal and non-normal subgroup variables. The reduced first-order ODEs in the non-normal subgroup variables (x,y) lose a point symmetry but have a hidden symmetry. A procedure for finding analytical expressions for these first-order ODEs in non-normal subgroup variables is presented and the results tabulated. The first-order ODEs tabulated are more general in form than were presented previously. For four cases a solution $y(x)$ is given in terms of integrals; for four other cases parametric solution for y and x are given in terms of integrals. The integrating factor for linear first-order ODEs and the solution for a Riccati equation when a particular solution is known emerge naturally from the hidden symmetry analysis.

The type II hidden symmetry, the Lie point symmetry that is gained when the order of the ODE is reduced, is described for energy-conserving equations. The tabulated results which have been presented elsewhere may be put in more general form as was done for the type I hidden symmetry. The extension of the indirect method to particular equations is described briefly.

Several extensions of the of the hidden symmetry analysis are possible. More extensive tables with a larger class of hidden symmetries should be prepared. Some preliminary analysis of type I hidden symmetries has been done for several of the 24 Lie algebras[16] in the plane both in a more recent paper[17] and in unpublished calculations. The preparation of such tables is feasible with symbolic computing both for the computation of the group symmetries and for the manipulations of equations. Nonlocal generators occur for type I hidden symmetries. In addition to the exponential type out by Olver [1] on page 184 one may find a linear type for three-parameter groups.[17] The form of these nonlocal group generators can be found by analysis of the Lie algebras and the nonlocal generator retains dependence on a canonical coordinate of a higher-order ODE. A thorough understanding of nonlocal group generators is required if a direct method for type I hidden symmetries is to be developed. Type II hidden symmetries are not well understood and the counterpart to the nonlocal group generator has not been obtained for the symmetry when hidden in the higher-order ODE. Partial differential equations (PDEs) possess hidden symmetries. The disappearance of symmetry groups when a PDE is reduced by another symmetry group has been reported.[4,18] The analysis of hidden symmetries of PDEs could be very important as many phenomenon in science and engineering are modeled by PDEs.

REFERENCES

1. P. J. Olver, *Applications of Lie Groups to Differential Equations*, Springer-Verlag, New York, 1986.

2. W. I Fushchich and A. G. Nikitin, *Symmetries of Maxwell's Equations*, D. Reidel. Co.Dordrecht, 1987.

3. G. Kalman, "Nonlinear Oscillations and Nonstationary flow in a Zero Temperature Plasma," I. Ann. Phys.(1960), 1-28.

4. G. W. Bluman and S. Kumei, *Symmetries and Differential Equations*, Appl. Math. Sci. No. 81, Springer-Verlag, New York, 1989.

5. G. W. Bluman and S. Kumei, "Symmetry-based algorithms to relate partial differential equations, I. Local symmetries," Euro. Jnl. of Appl. Math. **1** (1989), 189-216.

6. G. W. Bluman and S. Kumei, "Symmetry-based algorithms to relate partial differential equations:II. Linearization by local symmetries, Euro. Jnl. of Appl. Math. **1** (1989), 217-223.

7. A. M. Vinogradov and I. S. Krasil'shchik, "On the theory of nonlocal symmetries of nonlinear partial differential equations," Sov. Math. Dokl. **29** (1984), 337-341.

8. I. S. Krasil'shchik and A. M. Vinogradov, "Nonlocal symmetries and the theory of coverings: an addendum to A. M. Vinogradov's 'Local symmetries and conservation laws'," Acta Applic. Math. **2** (1984), 79- 95.

9. B. Abraham-Shrauner, "Lie transformation group solutions of the nonlinear one-dimensional Vlasov equation," J. Math. Phys. **26** (1985), 1425-1435.

10. H. Stephani, *Differential equations*, Cambridge University Press, Cambridge, 1989.

11. B. Abraham-Shrauner and A. Guo, "Hidden symmetries associated with the projective group of nonlinear first-order ordinary differential equations," J. Phys. A: **25** (1992) 5597-5608.

12. H. T. Davis, *Introduction to Nonlinear Differential and Integral Equations*, Dover, New York, 1962.

13. A. Guo and B. Abraham-Shrauner, "Hidden symmetries of energy conserving differential equations," IMA J. of Appl. Math. (accepted for publication).

14. E. Kamke, *Differentialgleichungen Losungsmethoden und Losungen*, Chelsea, New York, 1971.

15. B. Abraham-Shrauner and P. G. L. Leach, "Hidden symmetries of Nonlinear Ordinary Differential Equations," Proc. 1992 AMS-SIAM Summer Seminar, AMS *Lectures in Applied Mathematics* (to be published)

16. A. Gonzalez-Lopez, N. Kamran and P. J. Olver, "Quasi-exactly solvable Lie algebras differential operators in two complex variables, J. Phys. A: Math. Gen. **24**, (1991) 3995.

17. B. Abraham-Shrauner, and A. Guo, "Hidden and Nonlocal Symmetries of Nonlinear Differential Equations," Proceedings of the International Workshop, Modern Group analysis, Acireale, Italy, October, 1992 (submitted).

18. S. Coggeshall (private communication).

DEPARTMENT OF ELECTRICAL ENGINEERING, WASHINGTON UNIVERSITY, ST. LOUIS, MO 63130.
E-*mail* address: bas@wuee1.wustl.edu

Contemporary Mathematics
Volume **160**, 1994

ALGEBRAIC METHODS IN SCATTERING

Y. Alhassid

ABSTRACT. The use of algebraic methods in scattering problems is reviewed. Applications are presented.

1. Introduction

The scattering from composite particles such as nuclei and molecules where many channels are strongly coupled cannot be treated by perturbation methods (e.g. Born approximation) and is usually solved in the coupled channel formalism. The latter approach becomes prohibitively difficult as the number of channels increases. For long-range interactions, the number of partial waves that contribute to the scattering may also be very large.

When the internal motion of the colliding objects is described by an algebra, it is possible to take advantage of this algebraic structure. Under certain approximations (e.g. semi-classical or eikonal) the problem can be effectively solved to all orders in the interaction without a direct solution of the coupled channel equations. The techniques are reviewed in Section 2.

The above methods do not treat the relative motion of the colliding particles algebraically. An algebraic approach which includes the relative motion is discussed in Section 3 in the context of solvable potentials.

1991 Mathematics Subject Classification, Primary 17B, 22E.
Supported in part under U.S. DOE Grant DE-FG02-31ER40608
This paper is in final form and no version of it will be submitted for publication elsewhere.

Both bound and scattering states of certain classes of potentials are related to the unitary representations of certain non-compact groups. For such potentials the S-matrix can be calculated by pure algebraic techniques through the use of the Euclidean group that describes asymptotic behavior.

2. Scattering I

We first discuss the use of algebraic methods for scattering problems in which only the internal motion is described algebraically. We assume a structureless projectile with relative coordinate $\vec{r}$ and momentum $\vec{p}$ which is scattered from a target whose internal coordinates are denoted by ξ. The Hamiltonian is

$$H = \frac{\vec{p}^{\,2}}{2\mu} + H_0(\xi) + V(\vec{r},\xi) \quad , \tag{1}$$

where μ is the reduced mass, H_0 is the target Hamiltonian and V is the projectile-target interaction. H_0 is assumed to be described by an algebra $\mathcal{G}$, i.e. it can be written as a function of the generators of $\mathcal{G}$. Usually H_0 is a bilinear form in the generators. The bound eigenstates of H_0 span the carrier space of a representation of $\mathcal{G}$.

2.1. Semi-classical approximation

In the semi-classical approximation the relative motion is described by a classical trajectory $\vec{r}(t)$. The interaction is then time-dependent $V=V(\vec{r}(t))$ and the scattering problem reduces to the excitation of the target by a time-dependent perturbation.

The scattering operator is the evolution operator in the interaction picture between the times $-\infty$ and ∞:

$$S = \mathcal{T} \exp\left[- \frac{i}{\hbar} \int_{-\infty}^{\infty} V_I(\vec{r}(t))\, dt\right] \quad , \tag{2}$$

where

(3) $$V_I = e^{iH_0t} V e^{-iH_0t} ,$$

and $\mathcal{T}$ denotes time ordering.

In the sudden approximation where the excitation energy is small compared with $\hbar/\tau$ (τ is an average collision time) $[H_0,V]\approx 0$ and $V_I \approx V$. We can then replace V_I by V in Eq. (2).

The algebraic approach to S-matrix (2) was developed in Refs. 1 and 2. If V is linear in the generators of $\mathcal{G}$

(4) $$V = \sum_r v_n G_n ,$$

then the S-operator (2) is an element of the Lie group G that is generated by the algebra $\mathcal{G}$

(5) $$S = \exp(-i \sum \theta_n G_n) .$$

The parameters θ_n are determined by v_n in (4). The S-matrix is then a representation matrix of the group

(6) $$S_{fi} = < f \mid \exp\left[-i \sum \theta_n G_n\right] \mid i > ,$$

where the carrier space for this representation are the bound states of H_0.

2.2. Examples

A simple example is the dipole excitation of a rigid molecule. The rotor Hamiltonian is given by $H_0 = \vec{L}^2/2I$ where $\vec{L}$ is the angular momentum and I is the moment of inertia. Denoting by $\vec{D}$ the dipole operator of the rotor, the set $\{\vec{L},\vec{D}\}$ form an algebra, the Euclidean algebra in three dimensions e(3).

The dipole interaction is given by $V = -(e/r^2)\hat{r}\cdot\vec{D}$. Since V is linear in the generators of e(3), the S-matrix in the sudden limit is a representation matrix of the Euclidean group E(3). The representations of E(3) are denoted[3] by (d,K) where K is the angular momentum projection along the rotor symmetry axis and d is the intrinsic dipole moment. For K=0 and in a suitable coordinate frame

$$(7)\quad < K = 0\ L'\ M \mid S \mid K = 0\ L\ M > =$$

$$< L'\ M \mid \exp(-irD_z) \mid L\ M > = (-)^M [(2L+1)(2L'+1)]^{1/2}$$

$$\sum_\ell (-i)^\ell (2\ell+1) \begin{pmatrix} L' & \ell & L \\ 0 & 0 & 0 \end{pmatrix} \begin{pmatrix} L' & \ell & L \\ M & 0 & -M \end{pmatrix} j_\ell(rd) \quad ,$$

where j_ℓ are the spherical Bessel functions. Fig. 1 shows the excitation probabilities $|< L\ 0 \mid S \mid 0\ 0 >|^2$ versus x = rd for L = 0, 1, ... 5.

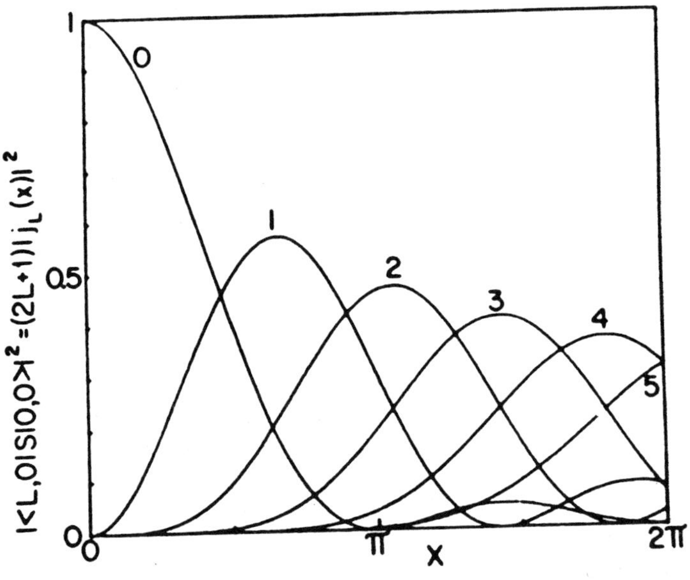

Fig. 1. Dipole excitation probabilities versus x=rd for L = 0,1,...,5.

Another example is the Coulomb excitation of a rotational nucleus. The relevant algebra is $\{\vec{L}, Q_\mu\}$ where Q_μ is the quadrupole operator. Two irreps appear in the case $K=0$; the states with even L and the states with odd L. They are obtained from the SU(3) representations (n,0) by a contraction in the limit $n \to \infty$. The S-matrix can be calculated in a closed form[2]. For an application to a reactive atom-molecule collision see in Ref. 2.

2.3. Eikonal Approximation

In the short wavelength (eikonal) approximation, the scattering amplitude (from i to f) at momentum transfer $\vec{q}$ is given by

$$f_{fi}(\vec{q}) = \frac{k}{2\pi i} \int d^2\vec{b} < f \mid U_I(b) - 1 \mid i > . \tag{8}$$

Here U_I is the evolution operator in the interaction picture for an impact parameter $\vec{b}$

$$U_I(\vec{b}) = \mathcal{T}_z \exp\left[- \frac{i\mu}{\hbar^2 k} \int_{-\infty}^{\infty} dz \; V_I(\vec{b}, z)\right] , \tag{9}$$

where $\mathcal{T}_z$ denotes z-ordering. Notice that $U_I(b)$ is just the S-operator (2) where $\vec{r}(t)$ is replaced by a straight line trajectory and $t \to mz/\hbar k$. Thus, in the sudden limit we replace $V_I \to V$ and for V linear in the generators of $\mathcal{G}$, $U_I(b)$ is a group element[4] of G and has the form (5). To calculate the parameters θ_n (with z-ordering) in (5) we write the evolution equation for $U = U(b,z)$(the sudden evolution operator between $-\infty$ and z)

$$i(\hbar^2 k/\mu) \frac{\partial U}{\partial z} U^{-1} = V . \tag{10}$$

Since U(b,z) is a group element (for any z) we can parametrized it in the form

$$U(b,z) = \exp\left[-i\sum_n \theta_n(b,z)\, G_n\right] \quad . \tag{11}$$

Eq. (10) is an algebraic equation which is valid in any irrep of G. Substituting (11) in (10) and using the fundamental irrep, we obtain first-order differential equations for $\theta_n(z)$ with the initial conditions $\theta_n(b,z=-\infty)=0$. The parameters θ_n in (5) are then given by $\theta_n=\theta_n(\infty)$.

In realistic applications we often have $V=V_0+V_1$ where V_0 is in the algebra but V_1 is in the enveloping algebra. Usually V_1 is a bilinear form in the generators.[5] If $V_1 \ll V_0$, we can expand in V_1 to obtain

$$U \approx U_0(\infty,-\infty)\left\{1 - \frac{i\mu}{\hbar^2 k}\int_{-\infty}^{\infty} dz\; U_0^{-1}(z,-\infty)V_1(z)U_0(z,-\infty)\right\} , \tag{12}$$

where U_0 is the sudden propagator for V_0. Notice that since U_0 is a group element, $U_0^{-1}G_nU_0$ is a rotation of a generator, so that for V_1 quadratic in G_n:

$$U(b) \approx U_0\left[1 + \sum_{m,n} \gamma_{m,n}(b)\, G_mG_n\right] \quad . \tag{13}$$

with $\gamma_{m,n}$ known analytically. The matrix elements of (13) are easily calculated algebraically.

The same technique can be used to calculate algebraically corrections to the sudden limit[5]. We have

$$V_I(z) \approx V(z) + \frac{i\mu}{\hbar^2 k}\, z\, \left[H_0, V(z)\right] + \dots \quad . \tag{14}$$

Notice that $[H_0,V]$ plays the role of V_1 above, and we can expand V in $[H_0,V]$ similarly to Eq. (12).

2.4. Examples

The algebraic-eikonal approach has been applied to electron-molecule scattering[4,5]. The diatomic molecule is described by the vibron model[6] where $\mathcal{G} = u(4)$. The building blocks of the model are one s boson and three p bosons. The algebra u(4) is formed by all sixteen bilinear combinations $\{s^\dagger s, s^\dagger \tilde{p}_\mu, p_\mu^\dagger s, p_\mu^\dagger \tilde{p}_{\mu'}\}$.

The Hamiltonian H_0 of the molecule is a scalar composed from one- and two-body terms in the bosons. Most diatomic molecules are close to an o(4) dynamical symmetry where o(4) is the algebra formed by $\vec{L}$ and the dipole operator $D_\mu = p_\mu s + s^\dagger \tilde{p}_\mu$. In the o(4) limit the eigenstates are $|\sigma\ j\ m\rangle$ where $\sigma(\sigma+2)$ is the eigenvalue of the o(4) Casimir invariant $\vec{L}^2+\vec{D}^2$, and j is the angular momentum. The vibrational quantum number is $v=(N-\sigma)/2$ with N being the total number of bosons.

The dominant interaction for a polar molecule is the dipole interaction $V=[-e/(r^2+R_0^2)]\hat{r}\cdot\vec{T}$. $\vec{T}$ is the dipole operator of the molecule and can be expanded in the u(4) generators

$$\vec{T} = d_0\vec{D} + \frac{1}{2}\, d_1(\hat{n}_p\vec{D} + \vec{D}\,\hat{n}_p) + \ldots \quad . \tag{15}$$

Here $\hat{n}_p$ is the number of p bosons. When $d_1=0$, we can only obtain rotational excitations (since $\vec{D}$ is in the o(4) algebra whose irreps are characterized by a fixed v). The S-matrix in the eikonal approximation is then an O(4) representation matrix. Using the isomorphic $O(4)\approx SU(2)\times SU(2)$ it is possible to express it as product of rotational Wigner matrices[5]. In practical applications it is important to take $d_1\neq 0$. This term is non-linear in the generators and can be dealt with as in Eqs. (12) and (13). An application to e+LiF is shown in Fig. 2.

Calculations which include quadrupole and polarization interactions have also been carried out in the algebraic approach[7].

The advantage of the algebraic-eikonal approach is that the S-matrix is calculated in closed form which effectively solves the coupled-channel problem to all orders in the interaction strength. Furthermore, both rotational and vibrational excitations are treated on equal footing. The simplicity of the algebraic approach

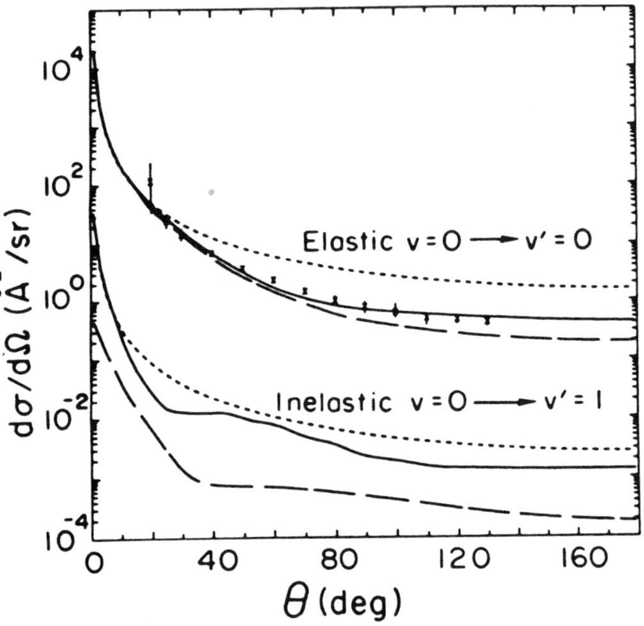

Fig. 2 The vibrational elastic (v=0) and inelastic (v'=1) diffenential cross-sections for electron scattering off LiF at E=5.44 eV. Solid lines: with the dipole operator given by Eq. (5). Long-dashed lines: the calculations of Ref. 4 with $d_1=0$. The short-dashed lines are the Born approximation.

makes it an attractive framework to be used in the scattering of electrons from polyatomic molecules. Applications to triatomic molecules were done in Ref. 8.

We note that the algebraic-eikonal was also applied to medium energy proton scattering off heavy nuclei[9].

3. Scattering II

A complete algebraic description of scattering requires an algebraic treatment of the relative motion.

In the following we shall describe such an algebraic approach to scattering problems with certain dynamical symmetries. The approach will be demonstrated for potential scattering in cases where there exists an exact algebraic description of the corresponding Schrödinger equation.

3.1 The potential algebra

Most potentials of practical interests in physics have a finite number of bound states, so that algebras used to describe bound systems are usually compact. To describe scattering states it is necessary to use non-compact groups[10]. Such groups possess unitary representations which are continuous and infinite dimensional and are thus suitable for the description of the continuum.

The simplest non-compact algebra that described one dimensional problems is su(1,1) ≈ so(2,1). Its commutation relations are

$$[J_x,J_y] = -iJ_z;\ [J_y,J_z] = iJ_x;\ [J_z,J_x] = iJ_y \quad . \tag{16}$$

The coset space so(2,1)/so(2) is an hyperboloid $x^2+y^2-z^2=1$ which is unbounded and reflects the non-compactness of the group.

Consider a problem with a dynamical symmetry su(1,1) ⊃ so(2). It can be diagnolized in a basis of states $|j,m>$

$$\begin{aligned} C\,|\,j,m > &= j(j+1)\,|\,j\ m > \\ J_z\,|\,j\ m > &= m\,|\,j\ m > \quad , \end{aligned} \tag{17}$$

where $C=-J_x{}^2-J_y{}^2+J_z{}^2$ is the quadratic Casimir invariant of su(1,1). The Hamiltonian is assumed to be a function of C; H=F(C).

The su(1,1) unitary irreps are of three types[11]:

(i) Discrete series D_j^+ and D_j^-. For the series D_j^+, j is a negative integer or half integer and m is unbounded from above

$$j = -1/2,\ -1,\ -3/2,\ -2,\ \ldots$$

(18)
$$m = -j, -j+1, -j+2, \dots \quad .$$

In a physical realization the states $|j,m\rangle$ describe bound states.

(ii) The continuous series C_k^0 and $C_k^{1/2}$ where

(19)
$$m = \begin{cases} 0, \quad \pm 1, \quad \pm 2, \quad \dots & \text{in } C_k^0 \\ \\ \pm 1/2, \quad \pm 3/2, \quad \pm 5/2, \quad \dots & \text{in } C_k^{1/2} \end{cases} \quad .$$

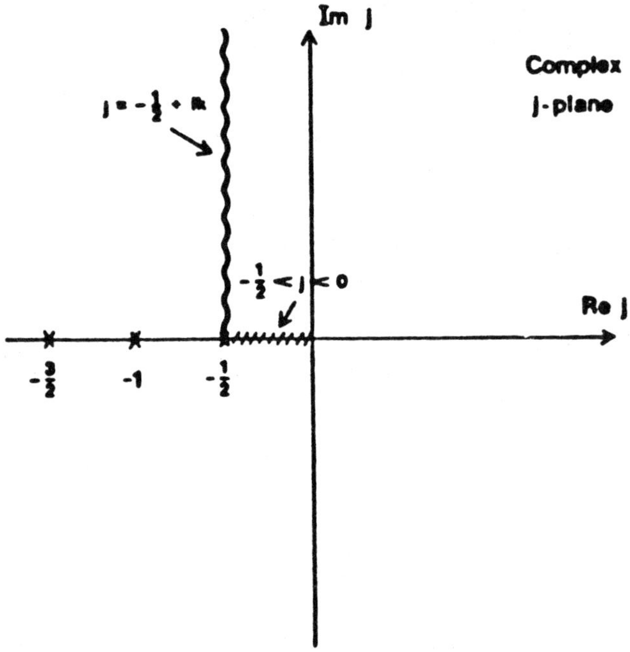

Fig. 3 The unitary representations of su(1,1) shown in the complex j-plane. The wavy line is the continuous series, the x-s belong to the discrete series and the shaded interval is supplementary series.

These states $|j,m\rangle$ describe the scattering states.
(iii) The supplementary series $-1/2<j<0$. These irreps are missing in the physical realization which lead to a Schrödinger equation. The discrete and continuous series already provide a complete set for functions on the hyperboloid. This is equivalent to the completeness of the discrete and continuous eigenfunctions of the Schrödinger Hamiltonian. The unitary irreps of su(1,1) are shown in Fig. 3 in the complex j plane.

3.2 Examples

We now give few examples of one-dimensional Schrödinger equations which are exactly described by the algebra su(1,1).

Introducing coordinates ρ,ϕ on the hyperboloid $x^2+y^2-z^2=1$, we arrive at the following realization for the hyperbolic angular momentum

$$(20)\qquad J_{\pm} = e^{\pm i\phi}\left[\pm\partial/\partial\rho + \tanh\rho\ (\pm 1/2 - i\partial/\partial\phi)\right]$$

$$J_z = -i\partial/\partial\phi \quad .$$

In this realization the basic states $|j,m\rangle$ are of the form

$$(21)\qquad |\ j,m\ \rangle = u^j_m(\rho)\ e^{im\phi} \quad ,$$

where $u^j_m(\rho)$ satisfies

$$(22)\qquad \left[-d^2/d\rho^2 - (m^2-1/4)/\cosh^2\rho\right]u^j_m(\rho) = -(j+1/2)^2 u^j_m(\rho).$$

This is a one-dimensional Schrödinger equation with a Pöschl-Teller potential $-(m^2-1/4)/\cosh^2\rho$.

The Hamiltonian is given by $H=-(C+1/4)$ so that the bound states are $E_j=-(j+1/2)^2$ and the scattering states have energy $E=k^2$. Notice that m appears in the strength

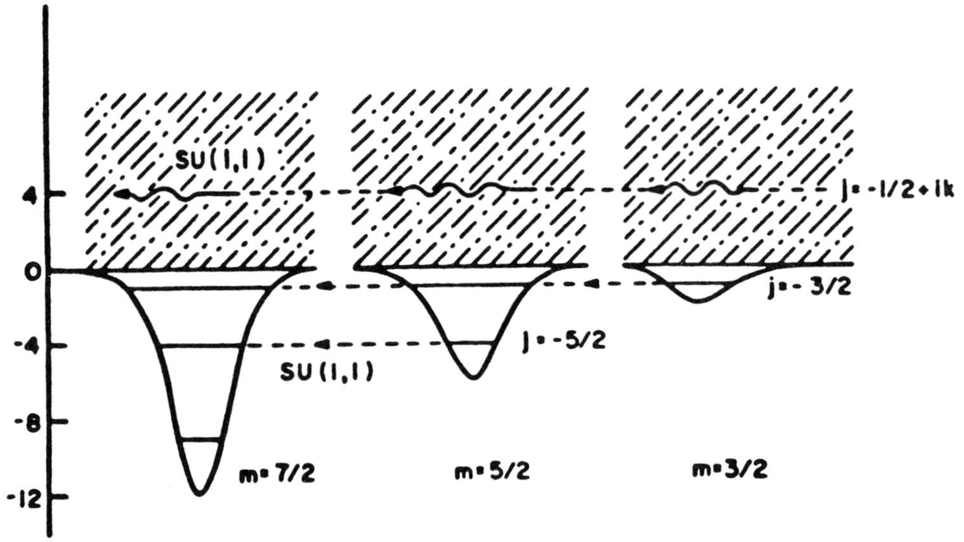

Fig. 4 Multiplets of the potential algebra su(1,1) (dashed horizontal lines) are shown for several Pöschl-Teller potentials.

of the potential. Therefore an irrep of su(1,1) describes a multiplet of levels with the same energy but belonging to potentials with various strengths. The corresponding algebra was therefore termed potential algebra. Fig. 4 illustrates the idea of the potential algebra.

A second example is provided by the realization

$$J_{\pm} = e^{\pm i\phi}(\mp\partial/\partial\rho - i\partial/\partial\phi - e^{-\rho} \pm 1/2)$$

(23)

$$J_z = - i\partial/\partial\rho \quad .$$

The basis states again has the form (21), but now the functions $u^j_m(\rho)$ satisfy

$$(24)\quad \left[-d^2/d\rho^2+(e^{-2\rho}-2me^{-\rho})\right]\ u^j_m(\rho) = -(j+1/2)^2\ u^j_m(\rho) \quad .$$

Using a translation $\rho\to\rho-\ell n m$, (24) becomes the Schrödinger equation with a Morse potential $m^2(e^{-2\rho}-2e^{-\rho})$. Here the Hamiltonian is also given by $H=-(C+1/4)$.

3.3 The Euclidean connection and the S-matrix

To find the S-matrix we have to inspect the asymptotic form of the scattering solution. In an algebraic formulation it is necessary to introduce an algebraic framework to describe aysmptotic behavior[13]. We write

$$(25)\qquad |\ j,m\ \rangle = A_m\ |\ -k,m\ \rangle - B_m\ |\ +k,m\ \rangle \quad ,$$

where $|\ \pm\ k,m\ \rangle = e^{\pm ik\rho}\ e^{im\rho}$ describe incoming (-k) and outgoing (+k) waves. They form (for a given k) a basis for irreps labeled -k and +k, respectively, of an Euclidean algebra $e(2) = \{P_x, P_y, L_z\}$ where

$$(26)\qquad [P_x,P_y] = 0 \quad , \quad [P_y,L_z] = iP_x, \quad [L_z, P_x] = iP_y \ .$$

The e(2) Casimir invariant is $P^2=P_x{}^2+P_y{}^2$. We have

$$(27)\qquad \begin{aligned} P^2\ |\ \pm\ k,m\ \rangle &= k^2\ |\ \pm\ k,m\ \rangle \\ L_z\ |\ \pm\ k,m\ \rangle &= m\ |\ \pm\ k,m\ \rangle \end{aligned} \qquad .$$

To calculate the S-matrix from (25) we need to determine a connection formula, where the su(1,1) generators are expanded in those of e(2). The general Euclidean connection for su(1,1) can be shown to be[12,13]

$$(28) \qquad J_+(\pm k) = \frac{e^{i\gamma_\pm(k)}}{(\pm k)} \left[(-1/2 \mp if(k))P_+ + L_z P_+\right] ,$$

where $\gamma_\pm$ are two real functions of k, and f(k) is determined from the relation between H and C using $C=-1/4+f^2(k)$. For the Pöschl-Teller and Morse potentials discussed above $f(k)=k$ and $\gamma_\pm=0$.

The connection formula can be used to derive recursion relations for the S-matrix in a purely algebraic way. This is accomplished by operating with (28) on Eq. (25). For the reflection amplitude $R_m=B_m/A_m$, we find

$$(29) \qquad R_{m+1}(k) = e^{i[\gamma_+(k)-\gamma_-(k)]} \frac{m+1/2-if(k)}{m+1/2+if(k)} R_m(k) \quad .$$

By solving (29) we can calculate $R_m(k)$ for a general su(1,1) scattering problem

$$(30) \quad R_m(k)=e^{i[\gamma_+-\gamma_-]m} \frac{\Gamma(m+1/2-if(k))\Gamma(m_0+1/2+if(k))}{\Gamma(m+1/2+if(k))\Gamma(m_0+1/2-if(k))} R_{m_0},$$

where R_m is an entire function of k.

Other families of exactly solvable potentials in one dimension can be described by the algebra so(2,2). These potentials and their S-matrices are discussed in detail in Ref. 14. For a systematic classification of algebraic potentials see Ref. 15. S-matrices that correspond to dynamical algebras so(m,n) in n dimensions were derived by using the Euclidean connection techniques. For example the partial S-matrix for the Coulomb potential can be derived based on its so(1,3) symmetry. The S-matrix for partial wave ℓ and momentum k for an so(1,3) symmetry is

$$S_\ell(k) = \frac{\Gamma(\ell+1+if(k))}{\Gamma(\ell+1-if(k))} e^{i\phi(k)} \quad , \tag{31}$$

where f(k) is again determined from the relation between H and the Casimir invariant of so(1,3). For the Coulomb problem $f(k) \propto 1/k$.

The methods can also be generalized to cases where there is an internal structure and several channels are coupled. S-matrix models which are based on so(1,3) were used in the description of heavy-ion reactions[16]. For applications to relativistic scattering see Ref. 17.

Conclusions

In this article we have reviewed algebraic methods in scattering problems. In one class of applications only the internal motion of the colliding particles is treated algebraically. Under certain conditions the S-matrix becomes a representation matrix of the group. In a complete algebraic approach the relative motion should also be treated algebraically. It was shown that for Lie algebraic potentials it is possible to calculate the S-matrix algebraically through the use of the Euclidean connection.

REFERENCES

1. Y. Alhassid and R.D. Levine, "Connection Between the Maximal Entropy and the Scattering Theoretic Analysis of Collision Processes", Phys. Rev. **A18** (1978) 89-116.
2. Y. Alhassid, Ph.D. Thesis (Jerusalem, 1979).
3. W. Miller, Jr., "Lie Theory and Special Functions", Academic, New York 1968.
4. R. Bijker and R.D. Amado "Algebraic-Eikonal Approach to Electron-Molecule Scattering: II. Rotational and Vibrational Excitations of LiF and KI", Phys Rev. **A34** (1986) 71-79.
5. Y. Alhassid and B. Shao, "Algebraic Eikonal Approach to Electron-Molecule Scattering: I. Generalized Formalism", Phys. Rev. **A46** (1992) 3978-3990.
6. F. Iachello, "Algebraic Methods for Molecular Rotation-Vibration Spectra", Chem. Phys. Lett. **79** (1981)

581-585; F. Iachello and R.D. Levine, "Algebraic Approach to Rotation-vibration Spectra: I. Diatomic Molecules", J. Chem. Phys. **79** (1983), 2515-2536.
7. Y. Alhassid and B. Shao, "Algebraic Eikonal Approach to Electron Molecule Scattering: II. Rovibrational Excitations", Phys. Rev. **A46**, (1992), 3991-4002.
8. R. Bijker and R.D. Amado, "Algebraic Eikonal Approach to Electron Molecule Scattering: III. Triatomic Molecules", Phys. Rev. **A37**, (1988), 1425-1437.
9. J.N. Ginocchio, T. Otsuka, R.D. Amado and D.A. Sparrow, "Medium Energy Probes and the Interacting Boson Model of Nuclei", Phys. Rev. C33 (1986) 247-259.
10. Y. Alhassid, F. Gürsey and F. Iachello, "Group Theory Approach to Scattering", Ann. Phys. (N.Y.) **148**, (1983) 346-380.
11. V. Bargman, Ann. Math. **48** (1947), 568;
G. Lindbland and B. Nagel, Ann. Inst. Henri Poincare **13**, (1970), 27.
12. Y. Alhassid, F. Gürsey and F. Iachello, "Group Theory Approach to Scattering: II. The Euclidean Connection", Ann. Phys. (N.Y.) 167 (1986) 181-200.
13. Y. Alhassid, J. Engel and J. Wu, "Algebraic Approach to the Scattering Matrix", Phys. Rev. Lett. **53**, (1984), 17-20.
14. J. Wu, Y. Alhassid and F. Gürsey, "Group Theory Approach to Scattering: IV. Solvable Potentials Associated with so(2,2)", Ann. Phys. (N.Y.) **196** (1989) 163-181.
15. N. Kamran and P.J. Olver, "Lie Algebras of Differential Operators and Lie Algebraic Potentials", J. Math and Anal. Appl. **145** (1990), 342-356.
16. Y. Alhassid and F. Iachello, "Algebraic Approach to Heavy-Ion Reactions", Nucl. Phys. **A501** (1989) 585-604.
17. Y. Alhassid, F. Gursey and F. Iachello, "Group Theory Approach to Relativistic Scattering", J. Phys. **A22** (1989) L947-951.

Current address: Center for Theoretical Physics, Sloane Physics Laboratory, Yale University, New Haven, Connecticut 06511, USA.

Contemporary Mathematics
Volume **160**, 1994

Exact Solutions to Operator Differential Equations

CARL M. BENDER

ABSTRACT. In this talk we consider the Heisenberg equations of motion $\dot{q} = -i[q, \mathcal{H}]$, $\dot{p} = -i[p, \mathcal{H}]$, for the quantum-mechanical Hamiltonian $\mathcal{H}(p, q)$ having one degree of freedom. It is a commonly held belief that such operator differential equations are intractable. However, a technique is presented here that allows one to obtain exact, closed-form solutions for huge classes of Hamiltonians. This technique, which is a generalization of the classical action-angle variable methods, allows us to solve, *albeit* formally and implicitly, the operator-differential equations of two anharmonic oscillators whose Hamiltonians are $\mathcal{H} = p^2/2 + q^4/4$ and $\mathcal{H} = p^4/4 + q^4/4$.

1. Introduction

The time-evolution equations

$$\dot{q} = \frac{\partial \mathcal{H}}{\partial p}, \qquad \dot{p} = -\frac{\partial \mathcal{H}}{\partial q}, \tag{1}$$

describe a classical dynamical system that evolves according to the Hamiltonian $\mathcal{H}(p, q)$. We assume that the solution to (1) satisfies the initial conditions $p(0) = p_0$, $q(0) = q_0$. Although (1) is equivalent to a single second-order differential equation, it is often possible to find a closed-form solution because the Hamiltonian $\mathcal{H}$ is a constant of the motion. Thus, we can in principle use the algebraic equation $\mathcal{H}(p, q) = E$ to solve for and eliminate one of the variables $p(t)$ or $q(t)$ in (1) and then to solve the resulting *first-order* differential equation for $q(t)$ or $p(t)$. For example, let us consider the Hamiltonian $\mathcal{H} = p^2/2 + V(q)$, for which Eq. (1) takes the form

$$\dot{q} = p, \qquad \dot{p} = -V'(q) \quad . \tag{2}$$

Subject Classification. Primary 81C05. Secondary 81D05, 81C10, 22E60.

Supported by US DOE Grant DE-FG02-91-ER40628.

This paper is in final form and no version of it will be submitted for publication elsewhere.

Now, solving $\mathcal{H}(p,q) = E$ for $p(t)$ and using the first equation of (2) gives $\dot{q} = \sqrt{2[E - V(q)]}$. This is a first-order separable equation whose implicit solution satisfying $q(0) = q_0$ is

$$F[q(t)] = t + F(q_0) \quad , \tag{3}$$

where

$$F(q) = \int^q \frac{dx}{\sqrt{2[E - V(x)]}} \quad . \tag{4}$$

The quantum equations of motion for the Hamiltonian $\mathcal{H}(p,q)$ are

$$\dot{q} = -i[q, \mathcal{H}] \,, \qquad \dot{p} = -i[p, \mathcal{H}] \,, \tag{5}$$

where the operators $q(t)$ and $p(t)$ satisfy the equal-time commutation relation $[q(t), p(t)] = i$. It is the noncommutivity of $q(t)$ and $p(t)$ that makes (5) difficult to solve if it is a nonlinear system. Only the special case of the harmonic oscillator $\mathcal{H} = p^2/2 + q^2/2$ gives rise to the easily solvable linear equations $\dot{q} = p$, $\dot{p} = -q$, whose exact solution is

$$q(t) = q_0 \cos t + p_0 \sin t \quad , \tag{6}$$

$$p(t) = p_0 \cos t - q_0 \sin t \quad , \tag{7}$$

We propose to solve operator equations that are nonlinear by constructing a quantum analogy to Eq. (3). Specifically, we will attempt to obtain a function $F(p,q)$ of the operators $p(t)$ and $q(t)$ that satisfies

$$-i[F(p,q), \mathcal{H}(p,q)] = \frac{d}{dt} F(p,q) = 1 \quad . \tag{8}$$

Note that F is not unique; $\bar{F} = F + \phi(\mathcal{H})$, where ϕ is an arbitrary function of the Hamiltonian, also satisfies (8). However, later on we will show how to find the simplest solution for F, which we will call the *minimal* solution. If such a function F can be found then of course the solution to (8),

$$F\left(p(t), q(t)\right) = t + F(p_0, q_0) \quad , \tag{9}$$

together with

$$\mathcal{H}\left(p(t), q(t)\right) = \mathcal{H}(p_0, q_0) \quad , \tag{10}$$

constitutes an exact *implicit* solution to the operator equations of motion (5). If we can then solve (9) and (10) simultaneously for $p(t)$ and $q(t)$ as functions of p_0, q_0, and t, we have an *explicit* solution to the equations of motion. Here are some simple examples:

Example 1 : $\mathcal{H} = (pe^q + e^q p)/2$. The operator equation of motion for q, $\dot{q} = e^q$, immediately suggests that $F(p,q)$ in (8) is actually a function of q only: $F(p,q) = -e^{-q(t)}$. The *explicit* solution to the operator equations of motion is

$$q(t) = -\log(e^{-q_0} - t)\,, \qquad p(t) = p_0 - t(p_0\, e^{q_0} + e^{q_0}\, p_0)/2\,.$$

Example 2 : $\mathcal{H} = pe^{aq}\,p$, where a is a constant. For this Hamiltonian, a function F satisfying (5) is

$$F = -\frac{1}{a}\frac{1}{\sqrt{\mathcal{H}}}\,p(t)\frac{1}{\sqrt{\mathcal{H}}} \quad .$$

The *explicit* solution to the operator equations of motion is

$$p(t) = -a\mathcal{H}(p_0, q_0)t + p_0\,, \quad q(t) = -\frac{1}{a}\ln[e^{-aq_0} - 2p_0 t + a^2\mathcal{H}(p_0, q_0)t^2]\,.$$

Example 3 : *Euler Hamiltonians.* We define an Euler Hamiltonian to be one in which the operator $p(t)$ is always accompanied by a multiple of $q(t)$; that is, $\mathcal{H} = \mathcal{H}(pq)$. In general, the operator equations of motion for Euler Hamiltonians can always be solved explicitly and in closed form.[1] This is true because the operator equations of motion for any Euler Hamiltonian can be written in the form

$$\dot{q} = q\sum_n \alpha_n (pq)^n\,, \qquad \dot{p} = -p\sum_n \alpha_n (qp)^n \quad .$$

The solution to these operator equations is

$$q = q_0 e^{\Sigma_n \alpha_n (p_0 q_0)^n t} \quad ,$$
$$p = p_0 e^{\Sigma_n \alpha_n (q_0 p_0)^n t} \quad .$$

For example, the operator equations for the Hamiltonian $\mathcal{H} = apq^2p + bqp^2q + c(pqpq + qpqp) + d(q^2p^2 + p^2q^2)$ have the form

$$\dot{q} = \alpha qpq\,, \qquad \dot{p} = -\alpha pqp \quad ,$$

where $\alpha = 2a + 2b + 4c + 4d$. The solution to these operator differential equations is

$$q(t) = q_0\, e^{\alpha p_0 q_0 t}\,, \qquad p(t) = p_0\, e^{-\alpha q_0 p_0 t} \quad .$$

A particularly interesting example of an Euler Hamiltonian is

$$\mathcal{H} = e^{c(pq+qp)}, \qquad (|c| < \pi/2) \quad .$$

The solution to the operator differential equations for this Hamiltonian is

$$q(t) = q_0\, e^{2t(\sin c)\, e^{2p_0 q_0 c}} \quad ,$$
$$p(t) = p_0\, e^{-2t(\sin c)\, e^{2q_0 p_0 c}} \quad .$$

Example 4 : $\mathcal{H} = p^\alpha q^{2\beta} p_\alpha$. For this Hamiltonian it is easy to see that a function F satisfying (5) is

$$F(p,q) = \frac{1}{4\alpha - 4\beta} \frac{1}{\sqrt{\mathcal{H}}}(pq + qp)\frac{1}{\sqrt{\mathcal{H}}} \quad . \tag{11}$$

Note that (11) ceases to exist when $\alpha = \beta$. However, the special case $\alpha = \beta$ gives an *Euler* Hamiltonian whose solution is discussed in Example 3 above. Solving $F(p,q) = t + F(p_0, q_0)$ simultaneously with $\mathcal{H}(p,q) = \mathcal{H}(p_0, q_0)$ can be complicated. However, a relatively simple case arises when $\alpha = 1$ and $\beta = N/2$. Now the explicit solution is

$$q(t) = \{[p_0 q_0 + (2-N)p_0 q_0^N\, p_0\, t]^{-1} p_0 q_0^N\, p_0 [q_0 p_0 + (2-N)p_0 q_0^N\, p_0 t]^{-1}\}^{1/(N-2)} ,$$
$$p(t) = \{[q(t)]^{-N/2} p_0 q_0^N\, p_0 [q(t)]^{-N/2} - \frac{1}{4}N(N-2)[q(t)]^{-2}\}^{1/2} .$$

Example 5 : $\mathcal{H} = \alpha p^\gamma + \beta q^{-\gamma}$. Here, the form of $F(p,q)$ is similar to that in Example 4:

$$F(p,q) = \frac{1}{2\gamma} \frac{1}{\sqrt{\mathcal{H}}}(pq + qp)\frac{1}{\sqrt{\mathcal{H}}} \quad .$$

2. Operator Basis Elements and Their Algebra

Our objective is now to describe a general procedure for obtaining an operator $F(p,q)$ that satisfies (8). To do so we introduce an operator basis. The basis elements $T_{m,n}$ are defined as the sum of all possible terms containing m factors of p and n factors of q multiplied by $m!n!/(m+n)!$. $T_{m,n}$ is thus a totally symmetric Hermitian object containing $(m+n)!/(m!n!)$ individual terms. For example,

$$T_{0,0} = 1 \quad ,$$
$$T_{0,3} = q^3 \quad ,$$
$$T_{1,1} = (pq + qp)/2 \quad ,$$
$$T_{2,1} = (p^2 q + pqp + qp^2)/3 \quad ,$$
$$T_{2,2} = (p^2q^2 + q^2p^2 + pqpq + qpqp + pq^2p + qp^2q)/6 \quad .$$

For two reasons, this seems to be a natural basis with which to express operators. First, $T_{m,n}$ contains positive powers of p and q, so it should be useful for constructing Taylor-like expansions of operators. Note that we can expand operators, regardless of whether they are symmetric. For example,

$$p^2q^3 = T_{2,3} - 3iT_{1,2} - 3T_{0,1}/2 \quad .$$

Second, $T_{m,n}$ satisfies an extremely useful set of commutation and anticommutation relations. Commuting with p or q has the effect of a lowering operator:

$$\begin{aligned} [T_{m,n}, p] &= in\, T_{m,n-1} \quad , \\ [T_{m,n}, q] &= -im\, T_{m-1,n} \quad . \end{aligned} \tag{12}$$

Anticommuting with p or q is analogous to applying a raising operator:

$$\begin{aligned} [T_{m,n}, p]_+ &= 2T_{m+1,n} \quad , \\ [T_{m,n}, q]_+ &= 2T_{m,n+1} \quad . \end{aligned} \tag{13}$$

The operator basis elements $T_{m,n}$ form a closed algebra under commutation and anticommutation; that is, the commutator or anticommutator of any two basis elements can be expressed as a sum of operator basis elements.

The basis elements have the crucial property that the totally symmetric operator $T_{m,n}$ can be completely reorganized using the commutation relation $[q, p] = i$ and recast in Weyl-ordered form

$$T_{m,n} = \frac{1}{2^m}\sum_{j=0}^{\infty}\binom{m}{j} p^j q^n p^{m-j} = \frac{1}{2^n}\sum_{k=0}^{\infty}\binom{n}{k} q^k p^m q^{n-k} \quad . \tag{14}$$

The operators $T_{m,n}$ have many more remarkable properties that have allowed them to play a central role in previous studies involving finite-element lattice approximations,[2] operator ordering,[3] and Hahn polynomials.[4] To summarize briefly, the operator $T_{m,m}$ can be expressed as a polynomial P of degree m in terms of the operator $T_{1,1}$:

$$T_{m,m} = P_m(T_{1,1}) \quad , \tag{15}$$

where

$$\begin{aligned} P_0(x) &= 1 \\ P_1(x) &= x \\ P_2(x) &= x^2 - 1 \quad , \\ P_3(x) &= x^3 - 5x \quad , \\ P_4(x) &= x^4 - 14x^2 + 9 \quad , \\ P_5(x) &= x^5 - 30x^3 + 89x \\ P_6(x) &= x^6 - 55x^4 + 439x^2 - 225 \quad . \end{aligned}$$

Equation (15) is reminiscent of the defining equation for Chebyshev polynomials $C_n(x)$: $\cos(n\theta) = C_n(\cos\theta)$.

The Hahn polynomials exhibit the following properties: First, they satisfy a three-term recursion relation:

$$P_n(x) = xP_{n-1}(x) - (n-1)^2 P_{n-2}(x) \quad .$$

Second, $P_n(x)$ satisfies a functional difference equation (rather than a differential equation):

$$(1-ix)P_n(x+2i) + (1+ix)P_n(x-2i) = (4n+2)P_n(x) \quad .$$

Third, the Hahn polynomials have a simple generating function $G(x)$:

$$\frac{e^{x\tan^{-1}t}}{\sqrt{1+t^2}} = \sum_{n=0}^{\infty} \frac{t^n}{n!} P_n(x) \quad .$$

Fourth, the Hahn polynomials are orthogonal on the line $-\infty < x < \infty$ with respect to the weight function $w(x) = sech(\pi x/2)/2$:

$$\int_{-\infty}^{\infty} dx\, w(x) P_m(x) P_n(x) = \delta_{m,n}(n!)^2 \quad .$$

It is easy to expand functions in terms of Hahn polynomials. For example,

$$e^{cx} = \sum_{n=0}^{\infty} \frac{(\tan c)^n}{n!\cos c} P_n(x) \qquad (|c| < \pi/2) \quad .$$

Finally, the Hahn polynomials can be expressed in terms of a generalized hypergeometric function:

$$P_n(x) = \sqrt{n!}\, i^n\, {}_3F_2(-n, n+1, (1-ix)/4; 1/2, 1; 1) \quad .$$

3. Minimal Solution of the Harmonic Oscillator

Now we return to the problem of obtaining a solution to (8). We represent $F(p,q)$ as an arbitrary sum of operator basis elements

$$F(p,q) = \sum_{m,n} \alpha_{m,n} T_{m,n} \quad , \tag{16}$$

where $\alpha_{m,n}$ are constants to be determined from the requirement in (8) that $-i[F(p,q), \mathcal{H}(p,q)] = 1$. To illustrate, we begin by finding $F(p,q)$ for the harmonic oscillator Hamiltonian $\mathcal{H}(p,q) = p^2/2 + q^2/2$. Equations (12) and (13) make the computation very easy:

$$\frac{1}{i}[T_{m,n}, \frac{1}{2}p^2] = \frac{1}{2i}[p, [T_{m,n}, p]]_+ = \frac{n}{2}[p, T_{m,n-1}]_+ = n\, T_{m+1,n-1} \quad . \tag{17}$$

Similarly,

$$\frac{1}{i}[T_{m,n}, \frac{1}{2}q^2] = -m\; T_{m-1,n+1} \quad . \tag{18}$$

Combining (16)-(18), we see that the commutation relation in (8) takes the form

$$\sum_{m,n} \alpha_{m,n}(nT_{m+1,n-1} - mT_{m-1,n+1}) = T_{0,0} \quad . \tag{19}$$

Hence, assuming completeness, we determine that the coefficients $\alpha_{m,n}$ satisfy the linear partial difference equation

$$(n+1)\alpha_{m-1,n+1} - (m+1)\alpha_{m+1,n-1} = \delta_{m,0}\,\delta_{n,0} \quad . \tag{20}$$

This partial difference relates pairs of coefficients $\alpha_{m,n}$. If we represent the coefficients as dots on an integer planar lattice then it is clear that next-to-nearest neighboring points on a diagonal whose slope is -1 are related by (20) (see Fig. 1). As we pointed out earlier, F is *not* uniquely determined. Thus, we are free to take the simplest particular solution to (20); namely, the one having the smallest possible number of nonvanishing coefficients $\alpha_{m,n}$. We call this the minimal solution. The minimal solution to (20) is given by

$$\alpha_{-2m-1,2m+1} = (-1)^m/(2m+1)\,, \qquad m = 0, 1, 2, 3, \ldots \quad ,$$

and $\alpha_{m,n} = 0$ for other values of m, n. Thus, the formula for $F(p,q)$ in (16) becomes

$$F(p,q) = \sum_{m=0}^{\infty} \frac{(-1)^m}{2m+1} T_{-2m-1,2m+1} \quad . \tag{21}$$

Observe that we are forced to generalize our initial assumption that, by analogy with Taylor series, our basis $T_{m,n}$ has $m, n \geq 0$. Apparently, a more accurate analogy is with Laurent series in which powers may be positive or negative. Fortunately, the formulas in (14) allow us to define $T_{m,n}$ when $m \geq 0$ and $n < 0$ and when $n \geq 0$ and $m < 0$. Moreover, the commutation and anticommutation relations in (12) and (13) continue to hold in this extended and singular basis.

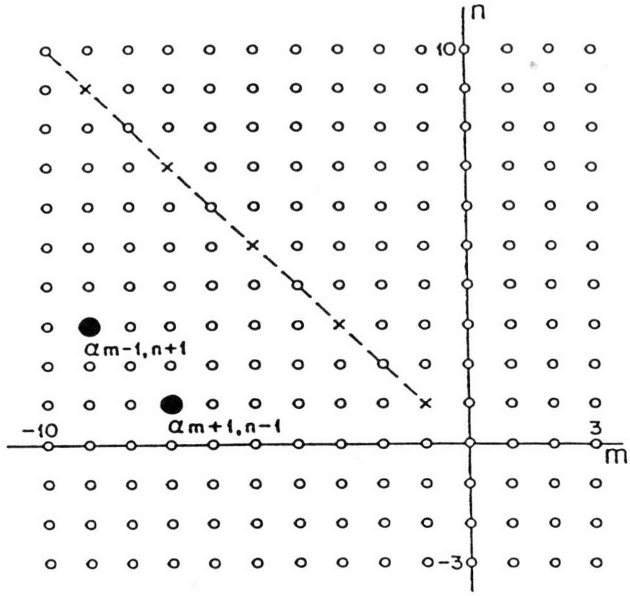

Figure 1: The partial difference equation (20) for the harmonic oscillator relates the coefficients $\delta_{m-1,n+1}$ and $\alpha_{m+1,n-1}$. the minimal solution consists of all $\alpha_{m,n}=0$ except $\alpha_{-2m-1,2m+1}$, $m=0,1,2,\ldots$, as indicated by crosses.

In order to understand the formal series in (21) we return to the exact solution to the harmonic oscillator in (6) and (7). We divide (6) by (7) and let $Z(t)=q(t)[p(t)]^{-1}$:

$$
\begin{aligned}
Z(t) &= (q_0 \cos t + p_0 \sin t)(p_0 \cos t - q_0\, sin\, t)^{-1} \\
&= (q_0 \cos t + p_0 \sin t)p_0^{-1}p_0(p_0 \cos t - q_0 \sin t)^{-1} \\
&= (Z_0 \cos t + \sin t)[(p_0 \cos t - q_0 \sin t)p_0^{-1}]^{-1} \\
&= \frac{Z_0 + \tan t}{1 - Z_0\, tan\, t} \quad . \qquad (22)
\end{aligned}
$$

Since $Z(t)$ is a function of Z_0, $Z(t)$ must commute with Z_0. Thus, we can treat (21) as a c-number algebraic equation and solve for tan t:
$\tan t = [Z(t) - Z_0]/[1 + Z(t)Z_0]$. Finally, taking the inverse tangent of this equation, we obtain

$$
arc \tan Z(t) = t + arc \tan Z_0 \quad . \qquad (23)
$$

This equation is an instance of (9), so we identify

$$F(p,q) = arc\ \tan\{q(t)[p(t)]^{-1}\} \quad . \tag{24}$$

Compare (21) and (24). Note that the coefficients α in (21) correspond exactly with the Taylor expansion of $arc\ \tan x = \Sigma_{n=0}^{\infty} \frac{(-1)^n}{2n+1} x^{2n+1}$. However, these two functions F are not the same even though they satisfy the same commutation relation in (8). As expected, they differ by a function of the Hamiltonian $\mathcal{H}$:

$$F_{\text{Eq. (21)}} - F_{\text{Eq. (24)}} = \frac{i}{2} \sum_{n=0}^{\infty} E_{2n} \mathcal{H}^{-2n-1} = \frac{i}{2\mathcal{H}} \int_0^{\infty} ds \frac{e^{-s}}{\cosh(\frac{s}{2\mathcal{H}})} ,$$

where E_2n are Euler numbers. (The first few Euler numbers are $E_0 = 1$, $E_2 = -1$, $E_4 = 5$, $E_6 = -61$, $E_8 = 1385$, and so on.)

Even more important, the physical meaning of $F(p,q)$ is now more evident: Classically, the harmonic oscillator describes an orbit in phase space which is a circle of radius $p_0^2 + q_0^2$. The singular quantity $arc\ \tan[q(t)/p(t)]$ is the *angle* θ of a point on this circle. Apparently, (21) is the generalization of the angle coordinate from a classical (c-number) theory to a quantum (operator) theory. This result clearly illustrates that the operator methods used here generalize the notion of classical action-angle variables to the realm of quantum mechanics.

4. Minimal Solution for Anharmonic Oscillator $\mathcal{H} = p^2/2 + q^4/4$

Next, we consider the anharmonic oscillator whose Hamiltonian is given by $\mathcal{H} = p^2/2 + q^4/4$. To find $F(p,q)$ we first evaluate the commutator

$$\begin{aligned} -i[T_{m,n}, q^4] &= -m(q^3 T_{m-1,n} + q^2 T_{m-1,n} q + q T_{m-1,n} q^2 + T_{m-1,n} q^3) \\ &= -2m(q^2 T_{m-1,n+1} + T_{m-1,n+1} q^2) \\ &= -m\,[q,[q,T_{m-1,n+1}]_+]_+ - m\,[q,[q,T_{m-1,n+1}]]_+ \\ &= -m(2[q,T_{m-1,n+2}]_+ + (m-1)i[q,T_{m-2,n+1}]) \\ &= -4m T_{m-1,n+3} + m(m-1)(m-2) T_{m-3,n+1} \,. \end{aligned} \tag{25}$$

Combining the result in (17) with that in (25), and assuming that the operators $T_{m,n}$ form a complete basis gives a recursion relation for $\alpha_{m,n}$

(26)

$$(n+1)\alpha_{m-1,n+1} - (m+1)\alpha_{m+1,n-3} + \frac{(m+3)(m+2)(m+1)}{4} \alpha_{m+3,n-1} = \delta_{m,0}\delta_{n,0} \,.$$

This partial difference equation relates triples of points on the integer lattice in Fig. 2 whose points correspond to coefficients $\alpha_{m,n}$. A careful analysis of (26) shows that a minimal solution exists consisting of certain nonzero values of $\alpha_{m,n}$. These nonzero values of $\alpha_{m,n}$ correspond to points in Fig. 2 lying in the quadrant $m < 0$, $n > 0$ and forming a triangular network as indicated in the figure.

A first step in solving (26) involves mapping the triangular network of nonvanishing values of $\alpha_{m,n}$ onto a triangular domain. On this domain the partial difference equation (26) works in much the same fashion as the difference equation that generates the binomial coefficients in Pascal's triangle. The relevant transformation of the independent variables is

$$M = -(n + 2m)/6 , \qquad N = (n - m)/6 \quad .$$

We define a new independent variable $A_{M,N}$ by

$$A_{M,N} = \alpha_{m-1,n+1} \quad ,$$

along with the constraint that $A_{M,N} = 0$ for $M < 0$, $N < 0$, and for $M > N$.

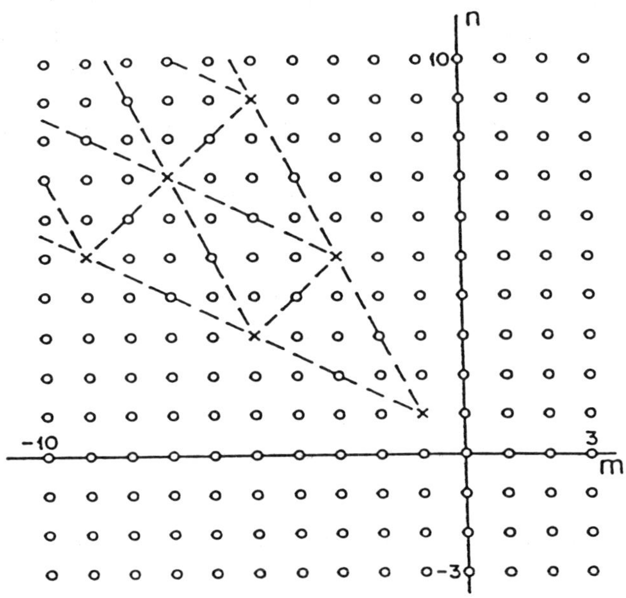

Figure 2: Triplets of points related by the partial difference equation (26) for the anharmonic oscillator with $\mathcal{H} = \frac{1}{2}p^2 + \frac{1}{4}q^4$. The minimal solution consists of nonvanishing values of $\alpha_{m,n}$ indicated by crosses; all other $\alpha_{m,n}$'s vanish.

The advantage of the M, N variables over the m, n variables is that the partial difference equation is first order:

(27)

$$(4N-2M+1)A_{M,N} + (2N+2M-1)A_{M,N-1}$$
$$-\frac{1}{4}(2N+2M-1)(2N+2M-2)(2N+2M-3)A_{M-1,N-1} = \delta_{M,0}\delta_{N,0}\,.$$

A further transformation, this time of the dependent variable, reduces the partial difference equation in (27) to one whose coefficients are linear functions of M and N. We define $B_{M,N} = 2^{-N}\Gamma(1/2)A_{M,N}/\Gamma(M+N+1/2)$ and the difference equation becomes

$$(4N - 2M + 1)B_{M,N} + B_{M,N-1} - (N + M - 1) = \delta_{M,0}\delta_{N,0}\,. \tag{28}$$

The simplest way to express a general element $B_{M,N}$ is in terms of a generating function; to wit, we define the generating function

$$g(x,y) = \sum_{N=0}^{\infty}\sum_{M=0}^{N} B_{M,N}x^M y^N \quad .$$

From (28) we can derive a linear first order partial differential equation satisfied by $g(x, y)$:

$$-x(2 + xy)g_x + y(4 - xy)g_y + (1 + y - xy)g = 1 \quad . \tag{29}$$

Using the method of characteristics we cna find the unique solution to this partial differential equation that satisfies the initial condition

$$g(0,y) = \sum_{N=0}^{\infty}(-y/4)^N\frac{\Gamma(5/4)}{\Gamma(N + 5/4)} = e^{-y/4}\int_0^1 du\, e^{u^4 y/4} \quad .$$

The solution is

(30)

$$g(x,y) = \int_0^1 \frac{du}{2\sqrt{(u)}\,(1 - xyu)}\exp\left(\frac{2}{3x^2y} - \frac{1}{x} + \frac{(1 - xy)^{3/2}(xyu - 2/3)}{x^2y(1 - xyu)^{3/2}}\right)\,.$$

In terms of the generating function $g(x, y)$,

$$B_{M,N} = \frac{1}{M!N!}\left(\frac{\partial}{\partial x}\right)^M\left(\frac{\partial}{\partial y}\right)^N g(x,y)\bigg|_{x=y=0}\,. \tag{31}$$

Using this expression for the coefficients $B_{M,N}$ in the formula (16) for $F(p,q)$ gives a complete and exact minimal solution to the Heisenberg operator differential equations for the anharmonic oscillator with the Hamiltonian $\mathcal{H} = p^2/2 + q^4/4$. The operator F is given by

$$F(p,q) = \sum_{N=0}^{\infty} \sum_{M=0}^{N} 2^N \frac{\Gamma(M+N+1/2)}{\Gamma(1/2)} B_{M,N} T_{-2N-2M-1,4N-2M+1} \, . \tag{32}$$

5. Minimal Solution for Anharmonic Oscillator $\mathcal{H} = p^4/4 + q^4/4$

Next, we consider the anharmonic oscillator whose Hamiltonian is given by $\mathcal{H} = p^4/4 + q^4/4$. This is the natural next step in our program of solving increasingly complex systems of quantum operator differential equations: $\mathcal{H} = p^2/2 = q^2/2$ gives rise to a two-term partial difference equation, $\mathcal{H} = p^2/2 = q^4/4$ gives rise to a three-term partial difference equation, and, as we will see, $\mathcal{H} = p^4/4 = q^4/4$ gives rise to a four-term partial difference equation. Specifically, the partial difference equation for the coefficients $\alpha_{m,n}$ in (16) has the form

$$\begin{aligned} &(n+1)\alpha_{m-3,n+1} + \frac{1}{4}(m+3)(m+2)(m+1)\alpha_{m+3,n-1} \\ &-(m+1)\alpha_{m+1,n-3} - \frac{1}{4}(n+3)(n+2)(n+1)\alpha_{m-1,n+3} = \delta_{m,0}\delta_{n,0} \, . \end{aligned} \tag{33}$$

This partial difference equation relates quartets of points, which lie at vertices of rectangles, on the integer lattice in Fig. 3 whose points correspond to coefficients $\alpha_{m,n}$. Examining (33) we see that a minimal solution exists with nonzero values of $\alpha_{m,n}$ lying in the quadrant $m < 0$, $n > 0$, and forming a wedge-shaped network as indicated in Fig. 3.

To solve (33) we use the transformation

$$N = (n-m)/8 \, , \qquad M = -(n+m)/4 \, ,$$

which maps points in the m,n plane for which the coefficient $\alpha_{m,n}$ is nonzero into a triangular lattice for which N runs from 0 to infinity and M runs from 0 to $2N$. We then define a new dependent variable $C_{M,N}$ by

$$C_{M,N} = \alpha_{m-3,n+1} \quad ,$$

along with the constraint that $C_{M,N} = 0$ for $M < 0$, $N < 0$, and $M > 2N$. $C_{M,N}$ satisfies a first order partial difference equation which can be simplified further by the transformation

$$D_{M,N} = 4^{-M}\frac{\Gamma(N-M/2+3/4)\Gamma(N-M/2+5/4)}{\Gamma(N+M/2+3/4)\Gamma(N+M/2+5/4)}C_{M,N}\,.$$

The equation satisfied by $D_{M,N}$ has linear coefficients:

(34)

$$\begin{aligned}&\left(N+\frac{M}{2}+\frac{1}{4}\right)D_{M,N}-\left(N-\frac{M}{2}+\frac{1}{2}\right)D_{M-1,N}\\+&\left(N-\frac{M}{2}-\frac{1}{4}\right)D_{M,N-1}-\left(N+\frac{M}{2}-\frac{1}{2}\right)D_{M-1,N-1}=\frac{1}{4}\delta_{M,0}\delta_{N,0}\,.\end{aligned}$$

A general element D_M,N in this array can be expressed in terms of a generating function $h(x,y)$, where

$$h(x,y)=\sum_{N=0}^{\infty}\sum_{M=0}^{2N}D_{M,N}x^My^N \quad .$$

From (34) we can derive a first order linear partial differential equation satisfied by $h(x,y)$:

$$\frac{x}{2}\frac{1+x}{1-x}h_x+y\frac{1+y}{1-y}h_y+\frac{1+3y-4xy}{4(1-x)(1-y)}h=\frac{1}{4(1-x)(1-y)}\,. \tag{35}$$

The unique solution to this partial differential equation that satisfies the initial condition

$$h(0,y)=\sum_{N=0}^{\infty}(-y)^N\frac{\Gamma(5/4)\Gamma(N+3/4)}{\Gamma(3/4)\Gamma(N+5/4)}=\int_0^1\frac{du}{\sqrt{(1+y)^2-4yu^4}}$$

is

$$h(x,y)=\int_0^1\frac{du}{\sqrt{(1+y)^2-4yu^4(1+x-xu^2)^2}} \quad . \tag{36}$$

In terms of this generating function,

$$D_{M,N}=\frac{1}{M!N!}\left(\frac{\partial}{\partial x}\right)^M\left(\frac{\partial}{\partial y}\right)^N h(x,y)\bigg|_{x=y=0} \quad .$$

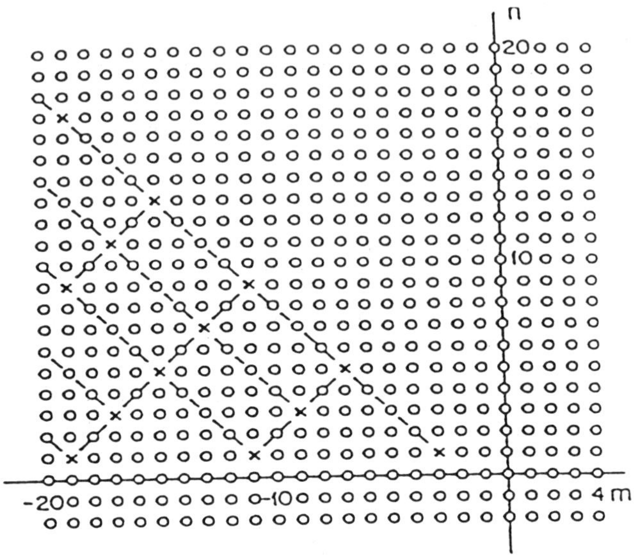

Figure 3: Quartets of points related by the partial difference equation (33) for the anharmonic oscillator with $\mathcal{H} = \frac{1}{4} + \frac{1}{4}q^4$. The minimal solution consists of nonvanishing values of $\alpha_{m,n}$ indicated by crosses; all other $\alpha_{m,n}$'s vanish.

Using this expression for the coefficients $D_{M,N}$ in the formula (16) for $F(p,q)$ gives a complete and exact minimal solution to the Heisenberg operator differential equations for the anharmonic oscillator whose Hamiltonian is $\mathcal{H} = p^4/4 + q^4/4$. The operator $F(p,q)$ satisfying (8) is

$$
\begin{aligned}
F(p,q) = \sum_{N=0}^{\infty} \sum_{M=0}^{2N} 4^M \frac{\Gamma(N+M/2+3/4)\Gamma(N+M/2+5/4)}{\Gamma(N-M/2+3/4)\Gamma(N-M/2+5/4)} \\
\times D_{M,N} T_{-2N-4M-3,4N-2M+1} \, .
\end{aligned}
\tag{37}
$$

6. Concluding Remarks

The progress we have reported here is formal because the operator $F(p,q)$ can be extremely singular as it contains in many cases arbitrary powers of $1/p(t)$ and/or $1/q(t)$. If $F(p,q)$ did exist as an operator in a Hilbert space, then by virtue of (8), $U \equiv \exp[i\lambda F(p,q)]$ would be an energy raising operator. That is, if $\mathcal{H}|E>= E|E>$, then $\mathcal{H}(U|E>) = (E+\lambda)(U|E>)$. Clearly, for Hamiltonians with discrete spectra, U must map states out of the Hilbert space except for special discrete values of λ. An obvious direction for future research is to find a way to identify these special values of λ and thereby to use the operator solution to a quantum theory to compute such quantities as eigenvalues and also unequal-time commutation relations. One should also consider applying the ideas presented here to systems having more than one degree of freedom and especially to models that exhibit quantum chaos. It was our sole intention here to point out that operator differential equations are not inaccessible to exact analytic methods and that in many cases they can actually be solved in closed form.

References

1. C. M. Bender and G. V. Dunne, Phys. Letters **B200**, 520 (1988).
2. C. M. Bender, L. R. Mead, and S. S. Pinsky, Phys. Rev. Lett. **56**, 2445 (1986).
3. C. M. Bender and G. V. Dunne, J. Math. Phys. **29**, 1727 (1988).
4. C. M. Bender, L. R. Mead, and S. S. Pinsky, J. Math. Phys. **28**, 509 (1987). For a review of the Hahn polynomials see W. Hahn, Math. Nachr. **2**, 4 (1949), N. M. Atakishiyev and S. K. Suslov, J. Phys. A **18**, 1583 (1985), and R. Askey, J. Phys. A **18** L1017 (1985).

Department of Physics, Washington University, St. Louis, MO 63130
Current Address: Department of Physics, Washington University, St. Louis, MO 63130
E-mail address: cmb@wuphys.wustl.edu

Contemporary Mathematics
Volume **160**, 1994

The Algebra of Tensor Operators for the Unitary Groups

L.C. Biedenharn

Abstract

The algebra of tensor operators for the unitary groups–a generalization of the concept of universal enveloping algebra–is defined and developed. The canonical classification of all irreducible unit-tensor operator components is discussed in terms of Gel'fand-Weyl and operator patterns, and also related to the concept of characteristic null space. Two open problems are sketched.

1 Introduction

The purpose of the present paper is to discuss the algebra of tensor operators associated to a Lie group. Our aim is to show that this concept is a far-reaching generalization of the concept of the enveloping algebra of a Lie group, capable of unifying and extending the representation theory of Lie groups and especially the physical applications of the theory.

The origin of this concept stems from quantum physics. The development and applications of quantum physics since its inception has been strongly influenced by symmetry considerations, and (directly

1991 *Mathematics Subject Classification.* Primary 20G05, Secondary 22E46, 81R05.

This paper is in final form and no version of it will be submitted for publication elsewhere.

or indirectly) by group theory. The paradigm for applying symmetry techniques in quantum physics has been the exploitation of rotational symmetry in the form of the quantal theory of angular momentum [5] (QTAM) involving the Lie group $SU(2)$. A key concept in such applications has been that of tensor operators [5], a concept developed specifically for QTAM. A basic problem has been *to extend*, as far as possible, *to an arbitrary Lie group the tensor operator concept* so decisive for QTAM. The developments in physics stimulated by this problem–originating in the work of Weyl, Heisenberg, Wigner and Racah–have led to results (in the physics literature) anticipating and/or independently obtaining similar results by mathematicians in Lie group theory (Harish-Chandra, Kostant, Weil, Borel). This points to a language barrier in the physics/mathematics discourse which, it is hoped, will not be too severe in the present discussion (necessarily based on the viewpoint of physics). For definiteness, the discussion will consider only the family of unitary groups $U(n)$, restricted to linear unitary integral representations.

2 The Concept of a Tensor Operator

(a) Let us consider the unitary group $SU(n)$ having the Lie algebra A_{n-1} in Cartan's notation. The elements λ in the set of all inequivalent unitary irreps (irreducible representations) can be uniquely labelled by a *Young frame*, denoted by $[m] \equiv [m_{1n}, m_{2n}, \cdots, m_{nn} = 0]$, where the m_{in} are integers obeying $m_{in} \geq m_{i+1\ n}$.

Let us now introduce a very special Hilbert space–called (by Gel'fand [10]) *model space*–for our system. This space is an abstraction from physical space and is defined to contain only symmetry degrees of freedom, suppressing all other degrees of freedom.

Let us denote by $V([m])$ the vector space carrying the irrep $[m]$.

Model space M is then defined as the direct sum:

$$M \equiv \bigoplus_{[m] \in \text{Irrep}} V([M]), \tag{2.1}$$

where each irrep $[m]$ occurs *precisely once.*

(At this point we introduce a simplifying technical device for $SU(n)$. Let us drop the unimodular restriction in model space (that is, $m_{nn} =$

0), and instead let m_{nn} be any integer. To return to $SU(n)$ we impose on the sum in (2.1) the equivalence relation: $[m_{in} + k] \simeq [m_{in}]$, restricting the sum to equivalence classes. This allows one to formally use $U(n)$ for various manipulations, but still remain in $SU(n)$ in actuality. Note, however, that $U(n)$ itself can be used as the symmetry group.)

(b) *Definition of a Tensor Operator*

A tensor operator is a linearly independent *set* of linear operators $\{T_i\}, T_i : M \to M$, with each T_i having an involution (Hermitian adjoint), $* : T_i \to T_i^*$, subject to the requirement that the set obeys the *equivariance condition:*

$$g(T_i) \equiv U_g^{-1} T_i U_g = \sum_j D_{j,i}(g) T_j. \tag{2.2}$$

Here $g(T_i)$ denotes an action by the group element $g \in G$ on the tensor operator $T_i; U_g$ is the unitary transformation induced by g on model space $M, g : M \to U_g M$, and D is a unitary representation of G. Expressed in words, the equivariance condition asserts that the action by g on the tensor operator T_i is conjugation by $U(g)$ such that the set $\{T_i\}$ carries a representation $D(g)$ of the symmetry group G. The individual T_i are called the *components* of the tensor operator.

An *irreducible tensor operator* is defined to be a tensor operator for which the representation in (2.2) is irreducible. Thus an irreducible tensor operator may be labelled *in part* by its associated irrep λ, that is, the set $\{T_i\}$ may be labelled $T^{(\lambda)}$.

Remarks:

(1) Tensor operators form an algebra: $\boldsymbol{T} \times \boldsymbol{T} \to \boldsymbol{T}$ with a composition ($\times$) that carries two tensor operators (sets) into a sum of tensor operators. This *algebra of tensor operators* is an associative, unital algebra with an involution.

(2) A positive, semi-definite, quasi-norm can be defined which, for each irreducible tensor operator $T^{(\lambda)}$ maps $T^{(\lambda)*} \times T^{(\lambda)} \to \mathcal{I} \equiv$ invariant positive operator.

(3) *Unit tensor operators,* $\hat{T}^{(\lambda)}$, can now be defined as a re-normalization of the tensor operator $T^{(\lambda)}$ such that: $\hat{T}^{(\lambda)*} \times \hat{T}^{(\lambda)} \to$ projection operator (eigenvalues 1 or 0 on irrep sub-spaces of M). Since unit tensor operators are bounded, it follows (from

the Gel'fand-Naimark theorem) that the components of unit tensor operators belong to a C^* algebra.

(4) The unit tensor operators form a linear basis for all tensor operators with invariant operators as scalar multipliers.

(5) The *elementary* unit tensor operators are an *algebraic* generating set for all tensor operators. (Elementary means an irrep having a Young frame of the form: $[\underbrace{1\cdots 1}_{k}\underbrace{0\cdots 0}_{n-k}], k = 1, 2, \cdots n-1$.) All elementary unit tensor operators for $SU(n), (n \geq 2)$ are explicitly constructible by an algorithm known as the pattern calculus [3].

(6) The *universal enveloping algebra* is a sub-algebra of the algebra of tensor operators, characterized by (a) an action on M that is diagonal (on the irrep labels) in M and (b) a quasi-norm that is polynomial in the (generalized) Casimir invariants.

(7) The tensor operator structure can be extended to quantum groups [6].

3 The Classification Problem for Tensor Operators

The problem of classifying all tensor operators is both fundamental and difficult. The equivariance condition (2.1) greatly helps in reducing the problem to more manageable proportions. For the irreducible tensor operators we have already seen that (2.2) allows a partial classification, the assignment of the irrep label λ.

For the unitary group $U(n)$, the components of the irreducible tensor operator $T^{(\lambda)}$ can be labelled by a *Gel'fand-Weyl pattern* which provides a canonical labelling for carrier space vectors in the irrep λ and hence a labelling for an individual operator (component) in the set $T^{(\lambda)}$. A Gel'fand-Weyl pattern (M) is a triangular pattern of integers:

$$(M) \equiv \begin{pmatrix} M_{1n} & M_{2n} & \cdots & M_{n-1,n} & M_{nn} \\ & M_{1,n-1} & \cdots & M_{n-1,n-1} & \\ & & \vdots & & \\ & & M_{11} & & \end{pmatrix} \tag{3.1}$$

where the integers M_{ij} obey the *betweenness conditions:* $M_{ij} \geq M_{i,j-1} \geq M_{i+1,j}, 1 \leq i,j \leq n$. (Here the irrep λ is specified by the top row which denotes the Young frame $[M] = [M_{1n} \cdots M_{nn}]$.) The Gel'fand-Weyl pattern (M) encodes the information that the component $T(M)$ of the tensor operator $T^{(\lambda)}$ is: (1) a component of a tensor operator belonging to the $U(n)$ irrep $[M_{1n} \cdots M_{nn}]$; (2) a component of a tensor operator belonging to the $U(n-1)$ irrep $[M_{1,n-1} \cdots M_{n-1,n-1}]$, acting in a sub-space of M; $\cdots$; , (n) a component of a tensor operator belonging to the $U(1)$ irrep M_{11} acting on a subspace of M. (That is, the elements of the tensor operator $T^{(\lambda)}$ are classified by irrep labels in the canonical sub-group chain $U(n) \supset U(n-1) \supset \cdots \supset U(1)$ using the Weyl branching theorem.)

Remark:

> The Gel'fand-Weyl pattern determines the *weights* (W_i) assigned to vectors carrying the irrep $[M]$, by the relation: $W_i = \Sigma_j M_{j,i} - \Sigma_j M_{j,i-1}$, where $M_{ij} \equiv 0$ if $i > j$. Note that the labelling by a Gel'fand-Weyl pattern uniquely and canonically splits all multiplicity in the weight space labelling.

This classification of tensor operators by Gel'fand-Weyl patterns is far from exhaustive. It is easily seen, for example, that multiplying all components of a tensor operator $T^{(\lambda)}$ by an invariant operator leaves the Gel'fand-Weyl patterns (M) of $T^{(\lambda)}$ unchanged. To eliminate this difficulty, one classifies *irreducible unit tensor operators* (defined in §2 Remark (3), §2 above, and denoted by $\hat{T}^{(\lambda)}$).

Even with this modification the classification is still not complete. To refine the classification further one introduces a labelling based on the *change* in the irrep labels of model space M induced by the action of the irreducible unit tensor operator $\hat{T}^{(\lambda)}$– these are the *shifts* Δ_i. To define the shifts Δ_i precisely, one uses the relation: $\Delta_i \equiv m_{in}^{final} - m_{in}^{initial}$, where $[m^{final}]$ and $[m^{initial}]$ are $U(n)$ irrep labels in M with the action $\hat{T}(M): \ V([m^{initial}]) \to V([m^{final}])$.

For a tensor operator belonging to the irrep $\lambda \equiv [M_{1n} \cdots M_{nn}]$, the shift values are constrained by group theory (the Kronecker product). It is well-known [8] that the *possible* (as opposed to actual) shift values for a tensor operator $T^{(\lambda)}$ acting on an arbitrary irrep in model space are *precisely the weights of the irrep* λ, *including multiplicity.* (The multiplicities are the Kostka numbers [14].) The *actual* shift values are determined by the triple of irreps, $[m^{initial}], \lambda$ and $[m^{final}]$, with multiplicities given by the Littlewood-Richardson numbers [14]. (This will be discussed in §5 below.)

The shifts $\Delta = \{\Delta_i\}$ are the operator space analog, for an irreducible tensor operator $T^{(\lambda)}$, of the weights of a representation in group theory. Just as the weights of a vector in an irrep space do not always characterize a unique vector, so too the shifts of the unit tensor operator component $\hat{T}(M)$ do not in general categorize a unique operator.

4 Operator Patterns

Significant progress toward fully resolving the classification problem of §3 can be made by exploiting further the analogy that exists between weight space $\{W_i\}$ and the shifts $\{\Delta_i\}$ of operator space. To do so, let us introduce the notion of an *operator pattern* (Γ).

An operator pattern (Γ) is a triangular array of integers Γ_{ij}:

$$(\Gamma) \equiv \begin{pmatrix} \Gamma_{1n} & \Gamma_{2n} & \cdots & \Gamma_{n-1,n} & \Gamma_{nn} \\ & \Gamma_{1,n-1} & \cdots & \Gamma_{n-1,n-1} & \\ & & \vdots & & \\ & & \Gamma_{11} & & \end{pmatrix} \tag{4.1}$$

where (a) the integers Γ_{ij} obey the *betweenness conditions*: $\Gamma_{i,j} \geq \Gamma_{i,j-1} \geq \Gamma_{i+1,j}$ and (b) for the unit tensor operator $\hat{T}^{(\lambda)}$ with $\lambda = [M_{1n}, \cdots, M_{nn}]$ we have: $\Gamma_{i,n} = M_{i,n}$.

Remarks:

(1) The shifts Δ_i associated to the unit tensor operator $\hat{T}(\Gamma)$ are given by:

$$\Delta_i = \sum_j \Gamma_{j,i} - \sum_j \Gamma_{j,i-1} \quad (\text{with } \Gamma_{ij} \equiv 0 \text{ if } i > j) \tag{4.2}$$

(This shift rule for operators patterns is in precise analogy to the weight rule for Gel'fand-Weyl patterns.)

(2) An operator pattern is structurally a Gel'fand-Weyl pattern of integers since both obey the betweenness conditions and determine 'weights', *but (unlike a Gel'fand-Weyl pattern) for an operator pattern there is no group-subgroup implication.* Put differently, there is *no analog* for an operator pattern of the Weyl branching theorem encoded in the structure of a Gel'fand-Weyl pattern.

(3) Note that for a unit tensor operator component of $\hat{T}^{(\lambda)}$, classified both by a Gel'fand-Weyl pattern (M) and an operator pattern (Γ), the irrep label $\lambda = [M_{1n}, M_{2n}, \cdots, M_{nn}] = [\Gamma_{1n}, \Gamma_{2n} \cdots, \Gamma_{nn}]$ with $\Gamma_{in} = M_{in}$. That is, (Γ) and (M) share a common top row (the Young frame λ). Since inverting a triangular pattern of integers leaves the betweenness relations unchanged, it is economical notationally to join the patterns (M) and (Γ) *inverted* as a single diamond-shaped pattern sharing a common largest row.

Summary of §3, §4: Irreducible unit tensor operator components for the simple Lie group $SU(n)$ can be labelled by three arrays of integers, $\hat{T}^{(\lambda)}((M), (\Gamma))$:

(a) an irrep (a Young frame): λ,

(b) a Gel'fand-Weyl pattern (M), and

(c) an operator pattern (Γ), with $[M_{1n} \cdots M_{nn} = 0] = [\Gamma_{in} \cdots \Gamma_{nn} = 0] = \lambda$.

Remarks:

(1) The labelling given in (a) and (b) is *canonical* (natural), that is, independent of any arbitrary choices (within equivalence). The labelling in (c) is canonical for the shifts $\Delta_i(\Gamma)$, and is conjectured (see below) to be canonical otherwise (that is, within a $\{\Delta_i\}$ multiplicity set).

(2) For an alternative approach to normalizing and labelling tensor operators see [9].

5 Characteristic Null Space

The invariant quasi-norm associated with a unit tensor operator $\hat{T}^{(\lambda)}$ is a projection operator on model space M. Let us define the *characteristic null space* of the unit tensor operator $\hat{T}^{(\lambda)}$ to be the set of $U(n)$ irreps μ such that all vectors in the irrep space $V([\mu])$ are annihilated by all components $\hat{T}_i^{(\lambda)}$ of $\hat{T}^{(\lambda)}$. It follows by construction that the characteristic null space of the unit tensor operator $\hat{T}^{(\lambda)}$ coincides with the null space of the invariant quasi-norm.

Consider now the set of distinct irreducible unit tensor operators $\hat{T}^{(\lambda)}$ having the same shift labels Δ. It can be shown [14] that the *characteristic null spaces* $\{\mathcal{N}_i\}$ *associated with this set are simply ordered by inclusion:* $\mathcal{N}_1 \supset \mathcal{N}_2 \supset \cdots$. If the number of distinct $\mathcal{N}_i$ equals the Kostka number for the multiplicity of the set $\{\hat{T}^{(\lambda)}(\Delta)\}$ then it follows that a labelling by characteristic null spaces canonically splits the multiplicity for the set $\{\hat{T}^{(\lambda)}(\Delta)\}$.

THEOREM [**2**]: *For irreducible unit tensor operators* $SU(3)$ *the characteristic null spaces are all distinct and ordered by inclusion. There exists a canonical splitting of all multiplicity in* $SU(3)$ *classifying uniquely all components of all irreducible unit tensor operators.*

This is a very satisfactory result for $SU(3)$, and *a posteriori* for $SU(2)$. An interesting extension is this:

THEOREM [**8**]: *The tensor operator algebra for* $SU(3)$ *can be embedded in the universal enveloping algebra of* $SO(6,2)$, *a non-compact realization of* D_4. *This algebra has no proper two-sided ideal.*

Unfortunately, the classification by characteristic null space cannot succeed in general

THEOREM [**1**]: *For unit tensor operators in* $SU(n)$, *with* $n \geq 4$, *there exist identical characteristic null spaces for distinct irreducible unit tensor operators in the same* $\{\hat{T}^{(\lambda)}(\Delta)\}$ *multiplicity set.*

6 Limit Relations

Important progress in the analysis of tensor operator algebras has been achieved in the last few years by exploiting a variant of the Borel-Weil construction [7] of irreps. The Borel-Weil construction induces an irreducible representation from a one-dimensional representation of the Cartan maximal toroid. The variant we shall exploit [11] uses the Borel-Weil procedure recursively, generating a $U(n)$ irrep from an irrep of the sub-group $U(n-1) \times U(1)$.

The irreps generated by this procedure are (to within an easily determined normalization) the tensor product of a totally symmetric (boson) $U(n-1)$ irrep and a generic $U(n-1) \times U(1)$ irrep (recursion hypothesis). The tensor product is effected by $U(n-1)$ $3j$-coefficients (matrix elements of $U(n-1)$ irreducible unit tensor operators).

The significance of this construction can be seen by remarking that (a) these $U(n)$ irreps are constructed algebraically *entirely by* $U(n-1)$ *tensor operator structures* and (b) the method is well-adapted to investigating, order-by-order, the limit $m_{nn} \to$ large negative integer.

THEOREM **[12]**: *Matrix elements of the $U(n)$ irreducible unit tensor operator $\hat{T}((M),(\Gamma))$ labelled by the Gel'fand pattern (M) and the operator pattern (Γ) obey the limit relation:*

$$\lim_{m_{nn}\to-\infty}\left(\hat{T}((M),(\Gamma))\right) = \prod_{i=1}^{n-1}\left(\delta_{\Gamma_{i,n-1}}^{m_{i,n-1}}\right) \cdot \begin{pmatrix}\text{Invariant}\\ \text{operator}\end{pmatrix} \cdot \hat{T}((M'),(\Gamma')) \tag{6.1}$$

where it is understood that (a) matrix elements are to be taken in model space (so that m_{nn} is well defined) (b) the invariant operator is uniquely defined, and (c) the (M') and (Γ') patterns have row n (top row) deleted.

Remark:

This result is essential to the classification problem since *it establishes recursively all operator pattern integers* Γ_{ij}. (Recall that the Gel'fand-Weyl pattern integers M_{ij} are eigen-values of well-defined operators [13].)

7 Open Problems

A canonical classification of *all* irreducible unit tensor operator components has, so far, been proven only for $U(2)$ and $U(3)$ but the result is conjectured to be true for all $U(n)$ and has been verified in a large number of individual cases. If this conjecture is valid, the canonical ("author independent") construction of $(3j), (6j), \cdots, (3nj)$ tensor operators [4],[5] can be carried out, developing extended tensor operator algebras [4], and for q-groups as well. This alone could generate a very large number of generalized hypergeometric and basic hypergeometric identities. A proof of this conjecture is our first open problem.

The space of operator patterns is strikingly similar to the space of Gel'fand-Weyl patterns, but, unlike Gel'fand-Weyl patterns, one can show that there is no underlying group structure. Something more than a group seems to be involved in which some group-like properties are restored in the limit of large m_{nn}. Although poorly posed this is our second open problem: what is the nature of operator pattern space, and what structure does this space carry?

8 Acknowledgement

We could like to thank Professor Daniel Flath for his critical reading of the manuscript and to the organizers, Professors Niki Kamran and P.J. Olver, for their help and encouragement.

References

[1] K. Baclawski, "A New Rule for Computing Clebsch-Gordan Series," *Adv. in Appl. Math.* **5** (1984) 416-432.

[2] L.C. Biedenharn, A. Giovannini, and J.D. Louck, "Canonical Definition of Wigner Coefficients in U_n", *J. Math. Phys.* **8** (1967) 691-700.

[3] L.C. Biedenharn and J.D. Louck, "A Pattern Calculus for Tensor Operators in the Unitary Groups," *Commun. Math. Phys.* **8** (1968) 89-131.

[4] L.C. Biedenharn and J.D. Louck, "The Racah-Wigner Algebra in Quantum Theory," Vol. 9 *Ency. of Math. and Appl.* (G.-C. Rota, Ed.), Addison-Wesley (Reading MA) 1981.

[5] L.C. Biedenharn and J.D. Louck, *Angular Momentum in Quantum Physics Theory and Application*, Vol. 8, *Ency. of Math. and Appl.* (G.-C. Rota, Ed.) Cambridge University Press (Cambridge, U.K.) 1989.

[6] L.C. Biedenharn and Marco Tarlini, "On q-Tensor Operators for Quantum Groups," *Lett. in Math. Phys.* **20** (1990) 271-278.

[7] A. Borel and A. Weil, *Representations linéaires et espaccs homogènes Kählerians des groups de Lie compacts, Seminaire Bourbaki,* May 1954, (exposé by J.-P. Serre).

[8] D. Flath and L.C. Biedenharn, "Beyond the Enveloping Algebra for sl_3", *Can. J. Math.* **37** (1985) 710-729.

[9] D. Flath, "Coherent Tensor Operators", to be published in the present volume.

[10] I.M. Gel'fand and A.V. Zelevinsky, *Societé Math. de France, Astérique, hors série* (1985) 117-128.

[11] K.T. Hecht, "The Vector Coherent State Method and Its Application to Problems of Higher Symmetries," *Lecture Notes in Physics* **190**, Springer-Verlag, Berlin, (1987).

[12] R. LeBlanc and L.C. Biedenharn, "Implementation of the $U(n)$ Tensor Operator Calculus in a Vector Bargmann Hilbert space," *J. Phys. A: Math. Gen* **22** (1989) 31-48.

[13] J.D. Louck and L.C. Biedenharn, "Canonical Unit Adjoint Tensor Operator in $U(n)$," *J. Math Phys.* (1970) 2368-2414.

[14] J.D. Louck and L.C. Biedenharn, "Some Properties of the Intertwining Number of the General Linear Group," in *Science and Computers, Adv. Math Supp. Studies*, **10** (1986) 265-311.

Center for Particle Physics
Department of Physics, University of Texas at Austin,
Austin, TX 78712

Contemporary Mathematics
Volume **160**, 1994

Lie Groups and Probability

PHILIP FEINSILVER

ABSTRACT. Using the example of sp(4), we illustrate the close connection between probability theory and representation theory. Of particular interest are the connections with Hermitian symmetric spaces, the systems of orthogonal polynomials which arise as bases for the representations, and the rôle played by matrix Riccati equations.

1. Introduction

Probability theory principally deals with measures, probability measures. A rich source of measures with particularly interesting properties is representation theory. A representation of a Lie group is given by an action of the group on a vector space, which is typically a Hilbert space, requisite for unitary representations. A realization of the Hilbert space as an L^2 space gives a measure. A study of representations of the Heisenberg algebra and sl(2) showed that the corresponding measures include the fundamental distributions of probability theory — binomial, negative binomial, exponential, Poisson, and Gaussian — as well as the 'continuous binomial' associated to an infinite-dimensional representation of su(1,1), which representation is well-known in representation theory, but the corresponding measure is not emphasized as important in probability theory. Rather, it is known to workers in special functions, as the orthogonal basis is given by Meixner-Pollaczek polynomials. (Further details on these basic results may be found in the book [3], part of a series written in collaboration with R. Schott.) It is important to note that this is *different* from the standard theory of stochastic processes on Lie groups as a generalization of the theory for $\mathbf{R}^n$. However, the techniques used here do apply in that context. The difference is mainly that here we are interested in getting measures on, say $\mathbf{R}^n$, *via* the representation theory. Another note is that the approach taken here allows one to develop a stochastic analysis — stochastic integrals and stochastic differential equations — for the corresponding processes (e.g. [6]).

In this article, we will illustrate via an example using sp(4). The approach presented here works for nilpotent and solvable as well as semisimple Lie groups. It gives a generalization of the theory for SL(2) and the Heisenberg group to an arbitrary number of dimensions. Particularly, the semisimple case deals with

---

1991 Mathematics Subject Classification. 17B15, 32M15, 33C50, 60B99.

This paper is in final form and no version of it will be submitted elsewhere.

Hermitian symmetric spaces of arbitrary rank. The basis states for the representations are realized as orthogonal polynomials. The measures for the L^2 spaces form convolution families, which correspond in the continuous parameter case to stochastic processes with independent increments. In the multivariable case, one gets special functions of matrix argument (see [7] for a point of view akin to ours). The representation theory is closely related to the 'vector coherent state' approach of Hecht, Le Blanc, Rowe, et al. (see [8]). For information concerning symmetric domains and Hermitian symmetric spaces, [9] and [12] are highly recommended. Especially, results given in the back of [12] were useful for the work presented here. For connections regarding symmetric spaces and Jordan algebras, see [10], and for further connections with quantum theory, the lectures by Upmeier [11] should be looked at, and [2] as well. The paper [1] contains calculations and ideas close to our point of view. There, what we call X, they use as a Hamiltonian to generate equations of motion on Hermitian symmetric spaces, which correspond to the flow for V discussed below.

2. Basic construction: Remarks

Geometrically speaking, we construct an L^2 bundle over the Lie algebra. The fiber over a typical element is an L^2 space in which that element is realized as a self-adjoint operator X_t — actually an operator-valued process (see below) — hence may be considered a quantum observable (for which reason this approach to probability theory is presently called 'quantum probability'). By looking at a family of representations, indexed by t (running through the positive reals or integers), related to a given element of the Lie algebra, the fiber may be considered as (L^2 of) a stochastic process, i.e., a measure on a function space, in which for each fixed time the element is realized as a self-adjoint operator X_t. The basis states for the representations are orthogonal polynomials, orthogonal with respect to the corresponding probability measures, in the variable X_t realized as multiplication by x via the spectral theorem. The measures corresponding to various representations are seen to form *naturally* a convolution family and hence correspond (in the continous-parameter case) to a stochastic process with independent increments.

The multivariable case deals with commuting elements, thus giving rise to a family of commuting observables, $(X_1, \ldots, X_n)$, say. And looking at families of representations, one has a multivariate stochastic process $(X_1(t), \ldots, X_n(t))$. The 'rank-one case' is presented in [5]. There, the basis states are Lauricella polynomials, which are a multivariable extension of the ${}_2F_1$ hypergeometric function.

3. Representation theory

One nice feature of this work is that it uses a synthesis of various approaches to the representation theory of Lie algebras and Lie groups. These involve:

1) Representations on the universal enveloping algebra and its quotients, i.e., induced representations
2) Matrix groups and factorizations
3) Realizations as vector fields or as (generalized) differential operators, equivalently, boson operator realizations

In all cases, we are interested in bases of polynomials, which for enveloping algebras of course involve noncommuting variables.

3.1. Representations on enveloping algebras. We take $\{\xi_1,\ldots,\xi_d\}$ as basis for the given Lie algebra, $\mathcal{G}$.
The Poincaré-Birkhoff-Witt theorem says that the universal enveloping algebra, $\mathcal{U}(\mathcal{G})$, has basis consisting of ordered monomials

$$\psi_n(\xi)=\xi^n=\xi_1^{n_1}\cdots\xi_d^{n_d}$$

where the multi-index $n=(n_1,\ldots,n_d)=\sum n_je_j$, $n_j\geq 0$, with e_j denoting the standard base vectors having a single 1 in the jth position. An action of $\mathcal{G}$ on $\mathcal{U}(\mathcal{G})$ is given by multiplication on the left, using Wick dots to denote the result,

$$\xi_j\xi^n =: \xi^n :_j = \xi^{n+e_j} + \text{commutation terms}$$

where the commutation terms arise from realizing the Lie bracket as a commutator

$$[\xi_j,\xi_k]=\xi_j\xi_k-\xi_k\xi_j=\sum_l c_{jk}^l\xi_l$$

for structure constants c_{jk}^l.

Let $\mathcal{K}$ be a Lie subalgebra of $\mathcal{G}$, with corresponding universal enveloping algebra $\mathcal{U}(\mathcal{K})$. Let E denote a representation space for $\mathcal{K}$. Then the corresponding *induced representation* is defined on $\mathcal{U}(\mathcal{G})\otimes_{\mathcal{U}(\mathcal{K})}E$, where the subscript means that elements of $\mathcal{U}(\mathcal{K})$ are pulled across the tensor sign to act on elements of E. The basis $\{\xi_1,\ldots,\xi_d\}$ is ordered so that $\{\xi_{m+1},\ldots,\xi_d\}$ span $\mathcal{K}$, $\dim\mathcal{K}=d-m$. Assuming that E has a cyclic vector η, with 1 denoting the monomial $\xi^0\in\mathcal{U}(\mathcal{G})$, we set $\Omega=1\otimes\eta$. Then

$$\psi_n(\xi)\Omega=\xi^n\Omega=\xi_1^{n_1}\xi_2^{n_2}\cdots\xi_m^{n_m}1\otimes\xi_{m+1}^{n_{m+1}}\cdots\xi_d^{n_d}\eta$$

The modules $\mathcal{U}(\mathcal{G})\otimes E$ have been utilized by Harish-Chandra, Verma, and others, as well as by mathematical physicists, notably B. Gruber, who introduced me to the method, to realize representations of $\mathcal{G}$ and those of the corresponding simply connected group G.

3.2. Realizations via differential operators. To get a realization in terms of differential operators, introduce canonical boson variables $\{\mathcal{R}_i\}$, $\{\mathcal{V}_i\}$, with $[\mathcal{V}_j,\mathcal{R}_i]=\delta_{ji}$, defined on the basis ψ_n by

$$\mathcal{R}_i\psi_n=\psi_{n+e_i},\qquad \mathcal{V}_i\psi_n=n_i\psi_{n-e_i}$$

These extend by linearity to $\mathcal{U}(\mathcal{G})$. Then the action of ξ_j on ψ_n can be expressed in terms of these variables. Thinking of $\mathcal{R}_i$ as multiplication by a variable y_i and $\mathcal{V}_i$ as the corresponding operator $\partial/\partial y_i$ yields a realization via differential operators. The term *generalized* differential operators is used when, e.g. the shift operator, sending $y\to y+1$ arises. This is realized as $\exp(\partial/\partial y)$, which is not a local operator, hence not a differential operator.

Next we introduce the *left dual representation*. The formal sum

$$g(A,\xi)=\sum_n c_n(A)\psi_n(\xi)=e^{A_1\xi_1}\cdots e^{A_d\xi_d}$$

where $A = (A_1, \ldots, A_d)$ are commuting variables, and
$c_n(A) = A_1^{n_1} \cdots A_d^{n_d}/n_1! \cdots n_d!$, gives a local group element, near the identity. The action of multiplication by ξ_j dualizes via

$$\xi_j g(A, \xi) = \sum_n c_n(A) \colon \psi_n(\xi) \colon_j = \sum_n \xi_j^\dagger c_n(A) \psi_n(\xi)$$

to an action on functions of A. The $\xi_j^\dagger$ can be calculated by acting on the exponential factors of $g(A, \xi)$ using the adjoint action

$$e^{t\xi_k} \xi_j e^{-t\xi_k} = e^{t \operatorname{ad} \xi_k} \xi_j \tag{3.2.1}$$

where $\operatorname{ad}\xi_k(\xi_j) = [\xi_k, \xi_j]$, to commute ξ_j to the jth position. Resulting terms of the form

$$\xi_i e^{A_i \xi_i}$$

are written as the differential operator $\partial/\partial A_i$ acting on g.

REMARK. As the left dual is adjoint to multiplication on the left, you get a Lie anti-homomorphism, since $(\xi_j \xi_k)^\dagger = \xi_k^\dagger \xi_j^\dagger$.

For example, for the Heisenberg algebra, with basis $\{\xi, h, \delta\}$, $[\delta, \xi] = h$, central, we have

$$g(A, B, C; \xi, h, \delta) = e^{A\xi} e^{Bh} e^{C\delta}$$

yielding

$$\xi^\dagger = \frac{\partial}{\partial A}, \quad h^\dagger = \frac{\partial}{\partial B}, \quad \delta^\dagger = \frac{\partial}{\partial C} + A \frac{\partial}{\partial B} \tag{3.2.2}$$

To get a Lie homomorphism, we form the *double dual*, i.e., interchanging

$$A_i \leftrightarrow \frac{\partial}{\partial A_i}$$

This is thus the action of g on the enveloping algebra. We denote the corresponding representation by $\hat{\xi}_j$, e.g., equation (3.2.2) yields for the Heisenberg algebra,

$$\hat{\xi} = A, \quad \hat{h} = B, \quad \hat{\delta} = C + B \frac{\partial}{\partial A}.$$

3.2.1 PROPOSITION. *The double dual is the boson realization of the Lie algebra acting on $\mathcal{U}(\mathcal{G})$, with the identification*

$$\mathcal{R}_i \leftrightarrow A_i, \qquad \mathcal{V}_i \leftrightarrow \frac{\partial}{\partial A_i}$$

3.2.2 COROLLARY. *When $\xi_j^\dagger$ are given as vector fields, the $\hat{\xi}_j$ are first-order in the A variables.*

REMARK. The (dual) action from the right gives a Lie homomorphism directly and can be used to find polynomial representations of the group. See [4] for this and related material.

3.3. Matrix groups and factorizations. A decomposition of the Lie algebra $\mathcal{G}$ into subalgebras, e.g., $\mathcal{G} = \mathcal{P}_1 + \mathcal{K} + \mathcal{P}_2$, exponentiates into a factorization of the Lie group: $G = P_1KP_2$. In terms of matrices, this corresponds to a factorization of $g \in G$ into matrices VCB, say, with $V \in P_1$, $C \in K$, $B \in P_2$. For example, with

$$\mathcal{P}_1 = \left\{\begin{pmatrix} 0 & v \\ 0 & 0 \end{pmatrix}\right\}, \quad \mathcal{K} = \left\{\begin{pmatrix} a & 0 \\ 0 & d \end{pmatrix}\right\}, \quad \mathcal{P}_2 = \left\{\begin{pmatrix} 0 & 0 \\ b & 0 \end{pmatrix}\right\}$$

we have for 2×2 matrices

$$g = \begin{pmatrix} 1 & V \\ 0 & 1 \end{pmatrix}\begin{pmatrix} A & 0 \\ 0 & D \end{pmatrix}\begin{pmatrix} 1 & 0 \\ B & 1 \end{pmatrix} = \begin{pmatrix} A + VDB & VD \\ DB & D \end{pmatrix}$$

This is a standard UDL factorization — upper triangular, diagonal, lower triangular — for the local Lie group.

Generally, the factorization is the exponentiation of the monomials in the enveloping algebra, where $\mathcal{G}$ is given according to the order: basis of $\mathcal{P}_1$, basis of $\mathcal{K}$, basis of $\mathcal{P}_2$. The action of $\mathcal{G}$ on $\mathcal{U}(\mathcal{G})$ corresponds to multiplication (on the left, say) by $X = \begin{pmatrix} a & v \\ b & d \end{pmatrix} \in \mathcal{G}$ on G:

$$Xg = \begin{pmatrix} aA + aVDB + vDB & aVD + vD \\ bA + bVDB + dDB & bVD + dD \end{pmatrix} \tag{3.3.1}$$

3.3.1 PROPOSITION. *The action on V corresponding to multiplication by $X \in \mathcal{G}$ is given by the fractional linear transformation*

$$V \to (aV + v)/(bV + d)$$

Proof: V is given by the ratio of the elements of the second column. Comparing with equation (3.3.1) gives the result. ∎

REMARK. Note that with entries above as block matrices, the same discussion holds.

3.4. Flows: Splitting lemma and Riccati equations. First we consider the 'converse problem', namely, given a group element how to factor it to a desired form corresponding to a decomposition of the Lie algebra. The idea is to look at one-parameter subgroups.

3.4.1 THEOREM. *Let $X = \sum a_i\xi_i \in \mathcal{G}$ generate the one-parameter subgroup $g(s) = \exp(sX) \in G$. The coefficients $A(s)$ correspond to the factorization*

$$g(s) = e^{A_1(s)\xi_1} \cdots e^{A_d(s)\xi_d} \tag{3.4.1}$$

Denote by $\xi_j^\dagger$ the corresponding left dual representation of $\mathcal{G}$, given by vector fields

$$\xi_j^\dagger = \sum_k \pi_{jk}(A)\,\frac{\partial}{\partial A_k}$$

Then $(A_1(s), \ldots, A_d(s))$ *are given by the flow*

$$\dot{A}_k = \sum_i a_i \pi_{ik}(A)$$

the dot denoting differentiation with respect to s, *with initial conditions* $A_i(0) = 0$, $1 \le i \le d$.

Proof: Differentiating with respect to s in equation (3.4.1) gives

$$\dot{g} = Xg = \sum_i a_i \xi_i \, g = \sum_k e^{A_1\xi_1} \cdots \dot{A}_k \xi_k e^{A_k \xi_k} \cdots e^{A_d \xi_d}$$

$$\sum_i a_i \xi_i^\dagger g = \sum_k \dot{A}_k \frac{\partial}{\partial A_k} g$$

Writing $\xi_i^\dagger = \sum \pi_{ik}(A)\partial/\partial A_k$ and comparing coefficients yields the result. ∎

REMARK. The multiplication on the right $\dot{g} = gX$ may be used as well, giving the same coordinates $A(s)$.

3.4.2 COROLLARY. *Given the flow* $\dot{A}$ *corresponding to the left action* $\dot{g} = Xg$, *one may deduce the left dual action.*

3.4.3 COROLLARY. *Corresponding to the flow* $\dot{A}_k = \sum_i a_i \pi_{ik}(A)$ *we have the double dual representation*

$$\hat{\xi}_i = \sum_k A_k \pi_{ik}\left(\frac{\partial}{\partial A}\right)$$

where $\dfrac{\partial}{\partial A} = \left(\dfrac{\partial}{\partial A_1}, \ldots, \dfrac{\partial}{\partial A_d}\right)$.

For the factorization VCB of the previous section, we have from equation (3.3.1),

$$\dot{g} = Xg = \begin{pmatrix} a & v \\ b & d \end{pmatrix} g = \begin{pmatrix} aA + aVDB + vDB & aVD + vD \\ bA + bVDB + dDB & bVD + dD \end{pmatrix}$$

Thus the flow for the variables (V, D, A, B) is given by

$$\begin{aligned} \dot{V} &= v + aV - Vd - VbV \\ \dot{D} &= (bV + d)D \\ \dot{A} &= (a - Vb)A \\ \dot{B} &= D^{-1}bA \end{aligned} \tag{3.4.2}$$

where we write these with order intact to include the case of block matrix form. Note that the flow for V is given by a (matrix) *Riccati equation.*

For the 2×2 case, write $X = v\xi_1 + d\xi_2 + a\xi_3 + b\xi_4$. Then we have the representation

$$\begin{aligned}
\hat{\xi}_1 &= V \\
\hat{\xi}_2 &= V\frac{\partial}{\partial V} + D\frac{\partial}{\partial D} \\
\hat{\xi}_3 &= V\frac{\partial}{\partial V} + A\frac{\partial}{\partial A} \\
\hat{\xi}_4 &= -V\frac{\partial^2}{\partial V^2} - A\frac{\partial}{\partial A}\frac{\partial}{\partial V} + D\frac{\partial}{\partial D}\frac{\partial}{\partial V} + B\frac{\partial}{\partial A}\left(\frac{\partial}{\partial D}\right)^{-1}
\end{aligned}$$

where $(\partial/\partial D)^{-1}D^n = D^{n+1}/(n+1)$.

3.5. Induced representations. For the induced representations, we reduce the basis as follows. Denote the variables of the inducing subalgebra by $\{\varepsilon, \delta, L\}$. We assume that m of the variables, the $\{L_i\}$, are *lowering operators,* satisfying $L_i\eta = 0$. The $\{\varepsilon_i\}$ annihilate η as well, $\varepsilon_i\eta = 0$. The $\{\delta_i\}$ give a nontrivial representation of $\mathcal{K}$. By the *scalar case,* considered below, we mean that the $\{\delta_i\}$ act as scalars on η: $\delta_i\eta = \lambda_i\eta$. From the double dual we see that η has a realization as common eigenfunctions of $\{\hat{\delta}_i\}$ and are annihilated by $\{\hat{\varepsilon}_i, \hat{L}_i\}$. The remaining m variables are the *raising operators,* $\{R_i\}$. Denote the subgroup with Lie algebra $\{\varepsilon_i\}$ by A, that with Lie algebra $\{\delta_i\}$ by D. The commuting variables dual to $\{R_i\}$ we denote by $\{V_i\}$, those dual to $\{L_i\}$ by $\{B_i\}$. Thus $g(s)$ has the form

$$g(s) = e^{sX} = e^{V(s)R}D_sA_se^{B(s)L} \tag{3.5.1}$$

where $e^{V(s)R} = e^{V_1(s)R_1}\cdots e^{V(s)_mR_m}$, and similarly for $e^{B(s)L}$

REMARK. With these assumptions, Ω is called a *vacuum state,* since it is annihilated by the lowering operators $\{L_i\}$.

The action of g on Ω takes the form

$$g(s)\Omega = e^{sX}\,\Omega = e^{V(s)R}\,1 \otimes D_s\eta \tag{3.5.2}$$

REMARK. The exponentials acting on Ω, $g(s)\Omega$ and $\exp(V(s)R)\Omega$ are often referred to as generalized *coherent states,* see [8]. In the scalar case, $D_s\eta = D(s)\eta$, with $D(s)$ a function, we have

$$e^{V(s)R}\,\Omega = D(s)^{-1}e^{sX}\,\Omega$$

Interestingly, the complete flow may be recovered from the action

$$Xe^{V(s)R} = e^{V(s)R}(e^{-V(s)R}Xe^{V(s)R})$$

3.5.1 PROPOSITION. *The action of X on $\exp(V(s)R)$ is given via*

$$e^{-V(s)R}Xe^{V(s)R} = \dot{V}R + (DA)^{\cdot}(DA)^{-1} + (DA)\dot{B}L(DA)^{-1} \tag{3.5.3}$$

Proof: From equation (3.5.1), we have

$$Xg = \dot{g} = e^{V(s)R}\dot{V}RD_sA_se^{B(s)L} + e^{V(s)R}(D_sA_s)^{\cdot}e^{B(s)L} + e^{V(s)R}D_sA_s\dot{B}Le^{B(s)L}$$

Thus,

$$e^{-VR}\dot{g}e^{-BL} = \dot{V}RDA + (DA)^{\cdot} + DA\dot{B}L$$

Which gives

$$e^{-VR}Xge^{-BL} = (e^{-VR}Xe^{VR})DA$$

Hence the result. ■

From equation (3.5.2), we have

3.5.2 PROPOSITION. *For the induced representation, we have*

$$Xe^{V(s)R}1 \otimes D_s\eta = e^{V(s)R}(\dot{V}R1 \otimes D_s\eta + 1 \otimes \dot{D}_s\eta)$$

Proof: Differentiate equation (3.5.2) with respect to s. ■

For example, for the VCB factorization above, we have

$$e^{-VR}Xe^{VR} = \begin{pmatrix} a - Vb & v + aV - Vd - VbV \\ b & bV + d \end{pmatrix} = \begin{pmatrix} \dot{A}A^{-1} & \dot{V} \\ D\dot{B}A^{-1} & \dot{D}D^{-1} \end{pmatrix}$$

in agreement with equation (3.4.2).

3.5.1. Cartan decomposition. We assume that the decomposition $\{R, \delta, \varepsilon, L\}$ corresponds to a *Cartan decomposition*

$$\mathcal{G} = \mathcal{P}_+ + \mathcal{K} + \mathcal{P}_-$$

such that

$$[\mathcal{K}, \mathcal{P}_-] \subset \mathcal{P}_-, \quad [\mathcal{K}, \mathcal{P}_+] \subset \mathcal{P}_+, \quad [\mathcal{P}_-, \mathcal{P}_+] \subset \mathcal{K}$$

We have $\mathcal{P}_+$ consisting of the R's, $\mathcal{P}_-$ the L's, and $\mathcal{K}$ the δ's and ε's. We take $\mathcal{P}_+$ and $\mathcal{P}_-$ to be abelian, i.e., $[R_i, R_j] = [L_i, L_j] = 0$. Then we have 'generically', with $[L, R] = K \in \mathcal{K}$:

$$\begin{aligned} [L, R^n] &= \sum R^{n-1-j}KR^j \\ &= \sum R^{n-1-j}\sum R^{j-1-i}\mathrm{r}\,R^i + nR^{n-1}K \\ &= n(n-1)R^{n-2}\mathrm{r} + nR^{n-1}K \end{aligned}$$

where $\mathrm{r} \in \mathcal{P}_+$. Thus the L's may be considered second-order differential operators with respect to the R variables.

And we get, cf. the action of X, Proposition 3.5.1 (here $\xi_i \in \mathcal{K}$):

$$\begin{aligned} e^{-VR}Xe^{VR} &= e^{-VR}(R + \sum_i \alpha_i\xi_i + L)e^{VR} \\ &= X + \sum_i \alpha_i\mathrm{r}_iV + KV + \mathrm{r}\,V^2/2 \end{aligned}$$

using the commutation properties of the Cartan decomposition and the formula for the adjoint action

$$e^{-VR}Xe^{VR} = e^{-V\,\mathrm{ad}\,R}X$$

4. Expectation

Now for the probability. We construct an inner product on the pre-Hilbert space with basis

$$J_n = R^n\Omega$$

REMARK. Here we will consider only the *scalar case*, where $D_s\eta = D(s)\eta$, with $D(s)$ a (scalar) function. That is we are inducing from a one-dimensional representation.

We define expected values relative to the vacuum state Ω:

$$\langle Y\rangle = \langle\Omega, Y\Omega\rangle$$

The main assumption is that R_i and L_i are adjoint. Then, since $L_i\Omega = 1\otimes L_i\eta = 0$, for all $1 \le i \le m$, we have

$$\langle\Omega, R^n\Omega\rangle = \langle L^n\Omega, \Omega\rangle = 0$$

for any non-zero multi-index n. With $Y\Omega = Y'1\otimes\eta + 1\otimes Y''\eta$, only the Y'' terms survive. Thus we have

$$\langle Y\rangle = \langle\eta, Y''\eta\rangle$$

For $g(s)$ we thus get from equation (3.5.2)

$$\langle g(s)\rangle = \langle e^{sX}\rangle = \langle\eta, D_s\eta\rangle$$

For $X\in\mathcal{G}$ a self-adjoint operator, we denote $p(dx)$ the corresponding spectral measure in the state Ω:

$$\langle e^{sX}\rangle = \langle\Omega, e^{sX}\,\Omega\rangle = \int_{-\infty}^{\infty} e^{sx}\,p(dx)$$

This is the probability measure we are looking for.

4.1. Variables. Canonically we take as variables self-adjoint operators in the form

$$X_i = R_i + \sum_j \alpha_{ij}\xi_j + b_i L_i$$

where $\xi_j\in\mathcal{K}$. The condition on the coefficients is that these commute: $[X_i, X_j] = 0$. Thus they are a commuting family of observables. With $X = \sum v_i X_i \in\mathcal{G}$, the expected value $\langle e^X\rangle$ is the *moment generating function* or *Fourier-Laplace transform* of the distribution of $(X_1,\ldots,X_d)$ as a family of random variables.

4.2. Orthogonal basis. Introduce the notation for the basis expressed in terms of the variables X (this is accomplished via recurrence relations, see the next section):

$$J_n(x) = R^n\Omega$$

The adjointness relation between R and L implies that these are an orthogonal basis. These will be a family of (in general) multivariate orthogonal polynomials. For example, in the case of the Heisenberg algebra, these will be Hermite polynomials. Equation (3.5.2) yields

$$\begin{aligned} e^{sX}\,\Omega &= e^{V(s)R}\,1\otimes D_s\eta \\ &= \sum_{n=0}^{\infty}\frac{V(s)^n}{n!}\,R^n 1\otimes D_s\eta \end{aligned} \tag{4.2.1}$$

where $n! = n_1!\cdots n_m!$. Then taking inner products in equation (4.2.1) with $J_n(x)$, we have

$$\begin{aligned} (V(s)^n/n!)^{-1}\,\langle J_n(X)e^{sX}\,\rangle &= \langle R^n 1\otimes\eta, R^n 1\otimes D_s\eta\rangle \\ &= \langle L^n R^n\Omega, 1\otimes D_s\eta\rangle \\ &= \langle 1\otimes K_n\eta, 1\otimes D_s\eta\rangle = \langle K_n\eta, D_s\eta\rangle \end{aligned}$$

with $K_n \in \mathcal{K}$. In the scalar case, denote the squared norms by

$$\gamma_n = \langle J_n(x), J_n(x)\rangle$$

Then

$$\langle J_n(X)e^{sX}\,\rangle = V(s)^n\,\frac{\gamma_n}{n!}\,\langle\eta, D_s\eta\rangle$$

Denote $\exp(H(s)) = \langle\eta, D_s\eta\rangle$. Setting

$$\phi_n(x) = \frac{n!}{\gamma_n}\,J_n(x)$$

we have

$$\langle\phi_n(X)e^{sX}\,\rangle = V(s)^n\,e^{H(s)} \tag{4.2.2}$$

In particular,

$$\langle e^{sX}\,\rangle = e^{H(s)} = \int_{-\infty}^{\infty} e^{sx}\,p(dx) \tag{4.2.3}$$

is the Fourier-Laplace transform of $p(dx)$. Thus the measure $p(dx)$ may be recovered by Fourier inversion.

4.3. Recurrence relations and Riccati equations. Consider X as a matrix variable, $X = (x_{kl})$, and correspondingly for S. We take (4.2.3) in the form

$$e^{H(S)} = \int e^{\langle S,X\rangle}\, p(dX)$$

Now (n) is a multi-index, a matrix of non-negative integers n_{ij} — with e_{ij} denoting the standard base elements having only one 1 as non-zero entry, we have $(n) = \sum_{ij} n_{ij}e_{ij}$ — and $\langle S, X\rangle$ denotes the usual matrix inner product $\operatorname{tr}(S^tX)$. Then (4.2.2) becomes, writing $V^{(n)} = \prod V_{ij}^{n_{ij}}$,

$$V^{(n)}e^{H(S)} = \int e^{\langle S,X\rangle}\, \phi_{(n)}(X)\, p(dX)$$

Note. In this discussion, the convention of summing over repeated *Greek* indices will be used.

Differentiating with respect to s_{kl} yields

$$(\sum_{i,j} n_{ij}V^{(n-e_{ij})}\frac{\partial V_{ij}}{\partial s_{kl}} + V^{(n)}\Gamma_{k\mu}V_{\mu l})\, e^{H(S)} = \int e^{\langle S,X\rangle}\, X_{kl}\phi_{(n)}(X)\, p(dX)$$

where $H(S)$ satisfies $\partial H/\partial s_{kl} = \Gamma_{k\mu}V_{\mu l}$.

Combining this with the Riccati equations, cf. (3.4.2),

$$\frac{\partial V}{\partial s_{kl}} = B_{kl} + A_{kl}V - VD_{kl} - VC_{kl}V$$

yields the recurrence

$$\begin{aligned} x_{kl}\phi_{(n)}(X) = &\\ &\sum_{i,j}[n_{ij}\phi_{(n-e_{ij})}B^{ij}_{kl} + n_{ij}\phi_{(n+e_{\mu j}-e_{ij})}A^{i\mu}_{kl} - n_{ij}\phi_{(n+e_{i\mu}-e_{ij})}D^{\mu j}_{kl}\\ &\qquad - n_{ij}C^{\lambda\mu}_{kl}\phi_{(n+e_{i\lambda}+e_{\mu j}-e_{ij})}] + \Gamma_{k\mu}\phi_{(n+e_{\mu l})} \end{aligned}$$

The recurrence relations effect the transition from the R variables to the X variables for generating the basis $\phi_{(n)}$. For the symmetric spaces, with the action of fractional linear mappings, choosing appropriate bases for the Lie algebra will give corresponding functions V, hence corresponding measures, via moment generating functions, and orthogonal bases consisting of special classes of polynomial functions $\phi_{(n)}$. Thus the theory for the 2×2 case extends to symmetric spaces in general.

5. An example from sp(4)

By sp(4) we mean the (real) Lie algebra of 4×4 matrices in the 2×2 block form

$$\begin{pmatrix} a & v \\ b & d\end{pmatrix} : \quad v^t = v,\ b^t = b,\ a = -d^t$$

the t denoting transpose (for more details, see [12], especially pp. 331–338). We have as basis (writing υ, τ for δ_1, δ_2), first R_1, R_2, R_3 given by the matrices

$$\begin{pmatrix} 0&0&1&0\\0&0&0&0\\0&0&0&0\\0&0&0&0 \end{pmatrix}, \begin{pmatrix} 0&0&0&1\\0&0&1&0\\0&0&0&0\\0&0&0&0 \end{pmatrix}, \begin{pmatrix} 0&0&0&0\\0&0&0&1\\0&0&0&0\\0&0&0&0 \end{pmatrix}$$

with $R_j = L_j^t$, $j = 1,2,3$. Thus, the bases for $\mathcal{P}_+$ and $\mathcal{P}_-$. For $\mathcal{K}$, we have the basis elements

$$\upsilon = \begin{pmatrix} 1&0&0&0\\0&0&0&0\\0&0&-1&0\\0&0&0&0 \end{pmatrix} \quad \tau = \begin{pmatrix} 0&0&0&0\\0&1&0&0\\0&0&0&0\\0&0&0&-1 \end{pmatrix}$$

$$\varepsilon_1 = \begin{pmatrix} 0&1&0&0\\0&0&0&0\\0&0&0&0\\0&0&-1&0 \end{pmatrix} \quad \varepsilon_2 = \begin{pmatrix} 0&0&0&0\\1&0&0&0\\0&0&0&-1\\0&0&0&0 \end{pmatrix}$$

The $\{ R_i \}$ span the set of 2×2 symmetric matrices. We see that $\dim \mathcal{G} = 10$, $\dim \mathcal{K} = 4$, $\dim \mathcal{P}_+ = \dim \mathcal{P}_- = 3$.

We want three commuting elements to be our variables. One possible choice for X_1, X_2, X_3 is

$$\begin{pmatrix} \alpha&0&1&0\\0&0&0&0\\-1&0&-\alpha&0\\0&0&0&0 \end{pmatrix}, \begin{pmatrix} 0&\alpha&0&1\\\alpha&0&1&0\\0&-1&0&-\alpha\\-1&0&-\alpha&0 \end{pmatrix}, \begin{pmatrix} 0&0&0&0\\0&\alpha&0&1\\0&0&0&0\\0&-1&0&-\alpha \end{pmatrix}$$

where $\alpha^2 = 1$. In terms of the basis $\{ R, \upsilon, \tau, \varepsilon, L \}$ we have

$$X_1 = R_1 + \alpha\upsilon - L_1,\ X_2 = R_2 + \alpha(\varepsilon_1 + \varepsilon_2) - L_2,\ X_3 = R_3 + \alpha\tau - L_3$$

Introduce dual boson variables $\{ V_i \}$ to $\{ R_i \}$: $[V_i, R_j] = \delta_{ij}$. Note that this is consistent with the $V_i(s)$:

$$V_i(s)e^{V(s)R} = \frac{\partial}{\partial R_i} e^{V(s)R} \tag{5.1}$$

5.1. Commutation relations. Re-introduce the boson variables $\{ \mathcal{R}_i \}$, $\{ \mathcal{V}_i \}$. We can write the basis in Jordan-Schwinger form, as it is commonly called in the physics literature, using these variables. This form is determined by the fact that the correspondence of matrices E_{ij} with $\mathcal{R}_i\mathcal{V}_j$ is a Lie isomorphism, where E_{ij} is the matrix with a single non-zero entry, a 1 in the ij-position. We thus have

$$R_1 = \mathcal{R}_1\mathcal{V}_3,\quad R_2 = \mathcal{R}_1\mathcal{V}_4 + \mathcal{R}_2\mathcal{V}_3,\quad R_3 = \mathcal{R}_2\mathcal{V}_4$$
$$L_1 = \mathcal{R}_3\mathcal{V}_1,\quad L_2 = \mathcal{R}_4\mathcal{V}_1 + \mathcal{R}_3\mathcal{V}_2,\quad L_3 = \mathcal{R}_4\mathcal{V}_2$$

and

$$\upsilon = \mathcal{R}_1\mathcal{V}_1 - \mathcal{R}_3\mathcal{V}_3,\ \tau = \mathcal{R}_2\mathcal{V}_2 - \mathcal{R}_4\mathcal{V}_4,\ \varepsilon_1 = \mathcal{R}_1\mathcal{V}_2 - \mathcal{R}_4\mathcal{V}_3,\ \varepsilon_2 = \mathcal{R}_2\mathcal{V}_1 - \mathcal{R}_3\mathcal{V}_4$$

Using the Jordan-Schwinger form, we readily find the commutation relations for the action on (R_1, R_2, R_3). We write only the non-zero commutators:

$$\begin{aligned} &[\upsilon, R_1] = 2R_1 && [\tau, R_2] = R_2 \\ &[\upsilon, R_2] = R_2 && [\tau, R_3] = 2R_3 \\ &[\varepsilon_1, R_2] = 2R_1 && [\varepsilon_2, R_1] = R_2 \\ &[\varepsilon_1, R_3] = R_2 && [\varepsilon_2, R_2] = 2R_3 \end{aligned}$$

and

$$\begin{gathered}[L_1, R_1] = -\upsilon,\ [L_2, R_1] = -\varepsilon_1,\ [L_3, R_2] = -\varepsilon_1 \\ [L_1, R_2] = -\varepsilon_2,\ [L_2, R_2] = -\upsilon - \tau,\ [L_3, R_3] = -\tau\end{gathered}$$

additionally, $[L_2, R_3] = -\varepsilon_2$.

5.2. Realization in R,V variables. For this case, it is not difficult to compute the double dual representation for the action of the Lie algebra on the induced module, with basis $R^n\Omega$. Using the commutation relations of the previous section, we calculate the action of the algebra on the group elements acting on Ω, exponential generating functions for the basis $R^n\Omega$. We only need the adjoint action, equation (3.2.1), cf. Propositions 3.5.1, 3.5.2. For example,

$$\begin{aligned} e^{-V_1R_1}\,\varepsilon_1 e^{V_1R_1} &= \varepsilon_1 \\ e^{-V_2R_2}\,\varepsilon_1 e^{V_2R_2} &= \varepsilon_1 + 2V_2R_1 \\ e^{-V_3R_3}\,\varepsilon_1 e^{V_3R_3} &= \varepsilon_1 + V_3R_2 \end{aligned}$$

This yields, with $\varepsilon_1\Omega = 0$,

$$\varepsilon_1 e^{V_1R_1}\, e^{V_2R_2}\, e^{V_3R_3}\,\Omega = (2R_1V_2 + R_2V_3)e^{VR}\,\Omega$$

Similarly, we have, identifying V_i with $\partial/\partial R_i$, as noted above, equation (5.1),

5.2.1 PROPOSITION. *The (double dual) representation on the basis $R^n\Omega$ may be given in terms of $\{R_i\}$, $\{V_i\}$, where $R_iJ_n = J_{n+e_i}$, $V_iJ_n = n_iJ_{n-e_i}$. We have, with $\hat{R}_j = R_j$, $j = 1, 2, 3$:*

$$\begin{aligned} \hat{\varepsilon}_1 &= 2R_1V_2 + R_2V_3 \\ \hat{\varepsilon}_2 &= 2R_3V_2 + R_2V_1 \end{aligned}, \quad \begin{aligned} \hat{\upsilon} &= \lambda + 2R_1V_1 + R_2V_2 \\ \hat{\tau} &= \lambda + R_2V_2 + 2R_3V_3 \end{aligned}.$$

and

$$\begin{aligned} \hat{L}_1 &= -\lambda V_1 - R_1V_1^2 - R_2V_1V_2 - R_3V_2^2 \\ \hat{L}_2 &= -2\lambda V_2 - R_2V_2^2 - R_2V_1V_3 \\ &\qquad - 2R_1V_1V_2 - 2R_3V_2V_3 \\ \hat{L}_3 &= -\lambda V_3 - R_3V_3^2 - R_2V_2V_3 - R_1V_2^2 \end{aligned} \tag{5.2.1}$$

We thus have the induced representation with

$$\hat{\upsilon}\Omega = \lambda\Omega, \quad \hat{\tau}\Omega = \lambda\Omega, \quad \hat{\varepsilon}_i\Omega = \hat{L}_j\Omega = 0$$

for $i = 1, 2$, $j = 1, 2, 3$.

The action of $g(s)$ on Ω is given by

$$e^{sX}\,\Omega = e^{V(s)R}\,1 \otimes D(s)\eta \tag{5.2.2}$$

where $D(s) = e^{\lambda H(s)}$. From Proposition 5.2.1,

5.2.2 PROPOSITION. *The variables* $\{X_1, X_2, X_3\}$ *act as*

$$\begin{aligned}
X_1 &= R_1 + \alpha(\lambda + 2R_1V_1 + R_2V_2) + \lambda V_1 \\
&\quad + R_1V_1^2 + R_2V_2V_1 + R_3V_2^2 \\
X_2 &= R_2 + \alpha(2R_1V_2 + R_2V_3 + 2R_3V_2 + R_2V_1) + 2\lambda V_2 \\
&\quad + R_2V_2^2 + 2R_1V_1V_2 + 2R_3V_3V_2 + R_2V_1V_3 \\
X_3 &= R_3 + \alpha(\lambda + R_2V_2 + 2R_3V_3) + \lambda V_3 \\
&\quad + R_3V_3^2 + R_2V_2V_3 + R_1V_2^2
\end{aligned}$$

These give recurrence formulas for multiplication by X_j, $j = 1, 2, 3$, on the basis $J_n(x)$.

Comparing with Proposition 3.5.2, we have

5.2.3 THEOREM. *Let* ∇ *denote* $(\partial/\partial s_1, \partial/\partial s_2, \partial/\partial s_3)^t$. *The functions* $H(s)$ *and* $V_j(s)$ *satisfy the differential equations*

$$\nabla H = \begin{pmatrix} \alpha + V_1 \\ 2V_2 \\ \alpha + V_3 \end{pmatrix}$$

And

$$\nabla V_1 = \begin{pmatrix} 1 + 2\alpha V_1 + V_1^2 \\ 2\alpha V_2 + 2V_1V_2 \\ V_2^2 \end{pmatrix}$$

$$\nabla V_2 = \begin{pmatrix} \alpha V_2 + V_1V_2 \\ 1 + \alpha V_1 + \alpha V_3 + V_1V_3 + V_2^2 \\ \alpha V_2 + V_2V_3 \end{pmatrix}$$

$$\nabla V_3 = \begin{pmatrix} V_2^2 \\ 2V_2(\alpha + V_3) \\ 1 + 2\alpha V_3 + V_3^2 \end{pmatrix}$$

Proof: As in Proposition 3.5.2, cf. equation (5.2.2), we have

$$\begin{aligned} X_j e^{sX} &= R_\sigma \frac{\partial V_\sigma}{\partial s_j} e^{V(s)R} 1 \otimes D(s)\eta + e^{V(s)R} 1 \otimes \lambda \frac{\partial H}{\partial s_j} D(s)\eta \\ &= \left(R_\sigma \frac{\partial V_\sigma}{\partial s_j} + \lambda \frac{\partial H}{\partial s_j} \right) D(s) e^{V(s)R} \Omega \end{aligned}$$

Comparing with Proposition 5.2.2, we see that we can read off the equations from the coefficients of R_j and λ. ■

Observe that effectively we have

$$\frac{\partial V_j}{\partial s_k} = [V_j, X_k]$$

with the relations $[V_j, R_i] = \delta_{ji}$.

Via the solution of these equations, one can find the distribution of the process X_1, X_2, X_3. Note that λ enters in the exponent, i.e. it corresponds to powers of the moment generating function $\exp(H(s))$ and thus to convolutions of the corresponding measures.

REMARK. Setting $s_3, s_2 = 0$ and $s_2, s_1 = 0$ in turn yields the marginal distributions of X_1 and X_3. With $\alpha = \pm 1$, the equation $V' = 1 + 2\alpha V + V^2$ has the form $V' = (1 + \alpha V)^2$. The solution is a fractional linear transformation of the variable s, either s_1 or s_3. Then integrating for $H(s)$ one finds that X_1 and X_3 have gamma distributions (convolution powers of the exponential distribution). Similarly, one finds that X_2 is distributed as the difference of two independent χ^2 variables. We have essentially recovered the Wishart distribution.

Acknowledgement. The author would like to thank J.-L. Clerc of Nancy and G. Letac and M. Casalis of Toulouse for valuable conversations.

References

1. S. Berceanu and A. Gheorghe, On equations of motion on compact Hermitian symmetric spaces, J. Math. Phys., **33** (1992), 998–1007.

2. F. Berezin, Quantization in complex symmetric spaces, Math. USSR Izvestia, **9** (1975), 341–379.

3. P. Feinsilver and R. Schott, *Representations and probability theory*, Kluwer Academic Press, Dordrecht, 1993.

4. P. Feinsilver and R. Schott, Appell systems on Lie groups, J. Theo. Prob., **5** (1992), 251–281.

5. P. Feinsilver, Orthogonal polynomials and coherent states, in *Symmetries in Science V*, Plenum Press, 1991, 159–172.

6. P. Feinsilver, Bernoulli fields, in *Quantum Probability and Applications IV*, Springer Lecture Notes, **1396** (1989), 158–179.

7. K.I. Gross and D. St. P. Richards, Special functions of matrix argument, I:Algebraic induction, zonal polynomials, and hypergeometric functions, Trans. Amer. Math. Soc., **301** (1987), 781–811.

8. K.T. Hecht, *The vector coherent state method and its application to problems of higher symmetries*, Springer Lecture Notes in Physics, **290**, 1987.

9. A.W. Knapp, Bounded symmetric domains and holomorphic discrete series, in *Symmetric Spaces*, W.M. Boothby & G.L. Weiss (eds.), Marcel Dekker, New York, 1972, 211-246.

10. I. Satake, *Algebraic structures of symmetric domains*, Princeton University Press, 1980.

11. H. Upmeier, *Jordan algebras in analysis, operator theory and quantum mechanics*, CBMS Regional Conference Series **67**, 1987.

12. J.A. Wolf, Fine structure of Hermitian symmetric spaces, in *Symmetric Spaces*, op.cit., 271–357.

DEPARTMENT OF MATHEMATICS
SOUTHERN ILLINOIS UNIVERSITY
CARBONDALE, ILLINOIS 62901

E-mail address: ga3613@siucvmb.siu.edu

Contemporary Mathematics
Volume **160**, 1994

Coherent Tensor Operators

DAN FLATH

ABSTRACT. Coherent tensor operators are certain recently defined linear endomorphisms of a model for all the representations of a connected compact lie group. A survey of the current state of their theory, including that of the algebra they generate, is given.

1. Introduction

The key to the successful application of the theory of group representations is often a detailed understanding of the decomposition of tensor (Kronecker) products of irreducible representations. Physicists long ago realized this, for the study of quantum angular momentum has become pretty much synonymous with the study of tensor products of representations of SU_2 (cf. [**4**], [**5**].) It should be emphasized at the outset that the question of interest is not merely which irreducible representations are contained in which tensor products and with what multiplicities - that can be taken as known - but rather how the irreducible subrepresentations fit into the tensor products.

Angular momentum results can be expressed in an elaborate and arcane formalism, sometimes called Racah-Wigner algebra, that seems little known to mathematicians but which physicists assert to be of the greatest utility for the representation theory of SU_2. For the past thirty years, beginning with [**3**], there has been an active but only partly successful search for an elaboration of the representation theory of SU_n (and other compact Lie groups), analogous to the Racah-Wigner algebra for SU_2, that might be useful for the study of higher symmetries of elementary particles ([**1**], [**14**].) The big stumbling block is that for the groups $K = SU_n, n \geq 3$, (and most other compact Lie groups) the tensor

1991 *Mathematics Subject Classification.* Primary 20G05; Secondary 22E46, 81R05.
This paper is in final form and no version of it will be submitted for publication elsewhere.

product of two irreducible representations of K can contain an irreducible representation with multiplicity greater than one. This situation does not arise for $K = SU_2$, a consequence of the famous Clebsch-Gordan formulas.

It has seemed best to many to formulate the hoped-for theory of SU_n in terms of "tensor operators" rather than "tensor products." This means that, given finite dimensional irreducible representations U and V of a group K, one should investigate the K-representation $\mathrm{Hom}_{\mathbf{C}}(U, V)$ rather than the K-representation $U \otimes_{\mathbf{C}} V$. We recall that the K-representation $\mathrm{Hom}_{\mathbf{C}}(U, V)$ is defined as follows: the elements T of $\mathrm{Hom}_{\mathbf{C}}(U, V)$ are linear transformations from U to V, and for $g \in K$ we define gT to be the linear transformation such that $(gT)(u) = g(T(g^{-1}u))$ for all $u \in U$. The difference between tensor operators and tensor products is partly superficial, since the representation $\mathrm{Hom}_{\mathbf{C}}(U, V)$ is isomorphic to $U^* \otimes_{\mathbf{C}} V$, where U^* is the vector space dual to U. The advantage of tensor operators over tensor products is that operators suggest the possibility of composition, a possibility that one wants to exploit to the fullest.

To bring in composition of operators, one starts with a *model* $W = \oplus_{U \in \mathrm{Irr}(K)} U$ of the finite dimensional representations of the group K. The vector space W is a representation of K that contains each isomorphism class of finite dimensional irreducible K-representations with multiplicity one. Note that for compact Lie groups K, the representation W will be infinite dimensional because K has an infinite number of irreducible representations. The truly enormous K-representation $\mathrm{End}_{\mathbf{C}}(W)$ of all linear transformations from the model W to itself contains as subspaces each of the representations $\mathrm{Hom}_{\mathbf{C}}(U, V)$ where U and V are irreducible representations of K. Moreover, elements of $\mathrm{End}_{\mathbf{C}}(W)$ can be composed. Thus $\mathrm{End}_{\mathbf{C}}(W)$ is an algebra

At first it might seem that the study of tensor operators would be identical with the study of the algebra $\mathrm{End}_{\mathbf{C}}(W)$. Certainly $\mathrm{End}_{\mathbf{C}}(W)$ is constructed by taking all the spaces $\mathrm{Hom}_{\mathbf{C}}(U, V)$ and fitting them together in the most obvious way. But for the groups we are interested in, connected compact Lie groups, there is a rich structure that suggests replacing $\mathrm{End}_{\mathbf{C}}(W)$ with a much smaller subalgebra. This was pointed out by Biedenharn at the very beginnings of the "theory of tensor operators of SU_n" in **[3]**, where he focused attention on an important stability result that we wish to state after the introduction of some notation.

Let K be a connected compact Lie group. Let P be the group of weights of K (with respect to some maximal torus), and let P_{++} be the submonoid of integral dominant weights (with respect to some ordering.) The weights in P_{++} index the irreducible representations of K; they are the weights that can be the highest weight of an irreducible representation of K. For $\alpha \in P_{++}$, write Λ^α for an irreducible representation of K with highest weight α. For nondominant weights $\alpha \in P$, we will write $\Lambda^\alpha = (0)$, the zero-dimensional representation. Thus the space $W = \oplus_{\alpha \in P_{++}} \Lambda^\alpha = \oplus_{\alpha \in P} \Lambda^\alpha$ is a model for the representations of K, and we are looking for a hint of an interesting subalgebra of $\mathrm{End}_{\mathbf{C}}(W)$. The hint is the following stability result, already alluded to, which links two different kinds of multiplicities.

THEOREM 1.1. *For α and μ in P_{++} and Δ in P, let* $\operatorname{mult}(\mu;\alpha,\Delta)$ *equal the multiplicity with which the representation Λ^μ appears as a subrepresentation of* $\operatorname{Hom}_{\mathbf{C}}(\Lambda^\alpha, \Lambda^{\alpha+\Delta})$*; and let* $\operatorname{mult}(\Delta;\mu)$ *equal the multiplicity with which the weight Δ occurs in the representation Λ^μ. Then*

$$\operatorname{mult}(\mu;\alpha,\Delta) \leq \operatorname{mult}(\Delta;\mu). \tag{1.1}$$

Moreover, for fixed μ and Δ, equality holds in (1.1) *for all "sufficiently large" weights α.*

Theorem 1.1 suggests that, for fixed μ and Δ, the set of operators in the μ-isotypic piece of $\operatorname{Hom}_{\mathbf{C}}(\Lambda^\alpha, \Lambda^{\alpha+\Delta})$ might be parameterized naturally and simultaneously for every α by the elements of a *single* vector space L_μ^Δ of dimension $\dim(\Lambda^\mu)\operatorname{mult}(\Delta;\mu)$ (with some collapsing in the case of "small" α, i.e., α near Weyl chamber walls.) An element of L_μ^Δ would give rise to a special "coherently" constructed set $\{T_\alpha\}_{\alpha\in P}$ of linear maps T_α from Λ^α to $\Lambda^{\alpha+\Delta}$. The sum $T = \sum_{\alpha\in P} T_\alpha$, which belongs to $\operatorname{End}_{\mathbf{C}}(W)$, could be be called a "coherent tensor operator of shift weight Δ and agent weight μ." Repeating the construction of coherent $T = \sum_{\alpha\in P} T_\alpha$ for all μ and Δ, one gets a large set of linear endomorphisms of W. The algebra A that is generated by the coherent tensor operators constructed in this way is the subalgebra of $\operatorname{End}_{\mathbf{C}}(W)$ we are seeking. It is somehow at the heart of decomposition questions.

This article is a survey of recent joint work of D.Flath and J.Towber on coherent tensor operators ([**9**], [**10**].) This work is part of a continuing effort to carry out, in a quite literal manner, the program suggested by the preceding paragraph for all connected compact Lie groups K. The method is an attempt to exploit the Cartan products $\Lambda^\alpha \otimes \Lambda^\beta \to \Lambda^{\alpha+\beta}$ to the fullest. The definitions are made and some of the basic properties have been established in complete generality. The main theorems, including the existence of "sufficiently many" coherent tensor operators, have been proved for the nontrivial case of $K = SU_3$, but they remain conjectural for $K = SU_n, n \geq 4$.

It should be mentioned that there can well be several different interesting subalgebras A of $\operatorname{End}_{\mathbf{C}}(W)$ for a given compact Lie group K. One might attempt a global construction of A with no attention to the coherence properties that suggest its existence. The problem then is to recover coherence by an analysis of the noncommutative algebra A. The classical choice for this approach is to take A to be the algebra of differential operators on G/U, where G is the complexification of K and U is the unipotent radical of a Borel subgroup of G. This was put forward in [**2**], where the proposed analysis has become famous as the Bernstein-Gelfand-Gelfand conjecture. Another choice for A, analogous to the classical Weyl algebra, has been proposed conjecturally for SU_n in [**6**], with proofs for SU_3 in [**8**] and for SU_4 in [**7**].

2. The main conjectures

We will work with the complexification G of a connected compact Lie group K, rather than with the compact group K itself, Weyl's unitarian trick in reverse. For example, we will discuss the complex connected reductive algebraic group $G = SL_n(\mathbf{C})$ in place of $K = SU_n$. The finite dimensional analytic representations of G are exactly the same as the finite dimensional continuous representations of K, and K is a maximal compact subgroup of G. The regular functions on G restrict to functions on K whose left and right translates by elements of K belong to a finite dimensional space, and all such functions on K extend uniquely to regular functions on G.

Let $W = \oplus_{\alpha \in P} \Lambda^\alpha$ be a model for the analytic representations of the complex connected reductive algebraic group G.

DEFINITION 2.1. *A linear operator $T : W \to W$ is a coherent tensor operator of shift weight Δ for G if and only if the following three conditions are satisfied.*

(i) *T is G-finite, meaning that the set $\{gT\}_{g \in G}$ lies in a finite dimensional space of operators.*

(ii) *$T(\Lambda^\alpha) \subset \Lambda^{\alpha+\Delta}$ for every $\alpha \in P$.*

(iii) *A certain strong relationship, called the coherence condition, made precise in (3.2), holds among the restrictions T_α of T to the subspaces $\Lambda^\alpha \subset W, \alpha \in P_{++}$.*

DEFINITION 2.2. *Let L_μ^Δ be the space of all coherent tensor operators T of shift weight Δ such that the linear span of $\{gT\}_{g \in G}$ is a sum of G-representations all isomorphic to Λ^μ. The elements of L_μ^Δ will be called coherent tensor operators of shift weight Δ and agent weight μ.*

CONJECTURE 1. (*Existence of coherent tensor operators*)

(i) *The set L_μ^Δ is a complex vector space of dimension* $\dim(\Lambda^\mu)\,\mathrm{mult}(\Delta;\mu)$. *As a representation of G, it is isomorphic to the direct sum of* $\mathrm{mult}(\Delta;\mu)$ *copies of Λ^μ.*

(ii) *For every $\Delta \in P$ and $\alpha \in P_{++}$, the map*

$$\oplus_{\mu \in P_{++}} L_\mu^\Delta \to \mathrm{Hom}_{\mathbf{C}}(\Lambda^\alpha, \Lambda^{\alpha+\Delta}), T \to T_\alpha$$

gotten by restricting T from W to $\Lambda^\alpha \subset W$ is surjective.

Conjecture 1 is proved for $G = SL_3(\mathbf{C})$ in **[9]**.

Let $L = \oplus_{\Delta,\mu} L_\mu^\Delta$ be the space of all finite linear combinations of coherent tensor operators of G, called the space of Biedenharn-Wigner operators in **[10]**. Let A be the subalgebra of $\mathrm{End}_{\mathbf{C}}(W)$ that is generated by all the operators in L.

We expect that the elements of L are naturally parameterized by the space $\mathbf{C}[G]$ of all regular functions on G in such a way that there is a kind of compatibility between composition of tensor operators and multiplication of functions in $\mathbf{C}[G]$, expressed as an algebra homomorphism from A to $\mathbf{C}[G]$. This is a bit surprising for two reasons. In the first place, the composition of coherent tensor operators need not be coherent. In the second place, composition of tensor operators is definitely not commutative, whereas multiplication of functions is commutative.

CONJECTURE 2. (*Parameterization of coherent tensor operators*)
There exists a vector space isomorphism $\pi : L \to \mathbf{C}[G]$ satisfying the following three conditions.

(i) *The map π is G-equivariant for the left regular representation on $\mathbf{C}[G]$: $\pi(gT) = g(\pi(T))$, where for $f \in \mathbf{C}[G]$ we set $(gf)(x) = f(g^{-1}x)$ for all g and x in G.*

(ii) *The subspace $\pi(L_\mu^\Delta)$ of $\mathbf{C}[G]$ is spanned by functions of the form $f_{v^*,v}(g) = \langle gv^*, v\rangle$, where $v \in \Lambda^\mu$, $v^* \in (\Lambda^\mu)^*$, and the vector v^* has weight Δ^*. (The map $\Delta \to \Delta^*$ on P is the dual involution that interchanges the highest weights of an irreducible representation and its dual. Thus $\Delta^* = -\omega_0(\Delta)$, where ω_0 is the element of the Weyl group that interchanges positive and negative roots.)*

(iii) *The map π extends to an algebra homomorphism $\pi : A \to \mathbf{C}[G]$. In particular, there is a direct sum decomposition $A = L \oplus I$ for which I, the kernel of π, is a two-sided ideal in A, and the quotient algebra A/I is commutative.*

Conjecture 2 is proved for $G = SL_3(\mathbf{C})$ in **[10]**.

A little more can be said about the space I, the kernel of π. An operator $S : W \to W$ is said to be a *scalar operator* if and only if there are scalars $c_\alpha \in \mathbf{C}$ such that $S(v) = c_\alpha v$ for every $\alpha \in P_{++}$ and every $v \in \Lambda^\alpha$. We say that S *vanishes at* ∞ if and only if $\lim_{\alpha\to\infty} c_\alpha = 0$. Then I is the set of finite linear combinations of operators of the form $T \circ S$, where T is a coherent tensor operator and S is a scalar operator that vanishes at ∞. Since I contains the commutators of all coherent tensor operators, one might say that coherent tensor operators "commute at ∞."

3. Definition of coherence

Let G be a complex connected reductive algebraic group. Let us choose once for all a maximal torus T of G, and a Borel sugroup B of G containing T. We let U be the unipotent radical of B. For example, for $G = SL_n(\mathbf{C})$ we will take T to be the subgroup of diagonal matrices and B to be the subgroup of upper triangular matrices, so that U will be the subgroup of upper triangular matrices with 1's on the main diagonal. The group P of weights and monoid P_{++} of dominant weights of G are determined by the choices of T and B.

The Borel-Weil theorem **[13]** provides the concrete model of the representations of G that we will use. As model we take the space $W = \mathbf{C}[G/U]$ of regular functions f on G such that $f(xu) = f(x)$ for all x in G and u in U. The action of G on W is by left translations: by definition, $(gf)(x) = f(g^{-1}x)$ for all x

and g in G. The irreducible representation of highest weight $\alpha \in P_{++}$ is the subrepresentation

$$\Lambda^\alpha = \{f \in \mathbf{C}[G/U] : f(xt) = \alpha^*(t)f(x) \text{ for } x \in G, t \in T\}.$$

For the classical groups, $\mathbf{C}[G/U]$ can be made explicit. For example, the regular functions on $G = SL_n(\mathbf{C})$ are the polynomials in the n^2 coordinate functions X_{ij}. The $\binom{n}{r}$ $r \times r$-minors of $(X) = (X_{ij})$ that can be formed on the first r columns of X, $1 \leq r \leq n-1$, belong to $W = \mathbf{C}[G/U]$, where U is the subgroup of upper triangular matrices in G with 1's on the diagonal. Finally $\mathbf{C}[G/U]$ consists of all functions on G that can be expressed as polynomials in these left-justified minors.

The key feature of the model $\mathbf{C}[G/U]$ is that it is a commutative algebra, sometimes called the shape algebra of G ([**15**].) Functions in it can be multiplied, and clearly $\Lambda^\alpha \cdot \Lambda^\beta \subset \Lambda^{\alpha+\beta}$.

There are three maps, m, d, and d^*, to be named.

(i)

$$m_{\alpha,\beta} : \Lambda^\alpha \otimes \Lambda^\beta \to \Lambda^{\alpha+\beta}$$

is multiplication of functions in the algebra $\mathbf{C}[G/U]$. We shall denote $m_{\alpha,\beta}(a \otimes b)$ simply by $a \cdot b$.

(ii)

$$d_{\alpha,\beta} : \Lambda^{\alpha+\beta} \to \Lambda^\alpha \otimes \Lambda^\beta$$

is the unique G-map such that $m_{\alpha,\beta} \circ d_{\alpha,\beta}$ is the identity on $\Lambda^{\alpha+\beta}$.

(iii)

$$d^*_{\alpha,\beta} : (\Lambda^\alpha)^* \otimes (\Lambda^\beta)^* \to (\Lambda^{\alpha+\beta})^*$$

is the transpose of $d_{\alpha,\beta}$ and provides a normalized Cartan product for the dual representations. We shall denote $d^*_{\alpha,\beta}(a^* \otimes b^*)$ simply by $a^* \cdot b^*$.

DEFINITION 3.1. *Let α,β, and λ be dominant integral weights of G. The coherence map*

$$\mathcal{C}^{\alpha,\beta}_\lambda : \mathrm{Hom}_{\mathbf{C}}(\Lambda^{\alpha+\lambda}, \Lambda^{\beta+\lambda}) \to \mathrm{Hom}_{\mathbf{C}}(\Lambda^\alpha, \Lambda^\beta)$$

is defined in the following way: Choose dual bases $\{e_{i,\beta}\}$ of Λ^β and $\{e^*_{i,\beta}\}$ of $(\Lambda^\beta)^*$; and choose dual bases $\{e_{j,\lambda}\}$ of Λ^λ and $\{e^*_{j,\lambda}\}$ of $(\Lambda^\lambda)^*$. For $T : \Lambda^{\alpha+\lambda} \to \Lambda^{\beta+\lambda}$ and $v \in \Lambda^\alpha$, then $(\mathcal{C}^{\alpha,\beta}_\lambda T)(v) \in \Lambda^\beta$ is given by

$$(\mathcal{C}^{\alpha,\beta}_\lambda T)(v) = \Gamma_{\beta,\lambda} \sum_{i,j} \langle T(v \cdot e_{j,\lambda}), e^*_{i,\beta} \cdot e^*_{j,\lambda} \rangle e_{i,\beta} \tag{3.1}$$

where

$$\Gamma_{\beta,\lambda} = \frac{\dim(\Lambda^\beta)}{\dim(\Lambda^{\beta+\lambda})}.$$

Note that the definition of $\mathcal{C}_\lambda^{\alpha,\beta}$ is independent of the choice of dual bases made. It is easy to show that $\mathcal{C}_\lambda^{\alpha,\beta}$ is G-equivariant. Also, the coherence maps are compatible under composition:

$$\mathcal{C}_\mu^{\alpha,\beta} \circ \mathcal{C}_\lambda^{\alpha+\mu,\beta+\mu} = \mathcal{C}_{\lambda+\mu}^{\alpha,\beta}.$$

We can now complete the definition of coherent tensor operators by giving the coherence condition for an operator $T : W \to W$ such that $T(\Lambda^\alpha) \subset \Lambda^{\alpha+\Delta}$ for every $\alpha \in P_{++}$. Writing T_α for the restriction of T to the subspace $\Lambda^\alpha \subset W$, the coherence condition is that

$$\mathcal{C}_\lambda^{\alpha,\alpha+\Delta}(T_{\alpha+\lambda}) = T_\alpha \quad \text{for every } \alpha \text{ and } \lambda \text{ in } P_{++}. \tag{3.2}$$

4. Construction of coherent tensor operators

What are the coherent tensor operators for a group G?

To begin with, there are the multiplications in the algebra $W = \mathbf{C}[G/U]$. For every $\beta \in P_{++}$ and every $u \in \Lambda^\beta \subset W$, let $M_u : W \to W$ be the linear operator defined by

$$M_u(v) = u \cdot v \in \Lambda^{\alpha+\beta}$$

for every $v \in \Lambda^\alpha$. The operator M_u is a coherent tensor operator of shift weight β.

Then there are the interior multiplications. For every $\beta \in P_{++}$ and every $u^* \in (\Lambda^\beta)^*$, let $D_{u^*} : W \to W$ be the linear operator defined by the conditions that

$$D_{u^*}(\Lambda^\alpha) \subset \Lambda^{\alpha-\beta} \quad \text{for all } \alpha \in P_{++}$$

and

$$\langle D_{u^*}(v), w^*\rangle = \frac{\dim(\Lambda^{\alpha-\beta})}{\dim(\Lambda^\alpha)}\langle v, u^* \cdot w^*\rangle$$

for all $v \in \Lambda^\alpha$ and $w^* \in (\Lambda^{\alpha-\beta})^*$. The operator D_{u^*} is a coherent tensor operator of shift weight $-\beta$.

THEOREM 4.1. *Every operator of the form*

$$(\prod_i D_{v_i^*}) \circ (\prod_j M_{u_j}) \tag{4.1}$$

is a coherent tensor operator.

CONJECTURE 3. *Every coherent tensor operator for G is a finite* **C**-*linear combination of coherent tensor operators of the form* (4.1).

Theorem 4.1 is proved in **[9]**, and Conjecture 3 is proved for $G = SL_3(\mathbf{C})$ in **[9]**. Conjecture 3 would follow in general from nonsingularity of the bilinear forms

$$F_{\alpha,\beta} : [(\Lambda^\alpha)^* \otimes \Lambda^\beta] \times [\Lambda^\alpha \otimes (\Lambda^\beta)^*] \to \mathbf{C}$$

defined by

$$F_{\alpha,\beta}(a^* \otimes b, a \otimes b^*) = \langle a \cdot b, a^* \cdot b^* \rangle.$$

We can now define the map π of Conjecture 2 that takes coherent tensor operators to functions on G. Since π is supposed to be linear and multiplicative, and since, by Conjecture 3, all coherent tensor operators can be constructed from multiplications and interior multiplications, it will suffice to explain how to assign functions on G to the operators M_u and D_{v^*}.

For $u \in \Lambda^\alpha \subset \mathbf{C}[G/U] \subset \mathbf{C}[G]$, we simply define $\pi(M_u) = u$.

For $\alpha \in P_{++}$, let θ_α be the unique lowest weight vector (i.e., of weight $-\alpha^*$) in Λ^α such that $\theta_\alpha(1_G) = 1$. For $v^* \in (\Lambda^\alpha)^*$ we define the function $\pi(D_{v^*}) \in \mathbf{C}[G]$ by the formula

$$(\pi(D_{v^*}))(g) = \langle g\theta_\alpha, v^* \rangle \quad \text{for all } g \text{ in } G.$$

We have defined the map π in such a way that $\pi(M_u) \in \mathbf{C}[G/U]$ and $\pi(D_{v^*}) \in \mathbf{C}[G/U^-]$, where U^- is the unipotent subgroup of G "opposite" to U. For example, for $G = SL_n(\mathbf{C})$ and U the subgroup of upper triangular matrices with 1's on the main diagonal, the group U^- will be the subgroup of lower triangular matrices with 1's on the main diagonal. One of the interesting problems that arises in attempting to prove Conjecture 2 is to determine a full set of relations between the two subalgebras $\mathbf{C}[G/U]$ and $\mathbf{C}[G/U^-]$ of $\mathbf{C}[G]$. This problem is solved for G any of the classical groups in **[11]**.

The proofs of Conjectures 1, 2, and 3 for $G = SL_3(\mathbf{C})$ in **[9]** and **[10]** rely crucially on remarkable explicit identities for certain of the commutators $M_u \circ D_{v^*} - D_{v^*} \circ M_u$, identities that are tabulated in Lemma 4.7 of **[9]**. It is ignorance of such identities for other groups that prevents us at this time from proving the conjectures in greater generality. A first step, as we hope, toward extension of the identities to $G = SL_n(\mathbf{C})$, in a different formalism, has been taken in **[12]**.

5. Example: $G = SL_2(\mathbf{C})$

Let $G = \{\left(\begin{smallmatrix} x & z \\ y & w \end{smallmatrix}\right) : xw - yz = 1\}$. Define the four functions $X, Y, Z, W \in \mathbf{C}[G]$ by the formulas

$$X(g) = x, Y(g) = y, Z(g) = Z, W(g) = w \quad \text{for } g = \left(\begin{smallmatrix} x & z \\ y & w \end{smallmatrix}\right) \in G$$

so that

$$\mathbf{C}[G] = \mathbf{C}[X, Y, Z, W]/(XW - YZ - 1).$$

Choose subgroups T, U, and U^- of G as follows:

$T = \{\left(\begin{smallmatrix} x & 0 \\ 0 & w \end{smallmatrix}\right) \in G\}$.

$U = \{\left(\begin{smallmatrix} 1 & z \\ 0 & 1 \end{smallmatrix}\right) \in G\}$ so that $\mathbf{C}[G/U] = \mathbf{C}[X, Y]$.

$U^- = \{\left(\begin{smallmatrix} 1 & 0 \\ y & 1 \end{smallmatrix}\right) \in G\}$ so that $\mathbf{C}[G/U^-] = \mathbf{C}[Z, W]$.

For $n \geq 0$, define $\alpha_n \in P_{++}$ by the formula $\alpha_n\left(\begin{smallmatrix} x & 0 \\ 0 & w \end{smallmatrix}\right) = x^n$. Then $\Lambda^{\alpha_n} \subset \mathbf{C}[G]$ is the space of homogeneous polynomials in X and Y of degree n.

Let $\{X^*, Y^*\}$ be the basis of $(\Lambda^{\alpha_1})^*$ that is dual to the basis $\{X, Y\}$ of Λ^{α_1}. We have:

$$\begin{aligned}
M_X(X^aY^b) &= X^{a+1}Y^b \\
M_Y(X^aY^b) &= X^aY^{b+1} \\
D_{X^*}(X^aY^b) &= \frac{a}{a+b+1}X^{a-1}Y^b \\
D_{Y^*}(X^aY^b) &= \frac{b}{a+b+1}X^aY^{b-1}
\end{aligned}$$

The map $\pi : L \to \mathbf{C}[G]$ is as follows:

$$\begin{aligned}
\pi(M_X) &= X \\
\pi(M_Y) &= Y \\
\pi(D_{X^*}) &= W \\
\pi(D_{Y^*}) &= -Z
\end{aligned}$$

The four nonvanishing commutators among the operators M_X, M_Y, D_{X^*}, and D_{Y^*} are as follows:

$$\begin{aligned}
M_X \circ D_{X^*} - D_{X^*} \circ M_X &= -D_{Y^*} \circ M_Y \circ S \\
M_Y \circ D_{Y^*} - D_{Y^*} \circ M_Y &= -D_{X^*} \circ M_X \circ S \\
M_X \circ D_{Y^*} - D_{Y^*} \circ M_X &= D_{Y^*} \circ M_X \circ S \\
M_Y \circ D_{X^*} - D_{X^*} \circ M_Y &= D_{X^*} \circ M_Y \circ S
\end{aligned}$$

where S is the scalar operator vanishing at ∞ defined by the formula

$$S(X^aY^b) = \frac{1}{a+b+1}X^aY^b.$$

Note finally that the map π applied to the identity

$$D_{X^*} \circ M_X + D_{Y^*} \circ M_Y = I$$

yields the determinant condition $WX - ZY = 1$.

References

1. G.E. Baird and L.C. Biedenharn, *On the representations of the semi-simple lie groups IV, a canonical classification for tensor operators in SU_3*, Journal of Mathematical Physics **5** (1964), 1730–1747.
2. I.N. Bernstein, I.M. Gelfand, and S.I.Gelfand, *Differential operators on the base affine space and a study of g-modules*, Lie Groups and their Representations (Proc. Summer School of the Bolyai Janos Mathematical Society, Budapest, 1971) (I.M. Gelfand, ed.), Halsted Press, New York, 1975, pp.21–64.
3. L.C. Biedenharn, *The generalization of the Wigner-Racah angular momentum calculus*, Physics Letters **3** (1963), 254–256.

4. L.C. Biedenharn and J.D. Louck, *Angular momentum in quantum physics*, Addison-Wesley, Reading, 1981.
5. L.C. Biedenharn and J.D. Louck, *The Racah-Wigner algebra in quantum theory*, Addison-Wesley, Reading, 1981.
6. D. Flath, *An analogue of the Weyl algebra*, Algebraic Geometry Seminar (Proc. of the Algebraic Geometry Seminar, Singapore, 1987) (M. Nagata and T.A. Peng, eds.), World Scientific, Singapore, 1988, pp.1–4.
7. D. Flath, *Remarks on tensor operators*, Journal of Mathematical Physics **31** (1990), 1605–1609.
8. D. Flath and L.C. Biedenharn, *Beyond the enveloping algebra of* $sl(3)$, Canadian Journal of Mathematics **37** (1985), 710–729.
9. D. Flath and J. Towber, *Tensor Operators I, the concept of a coherent tensor operator*, Communications in Algebra **18** (1990), 4047–4086.
10. D. Flath and J. Towber, *Tensor Operators II, the algebra of tensor operators*, Communications in Algebra **20** (1992), 2903–2917.
11. D. Flath and J. Towber, *Generators and relations for the affine rings of the classical groups*, Communications in Algebra **20** (1992), 2877–2902.
12. D. Flath and J. Towber, *Tensor Operators III, some fundamental tensor operator identities*, Journal of Mathematical Physics **34** (1993), 1523–1547.
13. A.W. Knapp, *Representation theory of semisimple groups*, Princeton University Press, Princeton, 1986.
14. J.D. Louck, *Recent progress toward a theory of tensor operators in the unitary groups*, American Journal of Physics **38** (1970), 3–42.
15. J. Towber, *Two new functors from modules to algebras*, Journal of Algebra **47** (1977), 80–104.

Department of Mathematics and Statistics, University of South Alabama, Mobile, Alabama 36688

E-mail address: F2KC@USOUTHAL.BITNET

Contemporary Mathematics
Volume **160**, 1994

$\mathcal{U}_q(sl(2))$ and q-Special Functions

ROBERTO FLOREANINI AND LUC VINET

ABSTRACT. The interpretation of the q-special functions based on the quantum algebra $\mathcal{U}_q(sl(2))$ is reviewed. A left-module $V^{(\lambda,m_0)}$ depending on two complex parameters is first considered. Matrix elements of q-exponentials in the generators are found to involve basic hypergeometric series. The q-analog of the metaplectic representation of the real form $\mathcal{U}(su(1,1))$ is examined next. It is shown to provide an algebraic framework for certain q-generalizations of the Gegenbauer polynomials.

1. Introduction

This paper is devoted to the application of quantum algebras to the theory of q-special functions. These functions are extensions to a base q of the standard special functions. Some of the best known examples are the basic hypergeometric series, [1]

$$
\begin{aligned}
&{}_r\phi_s(a_1,a_2,\ldots,a_r;b_1,\ldots,b_s;q;z) \\
&\quad = \sum_{n=0}^{\infty} \frac{(a_1;q)_n(a_2;q)_n\ldots(a_r;q)_n}{(q;q)_n(b_1;q)_n\ldots(b_s;q)_n} \left[(-1)^n q^{\frac{n(n-1)}{2}}\right]^{1+s-r} z^n ,
\end{aligned} \tag{1.1}
$$

which are the q-analogs of the hypergeometric series ${}_rF_s$. The q-shifted factorial $(a;q)_\alpha$ that enter the definition (1.1) is defined as follows for α complex:

$$
(a;q)_\alpha = \frac{(a;q)_\infty}{(aq^\alpha;q)_\infty} , \qquad (a;q)_\infty = \prod_{k=0}^{\infty}(1-aq^k) . \tag{1.2}
$$

When $0<|q|<1$, the series ${}_r\phi_s$ converges absolutely for all z if $r\le s$, and for $|z|<1$ if $r=s+1$; it terminates if one of the numerator parameters $\{a_i\}$ is of the q^{-m}, $m=0,1,2,\ldots$. Note that as far as convergence is concerned, a ${}_r\phi_s$ series with m numerator coefficients $\{a_i\}$ and n denominator coefficients $\{b_i\}$

1991 Mathematics Subject Classification. 33D80, 33D15, 33D25.

Supported in part by NSERC of Canada and FCAR of Québec.

This paper is in final form and no version of it will be submitted for publication elsewhere.

equal to zero, behaves as a ${}_{r-m}\phi_{s-n}$ series with no vanishing coefficients. In the following, we shall assume $0 < q < 1$.

The standard special functions are well-known to admit a Lie theoretic description.[2,3] A similar interpretation of q-special functions based on quantum algebras is now being developed.[4-16] We present here some of our results in this direction.

Quantum algebras are deformations as Hopf algebras of the universal enveloping algebras $\mathcal{U}(\mathcal{G})$ of Lie algebras $\mathcal{G}$.[17,18] If a single complex parameter q is involved, they are denoted $\mathcal{U}_q(\mathcal{G})$. In this paper, we shall focus on one of the simplest quantum algebra, $\mathcal{U}_q(sl(2))$, and shall discuss the q-special functions that are associated to its representations.

Our treatment is parallel to the standard Lie theoretic one. The following two q-exponentials play a central role in it

$$(1.3a) \qquad e_q(z) = \sum_{n=0}^{\infty} \frac{1}{(q;q)_n}\, z^n = \frac{1}{(z;q)_\infty}\ , \qquad |z| < 1\ ,$$

$$(1.3b) \qquad E_q(z) = \sum_{n=0}^{\infty} \frac{q^{\frac{1}{2}n(n-1)}}{(q;q)_n}\, z^n = (-z;q)_\infty\ ,$$

These functions satisfy $e_q(z)\, E_q(-z) = 1$, and $\lim_{q\to 1^-} e_q\big(z(1-q)\big) = \lim_{q\to 1^-} E_q\big(z(1-q)\big) = e^z$. They are furthermore eigenfunctions of the q-derivative operators. Indeed, if

$$(1.4) \qquad D_z^{\pm} = z^{-1}(1 - T_z^{\pm 1})\ ,$$

with T_z the q-dilatation operator, $T_z\, \varphi(z) = \varphi(qz)$, one has

$$(1.5a) \qquad D_z^{+}\, e_q(\lambda z) = \lambda e_q(\lambda z)\ ,$$

$$(1.5b) \qquad D_z^{-}\, E_q(\lambda z) = -q^{-1}\lambda E_q(\lambda z)\ ,$$

with λ a complex parameter.

The basic ides underlying our quantum algebra interpretation of q-special functions is to replace the exponential mappings from $\mathcal{G}$ into the corresponding Lie group by q-exponentials from $\mathcal{U}_q(\mathcal{G})$ into the completion of this algebra. We will thus consider products U of q-exponentials in quantum algebra generators to find that their matrix elements in representations of $\mathcal{U}_q(\mathcal{G})$ involve q-special functions.

Certain basic hypergeometric series will thus be connected to $\mathcal{U}_q(sl(2))$ in Section 2 by considering a left module $V^{(\lambda,m_0)}$ depending on two complex parameters. The real form $\mathcal{U}_q(su(1,1))$ of $\mathcal{U}_q(sl(2))$ will be the subject of Section 3. A q-analog of the $SU(1,1)$ metaplectic representation will be presented and shown to provide an algebraic setting for the discrete q-ultraspherical polynomials, which are certain q-generalizations of the Gegenbauer polynomials. The usefulness of such interpretations of q-functions to derive orthogonality relations, generating functions, addition formulas *etc.*, will be illustrated in Section 4 where various properties of the discrete q-ultraspherical polynomials will be obtained by exploiting the connection with $\mathcal{U}_q(su(1,1))$.

2. $\mathcal{U}_q(sl(2))$ and basic hypergeometric series

The quantum universal enveloping algebra $\mathcal{U}_q(sl(2))$ is the Hopf algebra generated by the elements k, k^{-1}, e and f satisfying the relations[17,18]

$$k\,e\,k^{-1} = q\,e\ , \qquad k\,f\,k^{-1} = q^{-1}f\ , \qquad [e,f] = \frac{k^2 - k^{-2}}{q - q^{-1}}\ , \tag{2.1}$$

and $k\,k^{-1} = k^{-1}k = 1$. The coproduct $\Delta : \mathcal{U}_q(sl(2)) \to \mathcal{U}_q(sl(2)) \otimes \mathcal{U}_q(sl(2))$, antipode $S : \mathcal{U}_q(sl(2)) \to \mathcal{U}_q(sl(2))$ and counit $\varepsilon : \mathcal{U}_q(sl(2)) \to \mathbf{C}$ are defined by:
(2.2)

$$\begin{array}{lll} \Delta(k) = k \otimes k\ , & \Delta(e) = e \otimes k + k^{-1} \otimes e\ , & \Delta(f) = f \otimes k + k^{-1} \otimes f\ , \\ S(k) = k^{-1}\ , & S(e) = -q\,e\ , & S(f) = -q^{-1}\,f\ , \\ \varepsilon(k) = 1\ , & \varepsilon(e) = 0\ , & \varepsilon(f) = 0\ . \end{array}$$

We now introduce a left $\mathcal{U}_q(sl(2))$-module $V^{(\lambda,m_0)} = \bigoplus_{j\in I} \mathbf{C}\xi_j$, where $I = \{i|\ i = m_0 + n,\ n \in \mathbf{Z}\}$, and λ, m_0 are complex numbers.[9] $V^{(\lambda,m_0)}$ is infinite dimensional, unless $m_0 + \lambda$ and $m_0 - \lambda$ are both integers. The corresponding representation is characterized by the following action of the generators on the basis vectors ξ_j, $j \in I$:

$$\begin{aligned} k\,\xi_j &= q^{-j}\,\xi_j\ , \\ e\,\xi_j &= q^{(1-2\lambda)/2}\,\frac{1 - q^{2(\lambda+j)}}{1-q}\,\xi_{j-1}\ , \\ f\,\xi_j &= q^{(1-2\lambda)/2}\,\frac{1 - q^{(\lambda-j)}}{1-q}\,\xi_{j+1}\ . \end{aligned} \tag{2.3}$$

Given any $a \in \mathcal{U}_q(sl(2))$, its matrix elements a_{ij} in this representation are defined by

$$a\,\xi_j = \sum_{i\in \mathbf{Z}+m_0} \xi_i\,a_{ij}\ . \tag{2.4}$$

In analogy with Lie theory, it is natural to consider elements in the completion of $\mathcal{U}_q(sl(2))$ that are of the form

$$U(\alpha,\beta) = \mathcal{E}_{q^2}(\alpha e)\ \mathcal{E}_{q^2}(\beta f)\ , \qquad \widetilde{U}(\alpha,\beta) = \mathcal{E}_{q^2}(\beta f)\ \mathcal{E}_{q^2}(\alpha e)\ , \tag{2.5}$$

with α, β complex parameters and $\mathcal{E}_{q^2}$ either a little, e_{q^2}, or a big, E_{q^2}, q-exponential. Indeed, once we set $k = q^{-h}$ and $q = e^{-\eta}$, in the limit $q \to 1^-$, $\eta \to 0^+$, the defining relations (2.1), become those of $sl(2)$: $[h,e] = -e$, $[h,f] = f$, $[e,f] = -2h$, and the quantities $U\big((1-q^2)\alpha,(1-q^2)\beta\big)$, $\widetilde{U}\big((1-q^2)\alpha,(1-q^2)\beta\big)$ go respectively into the $SL(2)$ group elements $e^{\alpha e}\,e^{\beta f}$ and $e^{\beta f}\,e^{\alpha e}$. Equation (2.5) clearly entails eight cases. Using the action (2.3) of the generators on the basis vectors and certain identities involving q-shifted factorials, it is immediate to show that the matrix elements of these $U(\alpha,\beta)$ and $\widetilde{U}(\alpha,\beta)$ in $V^{(\lambda,m_0)}$ are expressible in terms of q-hypergeometric series. One thus obtains the following results:

1. $U^{(e,e)}(\alpha,\beta) = e_{q^2}(\alpha\, e)\; e_{q^2}(\beta\, f)$

(2.6a)

$$U_{ij}^{(e,e)}(\alpha,\beta) = \left(\beta\,\frac{q^{(1-2\lambda)/2}}{1-q}\right)^{i-j}\frac{\left(q^{2(\lambda-i+1)};q^2\right)_{i-j}}{(q^2;q^2)_{i-j}}$$
$$\times\,{}_3\phi_1\left(q^{2(\lambda+i+1)},q^{2(i-\lambda)},q^2;q^{2(i-j+1)};q^2,\alpha\beta\,\frac{q^{1-2i}}{(1-q^2)^2}\right)\,,\quad \text{if } i-j\geq 0\,,$$

(2.6b)

$$U_{ij}^{(e,e)}(\alpha,\beta) = \left(\alpha\,\frac{q^{(1-2\lambda)/2}}{1-q^2}\right)^{j-i}\frac{\left(q^{2(\lambda+i+1)};q^2\right)_{j-i}}{(q^2;q^2)_{j-i}}$$
$$\times\,{}_3\phi_1\left(q^{2(\lambda+j+1)},q^{2(j-\lambda)},0;q^{2(j-i+1)};q^2,-\alpha\beta\,\frac{q^{1-2j}}{(1-q^2)^2}\right)\,,\quad \text{if } i-j\leq 0\,,$$

2. $U^{(e,E)}(\alpha,\beta) = e_{q^2}(\alpha\, e)\; E_{q^2}(\beta\, f)$

(2.7a)

$$U_{ij}^{(e,E)}(\alpha,\beta) = q^{(i-j)(i-j-1)}\left(\beta\,\frac{q^{(1-2\lambda)/2}}{1-q^2}\right)^{i-j}\frac{\left(q^{2(\lambda-i+1)};q^2\right)_{i-j}}{(q^2;q^2)_{i-j}}$$
$$\times\,{}_2\phi_1\left(q^{2(\lambda+i+1)},q^{2(i-\lambda)};q^{2(i-j+1)};q^2,-\alpha\beta\,\frac{q^{1-2j}}{(1-q^2)^2}\right)\,,\quad \text{if } i-j\geq 0\,,$$

(2.7b)

$$U_{ij}^{(e,E)}(\alpha,\beta) = \left(\alpha\,\frac{q^{(1-2\lambda)/2}}{1-q^2}\right)^{j-i}\frac{\left(q^{2(\lambda+i+1)};q^2\right)_{j-i}}{(q^2;q^2)_{j-i}}$$
$$\times\,{}_2\phi_1\left(q^{2(\lambda+j+1)},q^{2(j-\lambda)};q^{2(j-i+1)};q^2,-\alpha\beta\,\frac{q^{1-2j}}{(1-q^2)^2}\right)\,,\quad \text{if } i-j\leq 0\,,$$

3. $U^{(E,e)}(\alpha,\beta) = E_{q^2}(\alpha\, e)\; e_{q^2}(\beta\, f)$

(2.8a)

$$U_{ij}^{(E,e)}(\alpha,\beta) = \left(\beta\,\frac{q^{(1-2\lambda)/2}}{1-q^2}\right)^{i-j}\frac{\left(q^{2(\lambda-i+1)};q^2\right)_{i-j}}{(q^2;q^2)_{i-j}}$$
$$\times\,{}_2\phi_1\left(q^{2(\lambda+i+1)},q^{2(i-\lambda)};q^{2(i-j+1)};q^2,-\alpha\beta\,\frac{q^{1-2i}}{(1-q^2)^2}\right)\,,\quad \text{if } i-j\geq 0\,,$$

(2.8b)

$$U_{ij}^{(E,e)}(\alpha,\beta) = q^{(j-i)(j-i-1)}\left(\alpha\,\frac{q^{(1-2\lambda)/2}}{1-q^2}\right)^{j-i}\frac{\left(q^{2(\lambda+i+1)};q^2\right)_{j-i}}{(q^2;q^2)_{j-i}}$$
$$\times\,{}_2\phi_1\left(q^{2(\lambda+j+1)},q^{2(j-\lambda)};q^{2(j-i+1)};q^2,-\alpha\beta\,\frac{q^{1-2i}}{(1-q^2)^2}\right)\,,\quad \text{if } i-j\leq 0\,,$$

4. $U^{(E,E)}(\alpha,\beta) = E_{q^2}(\alpha\, e)\; E_{q^2}(\beta\, f)$

(2.9*a*)

$$U^{(E,E)}_{ij}(\alpha,\beta) = q^{(i-j)(i-j-1)} \left(\beta\,\frac{q^{(1-2\lambda)/2}}{1-q^2}\right)^{i-j} \frac{\left(q^{2(\lambda-i+1)};q^2\right)_{i-j}}{(q^2;q^2)_{i-j}}$$
$$\times\,{}_2\phi_2\left(q^{2(\lambda+i+1)},q^{2(i-\lambda)};q^{2(i-j+1)},0;q^2,\alpha\beta\,\frac{q^{1-2j}}{(1-q^2)^2}\right),\quad \text{if } i-j\geq 0\,,$$

(2.9*b*)

$$U^{(E,E)}_{ij}(\alpha,\beta) = q^{(j-i)(j-i+1)} \left(\alpha\,\frac{q^{(1-2\lambda)/2}}{1-q^2}\right)^{j-i} \frac{\left(q^{2(\lambda+i+1)};q^2\right)_{j-i}}{(q^2;q^2)_{j-i}}$$
$$\times\,{}_2\phi_2\left(q^{2(\lambda+j+1)},q^{2(j-\lambda)};q^{2(j-i+1)},0;q^2,\alpha\beta\,\frac{q^{1-2i}}{(1-q^2)^2}\right),\quad \text{if } i-j\leq 0\,,$$

5. $\widetilde{U}^{(e,e)}(\alpha,\beta) = e_{q^2}(\beta\, f)\; e_{q^2}(\alpha\, e)$

(2.10*a*)

$$\widetilde{U}^{(e,e)}_{ij}(\alpha,\beta) = \left(\beta\,\frac{q^{(1-2\lambda)/2}}{1-^2}\right)^{i-j} \frac{\left(q^{2(\lambda-i+1)};q^2\right)_{i-j}}{(q^2;q^2)_{i-j}}$$
$$\times\,{}_3\phi_1\left(q^{2(\lambda-j+1},q^{-2(\lambda+j)},0;q^{2(i-j+1)};q^2,\alpha\beta\,\frac{q^{1+2j}}{(1-q^2)^2}\right),\quad \text{if } i-j\geq 0\,,$$

(2.10*b*)

$$\widetilde{U}^{(e,e)}_{ij}(\alpha,\beta) = \left(\alpha\,\frac{q^{(1-2\lambda)/2}}{1-q^2}\right)^{j-i} \frac{\left(q^{2(\lambda+i+1)};q^2\right)_{j-i}}{(q^2;q^2)_{j-i}}$$
$$\times\,{}_3\phi_1\left(q^{2(\lambda-i+1)},q^{-2(\lambda+i)},0;q^{2(j-i+1)};q^2,\alpha\beta\,\frac{q^{1+2i}}{(1-q^2)^2}\right),\quad \text{if } i-j\leq 0\,,$$

6. $\widetilde{U}^{(e,E)}(\alpha,\beta) = e_{q^2}(\beta\, f)\; E_{q^2}(\alpha\, e)$

(2.11*a*)

$$\widetilde{U}^{(e,E)}_{ij}(\alpha,\beta) = \left(\beta\,\frac{q^{(1-2\lambda)/2}}{1-q^2}\right)^{i-j} \frac{\left(q^{2(\lambda-i+1)};q^2\right)_{i-j}}{(q^2;q^2)_{i-j}}$$
$$\times\,{}_2\phi_1\left(q^{2(\lambda-j+1)},q^{-2(\lambda+j)};q^{2(i-j+1)};q^2,-\alpha\beta\,\frac{q^{1+2j}}{(1-q^2)^2}\right),\quad \text{if } i-j\geq 0\,,$$

(2.11*b*)

$$\widetilde{U}^{(e,E)}_{ij}(\alpha,\beta) = q^{(j-i)(j-i-1)} \left(\alpha\,\frac{q^{(1-2\lambda)/2}}{1-q^2}\right)^{j-i} \frac{\left(q^{2(\lambda+i+1)};q^2\right)_{j-i}}{(q^2;q^2)_{j-i}}$$
$$\times\,{}_2\phi_1\left(q^{2(\lambda-i+1)},q^{-2(\lambda+i)};q^{2(j-i+1)};q^2,-\alpha\beta\,\frac{q^{1+2j}}{(1-q^2)^2}\right),\quad \text{if } i-j\leq 0\,,$$

7. $\widetilde{U}^{(E,e)}(\alpha,\beta) = E_{q^2}(\beta f)\, e_{q^2}(\alpha e)$

(2.12*a*)

$$\widetilde{U}^{(E,e)}_{ij}(\alpha,\beta) = q^{(i-j)(i-j-1)}\left(\beta\,\frac{q^{(1-2\lambda)/2}}{1-q^2}\right)^{i-j}\frac{\left(q^{2(\lambda-i+1)};q^2\right)_{i-j}}{(q^2;q^2)_{i-j}}$$
$$\times\,{}_2\phi_1\left(q^{2(\lambda-j+1)},q^{-2(\lambda+j)};q^{2(i-j+1)};q^2,-\alpha\beta\,\frac{q^{1+2i}}{(1-q^2)^2}\right),\quad \text{if } i-j\ge 0\,,$$

(2.12*b*)

$$\widetilde{U}^{(E,e)}_{ij}(\alpha,\beta) = \left(\alpha\,\frac{q^{(1-2\lambda)/2}}{1-q^2}\right)^{j-i}\frac{\left(q^{2(\lambda+i+1)};q^2\right)_{j-i}}{(q^2;q^2)_{j-i}}$$
$$\times\,{}_2\phi_1\left(q^{2(\lambda-i+1)},q^{-2(\lambda+i)};q^{2(j-i+1)};q^2,-\alpha\beta\,\frac{q^{1+2i}}{(1-q^2)^2}\right),\quad \text{if } i-j\le 0\,.$$

8. $\widetilde{U}^{(E,E)}(\alpha,\beta) = E_{q^2}(\beta f)\, E_{q^2}(\alpha e)$

(2.13*a*)

$$\widetilde{U}^{(E,E)}_{ij}(\alpha,\beta) = q^{(i-j)(i-j-1)}\left(\beta\,\frac{q^{(1-2\lambda)/2}}{1-q^2}\right)^{i-j}\frac{\left(q^{2(\lambda-i+1)};q^2\right)_{i-j}}{(q^2;q^2)_{i-j}}$$
$$\times\,{}_2\phi_2\left(q^{2(\lambda-j+1)},q^{-2(\lambda+j)};q^{2(i-j+1)},0;q^2,\alpha\beta\,\frac{q^{1+2i}}{(1-q^2)^2}\right),\quad \text{if } i-j\ge 0\,,$$

(2.13*b*)

$$\widetilde{U}^{(E,E)}_{ij}(\alpha,\beta) = q^{(j-i)(j-i-1)}\left(\alpha\,\frac{q^{(1-2\lambda)/2}}{1-q^2}\right)^{j-i}\frac{\left(q^{2(\lambda+i+1)};q^2\right)_{j-i}}{(q^2;q^2)_{j-i}}$$
$$\times\,{}_2\phi_2\left(q^{2(\lambda-i+1)},q^{-2(\lambda+i)};q^{2(j-i+1)},0;q^2,\alpha\beta\,\frac{q^{1+2j}}{(1-q^2)^2}\right),\quad \text{if } i-j\le 0\,.$$

The above formulas establish the connection between basic hypergeometric series and $\mathcal{U}_q(sl(2))$; they can now be used in various ways. Orthogonality or biorthogonality relations between basic series are derived by exploiting the fact that $\widetilde{U}^{(E,E)}(-\alpha,-\beta)\, U^{(e,e)}(\alpha,\beta) = \widetilde{U}^{(e,E)}(-\alpha,-\beta)\, U^{(e,E)}(\alpha,\beta) = \widetilde{U}^{(E,e)}(-\alpha,-\beta)\, U^{(E,e)}(\alpha,\beta) = \widetilde{U}^{(e,e)}(-\alpha,-\beta)\, U^{(E,E)}(\alpha,\beta) = 1$. Generating functions are obtained through the construction of a one-variable model for the module $V^{(\lambda,m_0)}$ and addition formulas are arrived at by observing that the matrix elements $U_{ij}(\alpha,\beta)$ and $\widetilde{U}_{ij}(\alpha,\beta)$ also provide models for $V^{(\lambda,m_0)}$. We describe in Section 4 how these applications are carried out, using as example the discrete q-ultraspherical polynomials that are introduced in the next section. For a specific treatment of th ${}_2\phi_1$ series the reader should consult Ref.[13].

Let us also note the following. In the dual of the quantum algebra $\mathcal{U}_q(sl(2))$ lies the Hopf algebra $A(SL_q(2))$ associated with the quantum group $SL_q(2)$. Its Hopf structure is inherited from $\mathcal{U}_q(sl(2))$; its generators obey quadratic relations and are referred to as the coordinate elements of $SL_q(2)$. The symbols

U_{ij} can be viewed as linear functionals on $\mathcal{U}_q(sl(2))$ and hence as elements of $A(SL_q(2))$, through the evaluation maps $U_{ij} \;:\; U(\alpha,\beta)k^\gamma \to U_{ij}(\alpha,\beta)\,q^{-j\gamma}$. From (2.4), these U_{ij} are seen to define a corepresentation of $SL_q(2)$. When expressed analytically in terms of the generators of $A(SL_q(2))$, the functionals U_{ij} are seen to involve q-hypergeometric functions with the coordinate elements as arguments.[19,20] These expressions are easily obtained in our approach. Since $U(\alpha,\beta)k^\gamma$ generates the complete Poincaré-Birkhoff-Witt basis of $\mathcal{U}_q(sl(2))$, the value $U_{ij}(\alpha,\beta)\,q^{-j\gamma}$, of U_{ij} on $U(\alpha,\beta)k^\gamma$ therefore completely determines U_{ij}. Its expression in terms of the coordinate elements of $SL_q(2)$ is then straightforwardly obtained from evaluation formulas. The reader is referred to [11] for details.

3. The metaplectic representation of $\mathcal{U}_q(su(1,1))$ and q-orthogonal polynomials

Let q be real and such that $-1 < q < 1$, $q \neq 0$. The $*$-structure that defines the real form $\mathcal{U}_q(su(1,1))$ of $\mathcal{U}_q(sl(2))$ is given by

$$k^* = k\;, \qquad e^* = -f\;, \qquad f^* = -e\;. \tag{3.1}$$

This anti-automorphism of $\mathcal{U}_q(sl(2))$ satisfy $*^2 = \text{id}$ and preserves the Hopf structure: $\Delta \circ * = (* \otimes *) \circ \Delta$, $(* \circ *)^2 = \text{id}$, $\varepsilon(a^*) = \overline{\varepsilon(a)}$, $a \in \mathcal{U}_q(sl(2))$.

It will be convenient to use a different set of generators $\tilde{k}$, $\tilde{k}^{-1}$, $\tilde{e}$ and $\tilde{f}$:

$$\tilde{k} = k^{-2} \qquad \tilde{k}^{-1} = k^2\;, \qquad \tilde{e} = (1-q^2)\,k^{-3}e\;, \qquad \tilde{f} = (q^2-1)\,kf\;; \tag{3.2}$$

in terms of these, the defining relations (2.1) become

$$\tilde{k}\,\tilde{e}\,\tilde{k}^{-1} = q^{-2}\tilde{e}\;, \quad \tilde{k}\,\tilde{f}\,\tilde{k}^{-1} = q^2\tilde{f}\;, \quad \tilde{e}\tilde{f} - q^2\,\tilde{f}\tilde{e} = (1-q^2)(1-\tilde{k}^2)\;. \tag{3.3}$$

The action of Δ, S and ε on $\tilde{k}$, $\tilde{e}$ and $\tilde{f}$ is

(3.4)
$$\begin{aligned}
&\Delta(\tilde{k}) = \tilde{k}\otimes\tilde{k}\;, && \Delta(\tilde{e}) = \tilde{e}\otimes\tilde{k} + \tilde{k}^2\otimes\tilde{e}\;, && \Delta(\tilde{f}) = \tilde{f}\otimes\tilde{k}^{-1} + \mathbf{1}\otimes f\;,\\
&S(\tilde{k}) = \tilde{k}^{-1}\;, && S(\tilde{e}) = -q^{-2}\,\tilde{k}^{-3}\,\tilde{e}\;, && S(\tilde{f}) = -q^{-2}\,\tilde{k}\,\tilde{f}\;,\\
&\varepsilon(\tilde{k}) = 1\;, && \varepsilon(\tilde{e}) = 0\;, && \varepsilon(\tilde{f}) = 0\;,
\end{aligned}$$

and that of $*$ is

$$\tilde{k}^* = \tilde{k}\;, \qquad \tilde{e}^* = q\,\tilde{f}\,\tilde{k}^2\;, \qquad \tilde{f}^* = q^3\tilde{e}\,\tilde{k}^{-2}\;. \tag{3.5}$$

We will consider in the following a special representation of $\mathcal{U}_q(su(1,1))$ which is a q-analog of the metaplectic representation of the Lie algebra $su(1,1)$. In order to define this reducible representation of $\mathcal{U}_q(su(1,1))$, we consider the direct sum $V = V^{(-1/4,1/4)} \oplus V^{(-1/4,3/4)}$ of two of the $\mathcal{U}_q(sl(2))$-modules $V^{(\lambda,m_0)}$ introduced in the last section. (Note that these vectors spaces are semi-infinite for the special choices $\lambda = -1/4$, $m_0 = 1/4, 3/4$.) One then takes as basis over the space V the set of vectors h_n, with $n = 0,1,2,\ldots$ given by

$$h_{2n} = q^{-3n^2/2}\,(q;q^2)_n\,\xi_{n+1/4}\;, \tag{3.6a}$$

$$h_{2n+1} = q^{-3n(n+1)/2}\,(q^2;q^2)_n\,\xi_{n+3/4}\;, \tag{3.6b}$$

in terms of the basis vectors ξ_j, $j \in \{1/4, 5/4, 9/4, \ldots\}$, respectively $j \in \{3/4, 7/4, 11/4, \ldots\}$, of $V^{(-1/4,1/4)}$, respectively $V^{(-1/4,3/4)}$.

The action of $\tilde{k}$, $\tilde{e}$ and $\tilde{f}$ on these h_n is readily determined from (2.3), (3.2) and (3.6):

$$\tilde{k}\, h_n = q^{n+1/2}\, h_n \ , \qquad \tilde{e}\, h_n = (1-q^n)(1-q^{n-1})\, h_{n-2} \ , \qquad \tilde{f}\, h_n = h_{n+2} \ . \tag{3.7}$$

By construction, this representation decomposes into two irreducible components, with the invariant subspaces $V^{(+)}$ and $V^{(-)}$ spanned by the h_n with n even and odd, respectively. Under an appropriate rescaling of $\tilde{e}$ and $\tilde{f}$, the representation (3.7) reduces to the metaplectic representation of the algebra $su(1,1)$ in the limit $q \to 1^-$. Furthermore, if we take the following positive definite Hermitian inner product on V:

$$\langle h_n, h_m \rangle = q^{-n^2/2} (q;q)_n\, \delta_{nm} \ , \tag{3.8}$$

the representation (3.7) is seen to be unitary, that is $\langle a\xi, \eta\rangle = \overline{\langle \xi, a^\star \eta\rangle}$, for any $\xi, \eta \in V$ and $a \in \mathcal{U}_q(sl(2))$.

A model for this representation can be constructed on the space $\mathcal{H} = \mathcal{H}^{(+)} \oplus \mathcal{H}^{(-)}$ of holomorphic functions of the complex variable z. A basis in $\mathcal{H}$ is provided by the monomials

$$h_n(z) = z^n \ , \qquad n = 0, 1, 2 \ldots \ , \tag{3.9}$$

$\mathcal{H}^{(+)} = \mathrm{span}\{z^{2n}\}$, $\mathcal{H}^{(-)} = \mathrm{span}\{z^{2n+1}\}$. It is easy to check that the relations (3.3) are realized under the identifications

$$\tilde{k} = q^{1/2}\, T_z \qquad \tilde{e} = (D_z^+)^2 \qquad \tilde{f} = z^2 \ , \tag{3.10}$$

and that these operators act on the functions (3.9) according to (3.7).

Let us now consider the following element in the completion of $\mathcal{U}_q(su(1,1))$:

$$\widetilde{U}(\alpha,\beta) = E_{q^2}(\alpha \tilde{f})\, E_{q^2}(\beta \tilde{e}) \ . \tag{3.11}$$

As $q \to 1^-$ this operator reduces to the element $e^{\alpha f}\, e^{\beta e}$ of the Lie group $SU(1,1)$. The matrix elements $\widetilde{U}^{(\pm)}_{kn}(\alpha,\beta)$ of the operator $\widetilde{U}(\alpha,\beta)$ in the spaces $\mathcal{H}^{(\pm)}$ are defined in the usual way

$$\widetilde{U}(\alpha,\beta)\ h_{2n} = \sum_{k=0}^{\infty} h_{2k}\ \widetilde{U}^{(+)}_{kn}(\alpha,\beta) \tag{3.12a}$$

$$\widetilde{U}(\alpha,\beta)\ h_{2n+1} = \sum_{k=0}^{\infty} h_{2k+1}\ \widetilde{U}^{(-)}_{kn}(\alpha,\beta) \ . \tag{3.12b}$$

They can be explicitly computed with the help of the definition (1.3) and of identities involving q-shifted factorials. One has

(3.13)
$$\widetilde{U}^{(+)}_{kn}(\alpha,\beta) = q^{n(n-1)+k(k-1)} \frac{(q;q^2)_n}{(q^2;q^2)_k}\, \alpha^k \beta^n\ {}_2\phi_1\Big(q^{-2k}, q^{-2n}; q; q^2, \frac{q^4}{\alpha\beta}\Big) \ ,$$
$$\widetilde{U}^{(-)}_{kn}(\alpha,\beta) = q^{n(n-1)+k(k-1)} \frac{(q^3;q^2)_n}{(q^2;q^2)_k}\, \alpha^k \beta^n\ {}_2\phi_1\Big(q^{-2k}, q^{-2n}; q^3; q^2, \frac{q^4}{\alpha\beta}\Big) \ .$$

This establishes the connection between terminating basic hypergeometric series ${}_2\phi_1$ and the algebra $\mathcal{U}_q(su(1,1))$.

In the case of

$$U(\alpha,\beta) = e_{q^2}(\beta\tilde{e})\, e_{q^2}(\alpha\tilde{f}) \ , \tag{3.14}$$

one similarly obtains

(3.15)

$$U^{(+)}_{kn}(\alpha,\beta) = q^{2kn}\frac{(q;q^2)_n}{(q^2;q^2)_k}\frac{\alpha^k\beta^n}{(\alpha\beta;q^2)_{k+n+1/2}}\ {}_2\phi_1\Big(q^{-2k},q^{-2n};q;q^2,\frac{q^2}{\alpha\beta}\Big)\ ,$$

$$U^{(-)}_{kn}(\alpha,\beta) = q^{2kn}\frac{(q^3;q^2)_n}{(q^2;q^2)_k}\frac{\alpha^k\beta^n}{(\alpha\beta;q^2)_{k+n+3/2}}\ {}_2\phi_1\Big(q^{-2k},q^{-2n};q^3;q^2,\frac{q^2}{\alpha\beta}\Big)\ .$$

In deriving these results, use have been made of the following transformation rule for the ${}_2\phi_1$ series:[1]

$${}_2\phi_1(q^a,q^b;q^c;q,z) = (q^{a+b-c}z;q)_{c-a-b}\ {}_2\phi_1(q^{c-a},q^{c-b};q^c;q,q^{a+b-c}z)\ . \tag{3.16}$$

A biorthogonality relation involving two terminating series ${}_2\phi_1$ is now easily derived by combining the results (3.13) and (3.15). One just needs to recall that $e_q(z)\,E_q(-z) = 1$, from where it follows that

$$U(\alpha,\beta)\,\widetilde{U}(-\alpha,-\beta) = \mathbf{1}\ . \tag{3.17}$$

Apply now both sides of this identity *e.g.* on h_{2n}, to find

$$\delta_{kn} = \sum_{l=0}^{\infty} U^{(+)}_{kl}(\alpha,\beta)\,\widetilde{U}^{(+)}_{ln}(-\alpha,-\beta)\ . \tag{3.18}$$

After some simplifications and the redefinition $w = q^2/\alpha\beta$, one arrives at the formula

(3.19)

$$\delta_{kn} = (-1)^k\, q^{k(k+1)} w^{-k}\frac{(q;q^2)_k}{(q^2;q^2)_k}\sum_{l=0}^{\infty}(-1)^l q^{l(l+1)+2ln}\frac{(q;q^2)_l}{(q^2;q^2)_l\,(q^2/w;q^2)_{n+l+1/2}}$$
$$\times\ {}_2\phi_1(q^{-2k},q^{-2l};q;q^2,q^2 w)\ {}_2\phi_1(q^{-2n},q^{-2l};q;q^2,w)\ .$$

Let us introduce now the following set of q-polynomials

$$\mathcal{C}^{(\lambda)}_k(q;z) \equiv \mathcal{C}_k(q;q^{2\lambda};z) = \sum_{l=0}^{[\![k/2]\!]}\frac{(-1)^l q^{l(l-1)}(q^{2\lambda};q^2)_{k-l}}{(q^2;q^2)_l\,(q;q)_{k-2l}}\, z^{k-2l}\ , \tag{3.20}$$

with $[\![x]\!]$ the integer part of x. They reduce to the usual Gegenbauer polynomials as $q\to 1^-$. The matrix elements (3.13) and (3.15) can be conveniently rewritten in terms of these $\mathcal{C}^{(\lambda)}_k(q;z)$; explicitly

(3.21)

$$\widetilde{U}^{(+)}_{kn}(\alpha,\beta) = q^{(k+n)^2+k-n}\frac{(q;q)_{2n}}{(q^2;q^2)_{n+k}}\,\alpha^k\beta^n\,\mathcal{C}^{(-n-k)}_{2k}\Big(q;\frac{q}{\sqrt{\alpha\beta}}\Big)\ ,$$

$$\widetilde{U}^{(-)}_{kn}(\alpha,\beta) = -q^{(k+n+1)^2+k-n}\frac{(q;q)_{2n+1}}{(q^2;q^2)_{n+k+1}}\,\alpha^k\beta^n\sqrt{\alpha\beta}\,\mathcal{C}^{(-n-k-1)}_{2k+1}\Big(q;\frac{q}{\sqrt{\alpha\beta}}\Big)\ ,$$

and
(3.22)

$$U_{kn}^{(+)}(\alpha,\beta) = q^{2k(2n+1)} \frac{(q;q)_{2n}}{(q^2;q^2)_{n+k}} \frac{\alpha^k \beta^n}{(\alpha\beta;q^2)_{k+n+1/2}} \, C_{2k}^{(-n-k)}\Big(q;\frac{1}{\sqrt{\alpha\beta}}\Big) ,$$

$$U_{kn}^{(-)}(\alpha,\beta) = -q^{2(2k+1)(n+1)} \frac{(q;q)_{2n+1}}{(q^2;q^2)_{n+k+1}} \frac{\alpha^k \beta^n}{(\alpha\beta;q^2)_{k+n+3/2}} \sqrt{\alpha\beta} \times C_{2k+1}^{(-n-k-1)}\Big(q;\frac{1}{\sqrt{\alpha\beta}}\Big) .$$

To get (3.22) from (3.15), one needs to use the following transformation formula for a terminating ${}_2\phi_1$ series:

(3.23)
$$ {}_2\phi_1(q^{-n},b;c;q,z) = (-1)^n q^{-n(n+1)/2} \frac{(b;q)_n}{(c;q)_n} z^n \times {}_2\phi_1(q^{-n},q^{1-n}/c;q^{1-n}/b;q,q^{n+1}c/bz) .$$

Also, note that: $q^{2n}\, C_{2n}^{(-n-k)} = q^{2k}\, C_{2k}^{(-n-k)}$ and $q^{2n}\, C_{2n+1}^{(-n-k-1)} = q^{2k}\, C_{2k+1}^{(-n-k-1)}$.

The polynomials (3.20) obey the three-term recurrence relation

(3.24)
$$(1-q^{k+1})\, C_{k+1}^{(\lambda)}(q;z) - (1-q^{2(\lambda+k)})\, z\, C_k^{(\lambda)}(q;z) + q^{k-1}(1-q^{2\lambda+k-1})\, C_{k-1}^{(\lambda)}(q;z) = 0 ,$$

and thus form a set of orthogonal polynomials. Indeed, they can be expressed in terms of q-Jacobi polynomials. One easily finds that

(3.25a)
$$C_{2k}^{(\lambda)}(q;z) = (-1)^k q^{k(k-1)} \frac{(q^{2\lambda};q^2)_k}{(q^2;q^2)_k}\, p_k(z^2;q^{-1},q^{2\lambda-1};q^2) ,$$

(3.25b)
$$C_{2k+1}^{(\lambda)}(q;z) = \frac{(-1)^k q^{k(k-1)}}{(1-q)} \frac{(q^{2\lambda};q^2)_{k+1}}{(q^2;q^2)_k}\, z\, p_k(z^2;q,q^{2\lambda-1};q^2) ,$$

where

(3.26)
$$p_n(x;a,b;q) = {}_2\phi_1(q^{-n},abq^{n+1};aq;q,qx) ,$$

are the little q-Jacobi polynomials [21,1]. The r.h.s. of (3.25) are actually proportional the big q-Jacobi polynomials $P_n^{(\lambda-1/2,\lambda-1/2)}(z;1,1:q)$. In fact, one also finds:

(3.27)
$$C_k^{(\lambda)}(q;z) = \frac{(q^2;q^2)_k (q^{2\lambda};q)_k}{(q;q)_k (q^{2\lambda+1};q^2)_k}\, P_k^{(\lambda-1/2,\lambda-1/2)}(z;1,1:q) ,$$

where in general[21]

(3.28)
$$P_n^{(\alpha,\beta)}(x;c,d:q) = c^n q^{-n(\alpha+1)} \frac{(q^{2\alpha+2};q^2)_n}{(q^2;q^2)_n} \times {}_3\phi_2(q^{-n},q^{n+\alpha+\beta+1},\frac{x}{c}q^{\alpha+1};q^{\alpha+1},-\frac{d}{c}q^{\alpha+1};q,q) .$$

From the orthogonality relation obeyed by $P_n^{(\alpha,\beta)}(x;c,d:q)$,[21] one derives that

$$\begin{aligned}\int_{-1}^{1} \mathcal{C}_k^{(\lambda)}(q;z)\,\mathcal{C}_n^{(\lambda)}(q;z)\;(q^2z^2;q^2)_{\lambda-1/2}\;d_qz \\ = 2\,q^{n(n-1)/2}\frac{(q^2;q^2)_{\lambda-1/2}}{(q^3;q^2)_{\lambda-1/2}}\,\frac{(q^{2\lambda};q)_n}{(q;q)_n}\,\frac{(1-q^{2\lambda})}{(1-q^{2(\lambda+n)})}\,\delta_{kn}\;,\end{aligned} \tag{3.29}$$

where the q-integral is defined in the usual way:[1]

$$\int_a^b f(t)\,d_qt = \int_0^b f(t)\,d_qt - \int_0^a f(t)\,d_qt\;, \tag{3.30}$$

with

$$\int_0^a f(t)\,d_qt = a(1-q)\sum_{n=0}^{\infty} f(aq^n)\,q^n\;. \tag{3.31}$$

4. Properties of the discrete q-ultraspherical polynomials

The definition of the discrete q-ultraspherical polynomials $\mathcal{C}_n^{(\lambda)}(q;z)$ given in the last section arose naturally in our analysis of the metaplectic representation of $\mathcal{U}_q(su(1,1))$. We shall now present properties of these polynomials that can be derived from this algebraic interpretation.

From formula (3.20), one directly obtains the q-difference equation that the polynomials $\mathcal{C}_n^{(\lambda)}(q;z)$ satisfy

$$\begin{aligned}\Big[(1-q^{2\lambda+1}\,z^2)\,D_z^+ + q^{2\lambda+1}(1-q^n)\,z\Big]\mathcal{C}_n^{(\lambda)}(q;z) \\ = (1-q^{2\lambda+n-1})\;\mathcal{C}_{n-1}^{(\lambda)}(q;z)\;,\end{aligned} \tag{4.1a}$$

$$\begin{aligned}\Big[(1-q^{2\lambda+1}\,z^2)\,D_z^+ - q^{1-n}(1-q^{2\lambda+n})\,z\Big]\mathcal{C}_n^{(\lambda)}(q;z) \\ = -q^{1-n}(1-q^{n-1})\;\mathcal{C}_{n+1}^{(\lambda)}(q;z)\;.\end{aligned} \tag{4.1b}$$

The one-variable model (3.9), (3.10) can now be used to obtain certain generating relations. Note that the special case $\mathcal{C}_k(q;0;z)$ coincides (up to a normalization factor) with the so called discrete q-Hermite polynomials[1]

$$\begin{aligned}H_k(q;z) &= (q;q)_k\;\mathcal{C}_k(q;0;z) \\ &= \sum_{l=0}^{[\![k/2]\!]} \frac{(-1)^l q^{l(l-1)}\,(q;q)_k}{(q^2;q^2)_l\,(q;q)_{k-2l}}\,z^{k-2l}\;.\end{aligned} \tag{4.2}$$

These polynomials allow to sum the l.h.s. of (3.12) in the holomorphic realization. In fact, quite in general one has

$$E_{q^2}\Big(-\eta\,(D_z^+)^2\Big)\,z^n = \eta^{n/2}\;H_n\big(q;z\,\eta^{-1/2}\big)\;. \tag{4.3}$$

For negative η, *i.e.* when the variable $z\,\eta^{-1/2}$ takes imaginary values, the polynomials H_n are still defined.[22] Inserting (4.3) in (3.12), together with the explicit expressions (3.21) of the matrix elements $\widetilde{U}^{(\pm)}_{kn}(\alpha,\beta)$, one obtains two additional relations connecting the discrete q-ultraspherical polynomials with the discrete q-Hermite polynomials. Introducing for convenience the new variables $w^2 = \alpha\, q^2\, z^2$, $x^2 = q^2/\alpha\beta$, one gets:

$$\begin{aligned} &E_{q^2}\Big(\frac{w^2}{q^2}\Big)\, H_{2n}\Big(q;\frac{iwx}{q^2}\Big) \\ &\qquad = (-1)^n \sum_{k=0}^{\infty} q^{(n+k)(n+k-1)}\, \frac{(q;q)_{2n}}{(q^2;q^2)_{n+k}}\, \mathcal{C}^{(-n-k)}_{2k}(q;x)\; w^{2k}\ , \end{aligned} \tag{4.4a}$$

$$\begin{aligned} &E_{q^2}\Big(\frac{w^2}{q^2}\Big)\, H_{2n+1}\Big(q;\frac{iwx}{q^2}\Big) \\ &\qquad = i(-1)^{n+1} \sum_{k=0}^{\infty} q^{(n+k+1)(n+k)}\, \frac{(q;q)_{2n+1}}{(q^2;q^2)_{n+k+1}}\, \mathcal{C}^{(-n-k-1)}_{2k+1}(q;x)\; w^{2k+1}\ . \end{aligned} \tag{4.4b}$$

If $k \geq n$, then $\mathcal{C}^{(-n-k)}_{2k}(q;x)$ and $\mathcal{C}^{(-n-k-1)}_{2k+1}(q;x)$ are polynomials in x of degree $2n$ and $2n+1$, respectively. In particular, for $n = 0$ the relations (4.4) imply $\mathcal{C}^{(-k)}_{2k}(q;x) = q^{-2k}$ and $\mathcal{C}^{(-k-1)}_{2k+1}(q;x) = -(1+q)[k+1]_q\, q^{-(3k+2)}\, x$, results that can also be directly derived from (3.20).

Further formulas involving the polynomials $\mathcal{C}^{(\lambda)}_n(q;z)$ can be obtained by observing that the matrix elements $U^{(\pm)}_{kn}(\alpha,\beta)$ and $\widetilde{U}^{(\pm)}_{kn}(\alpha,\beta)$ themselves define models for the metaplectic representation of $\mathcal{U}_q(su(1,1))$ on the spaces $V^{(\pm)}$. In fact, one can check that the following operators:

$$k = q^{2l+1/2}\, T_\alpha^{-2} T_\beta^2\ , \tag{4.5a}$$

$$e = -q^2\frac{1}{\beta}\big(1 - T_\beta^{-2}\big)\ , \tag{4.5b}$$

$$\begin{aligned} f = -q^2\frac{1}{\alpha}\big(1-T_\alpha^{-2}\big) + q^{2l}\beta\Big[&(1+q^{-1})\, T_\beta^2 \\ &- q^{2l-1}(1+q^2T_\beta^2)\, T_\alpha^{-2}\Big] T_\alpha^{-2}T_\beta^2\ , \end{aligned} \tag{4.5c}$$

acting on the basis functions

$$h^{(l)}_{2n}(\alpha,\beta) = \widetilde{U}^{(+)}_{ln}(\alpha,\beta)\ , \tag{4.6}$$

and

$$k = q^{2l+3/2}\, T_\alpha^{-2} T_\beta^2\ , \tag{4.7a}$$

$$e = -q^2\frac{1}{\beta}\big(1 - T_\beta^{-2}\big)\ , \tag{4.7b}$$

$$\begin{aligned} f = -q^2\frac{1}{\alpha}\big(1-T_\alpha^{-2}\big) + q^{2l}\beta\Big[&(1+q)\, T_\beta^2 \\ &- q^{2l+1}(1+q^2T_\beta^2)\, T_\alpha^{-2}\Big] T_\alpha^{-2}T_\beta^2\ , \end{aligned} \tag{4.7c}$$

acting on the basis functions

$$h^{(l)}_{2n+1}(\alpha,\beta) = \widetilde{U}^{(-)}_{ln}(\alpha,\beta)\ , \tag{4.8}$$

satisfy the relations (3.3). The expression of the operators in (4.5) and (4.7) can be easily derived using the properties (1.5) of the q-exponentials. For example, from (1.5b) one sees that

$$-q^2\frac{1}{\beta}(1 - T_\beta^{-2})\ \widetilde{U}(\alpha,\beta) = \widetilde{U}(\alpha,\beta)\ e\ . \tag{4.9}$$

This amounts to the following identities between matrix elements

$$(4.10a)\quad -q^2\frac{1}{\beta}(1 - T_\beta^{-2})\ \widetilde{U}^{(+)}_{ln}(\alpha,\beta) = (1-q^{2n})(1-q^{2n-1})\ \widetilde{U}^{(+)}_{ln-1}(\alpha,\beta)\ ,$$

$$(4.10b)\quad -q^2\frac{1}{\beta}(1 - T_\beta^{-2})\ \widetilde{U}^{(-)}_{ln}(\alpha,\beta) = (1-q^{2n+1})(1-q^{2n})\ \widetilde{U}^{(-)}_{ln-1}(\alpha,\beta)\ ,$$

which corresponds to the second relation in (3.7) under the identifications (4.6) and (4.8). To derive (4.5c) and (4.7c), one first acts with $-q^2\frac{1}{\alpha}(1-T_\alpha^{-2})$ on $\widetilde{U}(\alpha,\beta)$ to get $\widetilde{U}(\alpha,\beta)\ e_{q^2}(-\beta e)\ fE_{q^2}(\beta e)$, and then shows that

$$e_{q^2}(-\beta e)\ fE_{q^2}(\beta e) = f - \beta(1 - k^2 - fe) + \beta^2 k^2 e\ , \tag{4.11}$$

with the help of the following formula[5]

$$e_q(\zeta X)\ Y\ E_q(-\zeta X) = \sum_{n=0}^{\infty} \frac{\zeta^n}{(q;q)_n}\,[X,Y]_n\ , \tag{4.12}$$

where

$$[X,Y]_0 = Y, \qquad [X,Y]_{n+1} = X\,[X,Y]_n - q^n [X,Y]_n\, X, \quad \text{for } n = 1,2,\dots\ .$$

Similar results are obtained using $U^{(\pm)}_{ln}(\alpha,\beta)$.

From the realizations (4.5) and (4.7) of $\mathcal{U}_q(su(1,1))$ one can easily derive identities involving the polynomials $\mathcal{C}^{(\lambda)}_n(q;z)$. Insert (4.5$b$), (4.6) and (4.7$b$), (4.8) in the second relation of (3.7); after some simple manipulations one gets:

$$(1 - q^{2\lambda+n}\,T_z)\,\mathcal{C}^{(\lambda)}_n(q;z) = (1-q^{2\lambda})\,\mathcal{C}^{(\lambda+1)}_n(q;z)\ . \tag{4.13}$$

Similarly, inserting (4.5c), (4.6) and (4.7c), (4.8) in the third relation of (3.7), one obtains:

$$\begin{aligned}\Big[q^{2\lambda+1}z^2(q^n - T_z) - (1-T_z) - q(1-q^{1-2\lambda-n})\Big]\ \mathcal{C}^{(\lambda)}_n(q;z)\\ = q^{3-n}\frac{(1-q^{2\lambda+n-1})(1-q^{2\lambda+n-2})}{(1-q^{2\lambda-1})}\ \mathcal{C}^{(\lambda-1)}_n(q;z)\ .\end{aligned} \tag{4.14}$$

Since we have models for the representation (3.6), from the general definition of the matrix elements (3.12), one can write, *e.g.* using (4.6):

$$U(\alpha,\beta)\, h_{2n}^{(l)}(x,y) = \sum_{k=0}^{\infty} \widetilde{U}_{lk}^{(+)}(x,y)\; U_{kn}^{(+)}(\alpha,\beta)\ , \tag{4.15}$$

where the model independent matrix elements $U_{kn}^{(+)}(\alpha,\beta)$ are still given by the formula (3.15). This allows deriving further identities involving the polynomials $\mathcal{C}_n^{(\lambda)}(q;z)$. To this end, one needs to evaluate explicitly the l.h.s. of (4.15), *i.e.* to compute directly the action of $U(\alpha,\beta)$ on the basis functions (4.6), when the algebra generators e and f are realized as in (4.5), with α and β being replaced by x and y, respectively. We shall consider here only the simple case $\alpha = 0$, for which

$$U_{kn}^{(+)}(0,\beta) = \frac{\beta^{n-k}}{(q^2;q^2)_{n-k}} \frac{(q;q)_n}{(q;q)_{2k}}\ , \qquad n \geq k\ . \tag{4.16}$$

The action of $U(0,\beta) \equiv e_{q^2}\left(-q^2(\beta/y)(1-T_y^{-2})\right)$ on the basis function $h_{2n}^{(l)}(x,y)$ of (4.6) can be computed with the help of the following summation formula ($|\beta/y| < 1$)

$$\begin{aligned} e_{q^2}\left(-q^2(\beta/y)(1-T_y^{-2})\right)\, y^n &= y^n\, \frac{1}{(-q^2\beta/y;q^2)_{-n}} \\ &= q^{-n(n-1)}\,(-y/\beta;q^2)_n\; \beta^n\ . \end{aligned} \tag{4.17}$$

One finds

$$\begin{aligned} e_{q^2}\left(-q^2(\beta/y)(1-T_y^{-2})\right)\, h_{2n}^{(l)}(x,y) &= q^{k(k-1)}\,\frac{(q;q^2)_n}{(q^2;q^2)_k}\,(-y/\beta;q^2)_n \\ &\times x^k\beta^n\; {}_3\phi_2(q^{-2k},q^{-2n},0;q,-q^{2(1-n)}\beta/y;q^2,q^4/xy)\ . \end{aligned} \tag{4.18}$$

Insert now this result and the explicit expressions (3.21) and (4.16) for the matrix elements $\widetilde{U}_{lk}^{(+)}(x,y)$ and $U_{kn}^{(+)}(0,\beta)$ in (4.15); setting $u = q/\sqrt{xy}$ and $w = -q^{2(1-n)}\beta/y$, one finally gets, after suitable simplifications:

$$\begin{aligned} &\begin{bmatrix} k+n \\ n \end{bmatrix}_{q^2} (w;q^2)_n\; {}_3\phi_2(q^{-2k},q^{-2n},0;q,w;q^2,u^2) \\ &\qquad = \sum_{l=0}^{k+n} (-1)^l q^{l(l-1)+2k(n-l+1)} \begin{bmatrix} k+n \\ l \end{bmatrix}_{q^2} w^l\; \mathcal{C}_{2k}^{(-k-n+l)}(q;u)\ , \end{aligned} \tag{4.19}$$

where

$$\begin{bmatrix} m \\ n \end{bmatrix}_{q^2} = \frac{(q^2;q^2)_m}{(q^2;q^2)_n (q^2;q^2)_{m-n}}\ , \tag{4.20}$$

is the q-binomial coefficient. Similar results are obtained using the model (4.7). The details of the calculation are similar to the ones described before, we shall therefore only give the final formula:

(4.21)

$$\begin{bmatrix} k+n+1 \\ n \end{bmatrix}_{q^2} \frac{(1-q^{2(k+1)})}{(1-q)} (w;q^2)_n \ {}_3\phi_2(q^{-2k},q^{-2n},0;q^3,w;q^2,u^2)$$

$$= u \sum_{l=0}^{k+n+1} (-1)^l q^{l(l-1)+2[(k+1)(n-l+2)-1]} \begin{bmatrix} k+n+1 \\ l \end{bmatrix}_{q^2} w^l \ \mathcal{C}_{2k+1}^{(-k-n+l-1)}(q;u) \ .$$

These are just a few examples of the identities that can be obtained from the quantum algebra interpretation of the q-orthogonal polynomials $\mathcal{C}_n^{(\lambda)}(q;u)$. Further discussions on the connection between quantum algebras and q-special functions can be found in Refs.[6-16].

REFERENCES

1. Gasper, G. and Rahman, M., *Basic Hypergeometric Series*, (Cambridge University Press, Cambridge, 1990)
2. Miller, W., *Lie Theory and Special Functions*, (Academic Press, New York, 1968)
3. Vilenkin, N.Ya., *Special Functions and the Theory of Group Representations*, Amer. Math. Soc. Transl. of Math. Monographs **22**, (The American Mathematical Society, Providence, 1968)
4. Miller, W., Lie theory and q-difference equations, SIAM J. Math. Anal. **1**, 171-188 (1970)
5. Agarwal, A.K., Kalnins, E.G. and Miller, W., Canonical equations and symmetry techniques for q-series, SIAM J. Math. Anal. **18**, 1519-1538 (1987)
6. Floreanini, R. and Vinet, L., q-Orthogonal polynomials and the oscillator quantum group, Lett. Math. Phys. **22**, 45-54 (1991)
7. Floreanini, R. and Vinet, L., The metaplectic representation of $su_q(1,1)$ and the q-Gegenbauer polynomials, J. Math. Phys. **33**, 1358-1363 (1992)
8. Floreanini, R. and Vinet, L., q-Conformal quantum mechanics and q-special functions, Phys. Lett. B **277**, 442-446 (1992)
9. Floreanini, R. and Vinet, L., Quantum algebras and q-special functions, Ann. of Phys. **221**, 53-70 (1993)
10. Floreanini, R. and Vinet, L., Addition formulas for q-Bessel functions, J. Math. Phys. **33**, 2984-2988 (1992)
11. Floreanini, R. and Vinet, L., On the quantum group and quantum algebra approach to q-special functions, Lett. Math Phys. **27**, 179-190 (1993)
12. Floreanini, R. and Vinet, L., Generalized q-Bessel functions, University of Montreal-preprint, UdeM-LPN-TH87, 1992
13. Floreanini, R. and Vinet, L., Using quantum algebras in q-special function theory, Phys. Lett. A **170**, 21-28 (1992)
14. Floreanini, R. and Vinet, L., An algebraic interpretation of the q-hypergeometric functions, Journal of Group Theory in Physics **1**, to appear
15. Kalnins, E.G., Manocha, H.L. and Miller, W., Models of q-algebra representations: I. Tensor products of special unitary and oscillator algebras, J. Math. Phys. **33**, 2365-2383 (1992)

16. Kalnins, E.G, Miller, W., and Mukherjee, S., Models of q-algebra representations: the group of plane motions, University of Minnesota preprint, 1992

17. Drinfel'd, V.G., Quantum groups, in: *Proceedings of the International Congress of Mathematicians*, Berkeley (1986), vol. **1**, 798-820, (The American Mathematical Society, Providence, 1987)

18. Jimbo, M., A q-difference analogue of $U(g)$ and the Yang-Baxter equation, Lett. Math. Phys. **10**, 63-69 (1985); A q-analogue of $U(gl(N+1))$, Hecke algebra and the Yang-Baxter equation, *ibid.* **11**, 247-252 (1986)

19. Masuda, T., Mimachi, K., Nakagami, Y., Noumi, M., Saburi, Y. and Ueno, K., Unitary representations of the quantum group $SU_q(1,1)$: structure of the dual space of $\mathcal{U}_q(sl(2))$, Lett. Math. Phys. **19**, 187-194 (1990); Unitary representations of the quantum group $SU_q(1,1)$: II-matrix elements of unitary representations and the basic hypergeometric functions, *ibid.* **19**, 195-204 (1990)

20. Koornwinder, T.H., Orthogonal polynomials in connection with quantum groups, in *Orthogonal Polynomials*, Nevai, P., ed., 257-292 (Kluwer, 1990)

21. Andrews, G.E. and Askey, R. , Classical orthogonal polynomials, in *Polynômes orthogonaux et applications*, Lect. Notes in Math. **1171**, 36-62 (Springer-Verlag, Berlin, 1985)

22. Askey, R., Continuous q-Hermite polynomials when $q > 1$, in *q-Series and Partitions*, Stanton, D., ed., (Springer-Verlag, New York, 1989)

Istituto Nazionale di Fisica Nucleare, Sezione di Trieste, Dipartimento di Fisica Teorica, Università di Trieste, Strada Costiera 11, 34014 Trieste, Italy

Laboratoire de Physique Nucléaire, Université de Montréal, Montréal, Canada H3C 3J7

Contemporary Mathematics
Volume **160**, 1994

The Group Representation Matrix in Quantum Mechanical Scattering

JOSEPH N. GINOCCHIO

ABSTRACT. We derive a general expression for the group representation matrix elements for a vibrator and a displaced harmonic oscillator in an arbitrary number of dimensions. These matrix elements arise naturally in an eikonal treatment of scattering from complex systems and contain the results derived for proton-nucleus and electron-molecule scattering as special cases.

1. Introduction

Group theory and symmetries have played an important role in Physics in determining allowed quantum numbers of quantum states, transition selection rules and rates between states, and energy levels for the states. Generally these properties require a knowledge of the matrix elements of the generators of the groups, but not the matrix elements of the group representation matrix. In this paper we shall discuss the application of the group representation matrix to the scattering and decay of particles from complex composite systems which are dominated by one multiple variable $\alpha_{\lambda\mu}$. Examples are collective states in nuclei which arise from quadrupole oscillations ($\lambda = 2$) [1] and polar molecules which are dominated by a dipole degree of freedom ($\lambda = 1$) [2] and string models of hadrons [3]. For nuclear targets the scattering theory has been applied to medium-energy nucleon scattering [4], Coulomb excitation [5], and sub-barrier fusion [6]. For molecular targets this theory has been applied to medium-energy electron scattering [7]. For hadrons this formalism has been to applied to the decay of mesons and baryons [8].

1991 Mathematics Subject Classification, Primary 22E70, 81R05.

Supported by U.S. Department of Energy

This paper is in final form and no version of it will be submitted for publication elsewhere.

2. Schrödinger Equation

The dynamics of the system is governed by a Schrödinger equation in $2\lambda+1$ dimensions. The Hamiltonian can be written as

$$H = -\frac{\hbar^2}{2B}\left[\frac{1}{\beta^{2\lambda}}\frac{\partial}{\partial\beta}\left(\beta^{2\lambda}\frac{\partial}{\partial\beta}\right) - \frac{C(\Omega)}{\beta^2}\right] + V\ , \tag{1}$$

where we have introduced spherical coordinates: one radial variable β with $\beta^2 = \sum_\mu \alpha^*_{\lambda\mu}\alpha_{\lambda\mu}$ and 2λ angular variables, collectively denoted by Ω. The first term is the kinetic energy in the radial variable and the second term proportional to $C(\Omega)$ denotes the 'rotational' or 'angular' kinetic energy [9]. We shall consider a system for which the potential energy V only depends on the radial variable, $V = V(\beta)$. Two examples of such a system are the anharmonic vibrator [10] and the displaced harmonic oscillator [11] in $2\lambda+1$ dimensions.

In this paper we use a realization in terms of N interacting bosons with angular momentum $l=0$ ($s^\dagger$) and $l=\lambda$ ($b^\dagger_{\lambda\mu}$), as has been done in many quantum mechanical problems [12, 13]. These bosons span the $2\lambda+2$ dimensional space of the group $U(2\lambda+2)$. The connection between the so-called algebraic description and the geometric shape variables was studied in [14, 15].

3. Eikonal Scattering

First we discuss briefly the eikonal scattering operator and next we derive the scattering matrix elements for the anharmonic vibrator and the displaced harmonic oscillator.

In the eikonal approximation the scattering operator is expressed as an exponentiated multipole operator,

$$\hat{U}(\epsilon) = \exp[i\epsilon G_0] = \exp[i\epsilon(b^\dagger_{\lambda 0}s + s^\dagger b_{\lambda 0})]\ , \tag{2}$$

where $G_{\mu=0}$ represents the interaction of the projectile with the target system. The matrix elements of the exponentiated operator can be derived from the transformation properties of a single boson

$$\begin{aligned}
\hat{U}(\epsilon)\, s^\dagger\, \hat{U}(-\epsilon) &= \cos\epsilon\; s^\dagger + i\,\sin\epsilon\; b^\dagger_{\lambda 0}\;, \\
\hat{U}(\epsilon)\, b^\dagger_{\lambda 0}\, \hat{U}(-\epsilon) &= i\,\sin\epsilon\; s^\dagger + \cos\epsilon\; b^\dagger_{\lambda 0}\;, \\
\hat{U}(\epsilon)\, b^\dagger_{\lambda\mu}\, \hat{U}(-\epsilon) &= b^\dagger_{\lambda\mu} \qquad (\mu \neq 0)\;.
\end{aligned} \tag{3}$$

Both for the anharmonic vibrator and the displaced oscillator these matrix elements can be derived in closed form.

3.1 Anharmonic Vibrator. The anharmonic vibrator corresponds to the group chain

$$\begin{aligned}
U(2\lambda+2) \supset U(2\lambda+1) \supset SO(2\lambda+1) \supset \\
\ldots \supset SO(3) \supset SO(2)\;,
\end{aligned} \tag{4}$$

which yields a set of basis states $|[N]n\tau\alpha LM\rangle$. Since the degrees of freedom are bosonic the states are classified according to the symmetric representation $[N]$ of $U(2\lambda+2)$, where N is the total number of bosons. Further, n is the number of λ-bosons, τ is a boson seniority label denoting the number of λ-bosons that are not coupled pairwise to angular momentum zero, L is the angular momentum and M its z-component. Finally, α is a shorthand notation for all additional labels that are needed for a unique classification of the basis states. The allowed values of the number of λ-bosons are $n = 0, 1, \ldots, N$. The boson seniority can have values $\tau = n, n-2, \ldots, 1$ or 0 for n odd or even, respectively. The wave functions for the anharmonic vibrator are

$$|[N]n\tau\alpha LM\rangle = \frac{B_{n\tau}}{\sqrt{(N-n)!}}(s^\dagger)^{N-n}\,(b^\dagger_\lambda \cdot b^\dagger_\lambda)^{(n-\tau)/2}|[\tau]\tau\tau\alpha LM\rangle\;, \tag{5}$$

where the normalization coefficient is given by

$$B_{n\tau} = \sqrt{\frac{(2\tau+2\lambda-1)!!}{(n+\tau+2\lambda-1)!!(n-\tau)!!}}\;. \tag{6}$$

The matrix elements for the excitation from the ground state ($n = \tau = L = 0$) can now be derived simply by using the transformation properties of (3),

$$\begin{aligned} U_{n\tau\alpha LM}(\epsilon) &= \langle [N] n\tau\alpha LM | \hat{U}(\epsilon) | [N]00000 \rangle \\ &= \delta_{M0}\, B_{n\tau}\, A_{\tau\alpha L}\, \sqrt{N!/(N-n)!}\, (i \sin\epsilon)^n (\cos\epsilon)^{N-n} , \end{aligned} \tag{7}$$

with

$$A_{\tau\alpha L} = \langle [\tau]\tau\tau\alpha L0 | (b^\dagger_{\lambda 0})^\tau | [0]00000 \rangle / \tau! \, . \tag{8}$$

As far as we know it is not possible to derive this matrix element in general, with the exception of the case $\lambda = 1$ for which we have $A_{\tau=L} = \sqrt{(2\tau+1)/(2\tau+1)!!}$. For $\lambda = 2$ these matrix elements are tabulated in [4] for small values of τ. The coefficients $A_{\tau\alpha L}$ satisfy the sum rule

$$\sum_{\alpha L} A^2_{\tau\alpha L} = \frac{(2\tau+2\lambda-1)(\tau+2\lambda-2)!}{\tau!(2\lambda-2)!!(2\tau+2\lambda-1)!!} \, . \tag{9}$$

In the large N limit the energy spectrum in the $U(2\lambda+1)$ limit corresponds to that of a $(2\lambda+1)$-dimensional vibrator. The coefficient ϵ is usually normalized by the reduced matrix element of G between the ground state ($n = \tau = L = 0$) and the first excited state ($n = \tau = 1$, $L = \lambda$): $\epsilon = \eta/\sqrt{(2\lambda+1)N}$. In the large N limit, which is such that $n \ll N$ and η remains finite, the transition matrix element reduces to

$$U_{n\tau\alpha LM}(\epsilon) \to \delta_{M0}\, B_{n\tau}\, A_{\tau\alpha L} \left(\frac{i\,\eta}{\sqrt{2\lambda+1)}} \right)^n e^{-\eta^2/2(2\lambda+1)} \, . \tag{10}$$

It is easy to show that for $\lambda = 2$ the results obtained for the quadrupole vibrator limit [4] of the Interacting Boson Model [12] are recovered and for $\lambda = 1$ those derived for the dipole vibrator limit [7] of the vibron model [13].

The probability that a state belonging to the n-phonon multiplet can be excited from the ground state ($n = 0$) is given by

$$P_{0\to n}(\epsilon) = \sum_{\tau\alpha LM} | U_{n\tau\alpha LM}(\epsilon) |^2 = \binom{N}{n} (\sin^2\epsilon)^n (\cos^2\epsilon)^{N-n} \, . \tag{11}$$

For $\epsilon = 0$ only the ground state is excited. With increasing values of ϵ all higher phonon multiplets are successively excited until for $\epsilon = \pi/2$ all strength is concentrated in the highest vibrational multiplet with $n = N$. The excitation probability is symmetric around $\epsilon = \pi/2$ and is a periodic function with period π, $P_{0\to n}(\epsilon) = P_{0\to n}(\pi \pm \epsilon)$. This implies that for $\epsilon = \pi$ all strength is again concentrated in the ground state. This behavior is an artifact of the finiteness of the model space.

In physical applications the excitation probability is determined by the coupling between the projectile and the target. For example, electron scattering from polar molecules is largely determined by the static dipole moment of the molecule, which can be used to normalize the coefficient $\eta \equiv d_{exp}$. Similarly, proton-nucleus scattering is dominated by the quadrupole coupling strength. Therefore, in order to study the scattering process as a function of the physical coupling strength we present in Figure 1 the excitation probability as a function of η. The number of bosons is $N \geq 50$ for diatomic molecules ($\lambda = 1$) or $N \sim 15$ for deformed nuclei ($\lambda = 2$). The number of bosons we take in Figure 1 is $N = 20$. For these typical values of N it is interesting to compare the exact results with those in the large N limit, for which the excitation probability $P_{0\to n}(\epsilon)$ reduces to the familiar Poisson distribution for the harmonic oscillator

$$P_{0\to n}(\epsilon) \to \frac{1}{n!}\left(\frac{\eta^2}{2\lambda+1}\right)^n e^{-\eta^2/(2\lambda+1)} . \tag{12}$$

Figure 1 shows that for small values of η and n, the large N results (dashed lines) provide a good approximation to the exact results calculated with (11) (solid lines) with $N = 20$.

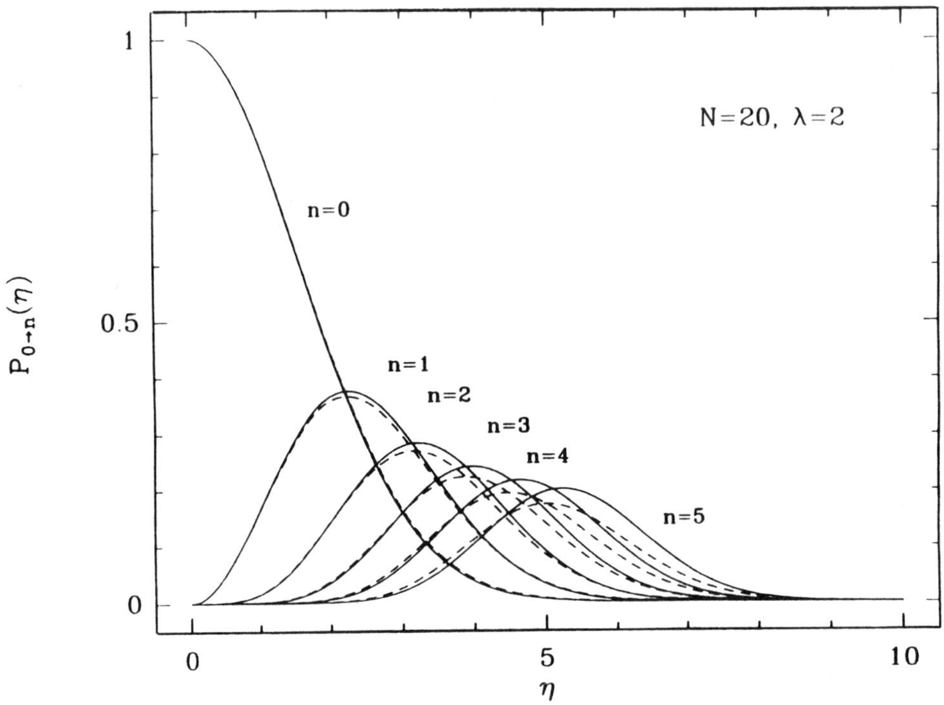

Figure 1. Excitation probability for the five-dimensional anharmonic vibrator ($U(5)$ limit), calculated for $N = 20$ (solid lines) and in the large N limit (dashed lines).

3.2 Displaced Harmonic Oscillation.

A second example of interest is that of a displaced harmonic oscillator in $2\lambda + 1$ dimensions. Its group chain

$$U(2\lambda+2) \supset SO(2\lambda+2) \supset SO(2\lambda+1) \supset \ldots \supset SO(3) \supset SO(2) \ , \tag{13}$$

yields a set of basis states $|[N]\sigma\tau\alpha LM\rangle$, where σ counts the number of bosons not paired to angular momentum zero (note that in this case the pairs are a superposition of s- and λ-bosons). The remaining labels are the same as for the anharmonic vibrator. The allowed values of σ are $\sigma = N, N-2, \ldots, 1$ or 0 for N odd or even, respectively, and those of τ are $\tau = 0, 1, \ldots, \sigma$. Using the quasi-spin group $SU(1,1)$ for a system of bosons, the wave functions for the displaced oscillator can be written as

$$|[N]\sigma\tau\alpha LM\rangle = B_{N\sigma}(I^\dagger)^{(N-\sigma)/2}\,|[\sigma]\sigma\tau\alpha LM\rangle\ , \tag{14}$$

where $I^\dagger$ is the pair creation operator in the boson space, $I^\dagger = s^\dagger s^\dagger - b_\lambda^\dagger \cdot b_\lambda^\dagger$, and $B_{N\sigma}$ is a normalization coefficient

$$B_{N\sigma} = \sqrt{\frac{(2\sigma + 2\lambda)!!}{(N + \sigma + 2\lambda)!!(N - \sigma)!!}}\ . \tag{15}$$

Since G_0 is a generator of $SO(2\lambda + 2)$ it is diagonal in the quantum number σ and it commutes with the pair creation operator, $[G_0, I^\dagger] = 0$. Therefore its matrix elements do not depend on the total number of bosons

$$\begin{aligned}&\langle[N]\sigma\tau\alpha LM|\hat{U}(\epsilon)|[N]\sigma'\tau'\alpha'L'M'\rangle = \\ &\qquad \delta_{\sigma'\sigma}\delta_{M'M}\langle[\sigma]\sigma\tau\alpha LM|\hat{U}(\epsilon)|[\sigma]\sigma\tau'\alpha'L'M\rangle\ .\end{aligned} \tag{16}$$

The wave function with $N = \sigma$ can be written as

$$|[\sigma]\sigma\tau\alpha LM\rangle = \sum_{k=0}^{[(\sigma-\tau)/2]} D_k(\sigma,\tau)(s^\dagger)^{\sigma-\tau-2k}(I^\dagger)^k|[\tau]\tau\tau\alpha LM\rangle\ , \tag{17}$$

with

$$\begin{aligned}D_k(\sigma,\tau) = &\sqrt{\frac{2^{\sigma+\lambda-1}(\sigma-\tau)!(2\tau+2\lambda-1)!!}{(\sigma+\lambda-1)!(\sigma+\tau+2\lambda-1)!}} \\ &\left(-\frac{1}{4}\right)^k \frac{(\sigma+\lambda-1-k)!}{(\sigma-\tau-2k)!k!}\ .\end{aligned} \tag{18}$$

The matrix element for the excitation from the ground state ($\tau = L = 0$) can be interpreted as $SO(2\lambda + 2)$ representation matrix element and hence can be expressed in terms of a Gegenbauer polynomial

$$\begin{aligned}U_{\sigma\tau\alpha LM}(\epsilon) &= \delta_{M0}\sqrt{\frac{(2\lambda-1)!!(2\tau+2\lambda-1)!!(\sigma-\tau)!\sigma!}{(\sigma+2\lambda-1)!(\sigma+\tau+2\lambda-1)!}} \\ &\quad (2\tau+2\lambda-2)!!\ A_{\tau\alpha L}\ (i\sin\epsilon)^\tau C_{\sigma-\tau}^{(\tau+\lambda)}(\cos\epsilon)\ .\end{aligned} \tag{19}$$

In the large σ $(=N)$ limit the energy spectrum in the $SO(2\lambda+2)$ limit corresponds to that of a $(2\lambda+1)$-dimensional displaced oscillator. The coefficient ϵ is again normalized by the reduced matrix element of G between the ground state $(\tau = L = 0)$ and the first excited state $(\tau = 1,\ L = \lambda)$: $\epsilon = \eta/\sqrt{\sigma(\sigma+2\lambda)}$. In the large σ limit $(\lambda, \tau \ll \sigma$ and η finite) the transition matrix element can be expressed in terms of a spherical Bessel function

$$U_{\sigma\tau\alpha LM}(\epsilon) \to \delta_{M0}\, A_{\tau\alpha L}\sqrt{(2\lambda-1)!!(2\tau+2\lambda-1)!!}\ \frac{i^{\tau}\, j_{\tau+\lambda-1}(\eta)}{\eta^{\lambda-1}}\ . \tag{20}$$

It is easy to show that for $\lambda = 2$ and $\lambda = 1$ the results obtained for the γ-unstable rotor limit of the IBM and the dipole rotor limit of the vibron model are recovered [4, 7].

The probability that a state belonging to the τ-multiplet of the ground state band $\sigma = N$ can be exited from the ground state with $\tau = 0$ is given by

$$P_{0\to\tau}(\epsilon) = \sum_{\alpha LM} \mid U_{\sigma\tau\alpha LM}(\epsilon) \mid^2 . \tag{21}$$

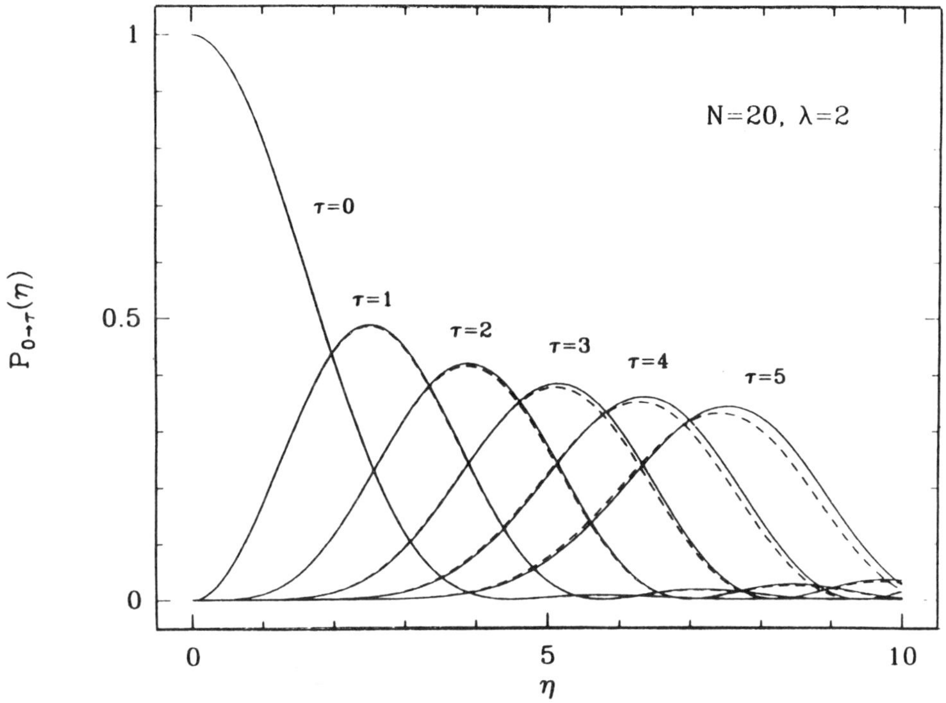

Figure 2. Excitation probability for the five-dimensional displaced oscillator ($SO(6)$ limit), calculated for $N = 20$ (solid lines) and in the large N limit (dashed lines).

Just as for the anharmonic vibrator we see that with increasing ϵ all multiplets of the ground state band are populated successively until for $\epsilon = \pi/2$ the maximum strength is contained in the highest multiplet with $\tau = \sigma$. The probability $P_{0\to\tau}(\epsilon)$ has the same symmetry properties as in the case of the anharmonic vibrator. In the large σ limit the excitation probability reduces to

$$P_{0\to\tau}(\epsilon) \to \frac{(2\tau+2\lambda-1)(\tau+2\lambda-2)!(2\lambda-1)!!}{\tau!(2\lambda-2)!!} \left(\frac{j_{\tau+\lambda-1}(\eta)}{\eta^{\lambda-1}}\right)^2 . \tag{22}$$

As an illustration we show in Figure 2 the excitation probability as a function of the physical coupling strength in the scattering process, η both for finite $N = \sigma = 20$ and in the large $N = \sigma$ limit.

4. Summary

The generalization of the transition matrix elements for quantum mechanical scattering from a target dominated by quadrupole and dipole degrees of freedom considered in [4] and [7] to an arbitrary multipole λ is straightforward and is given in (7) and (19) for finite N systems and (10) and (20) for infinite N (or classical) systems. From these results we see that there are common features for all multipoles. For the anharmonic vibrator the excitation probability of the vibrational multiplet with n λ-bosons is determined by the coupling strength ϵ and does not depend on the multipole. All multiplets are excited successively with increasing values of ϵ and then fall off exponentially. However, the relative transition matrix elements to states with the same number of λ bosons depend only on a geometric factor $A_{\tau\alpha L}$ (and hence on λ) and not on the strength of the interaction of the projectile with the target. Likewise for the displaced oscillator the interaction strength determines the transition to a multiplet of states with the same quantum number τ, but the relative transition matrix elements to states within the same multiplet are determined only by the geometric factor and are independent of the interaction of the projectile with the system.

REFERENCES

1. A. Bohr and B. R. Mottelson, 'Nuclear Structure', Volume 2 (Benjamin, Reading, MA 1975).

2. G. Herzberg, 'Molecular Spectra and Molecular Structure I: Spectra of Diatomic Molecules' (Van Nostrand Rheinhold, Princeton, NJ 1950).

3. F. Iachello, Phys. Rev. Lett. **62** (1989), 2440.

4. J.N. Ginocchio, T. Otsuka, R.D. Amado and D.A. Sparrow, Phys. Rev. **C33** (1986), 247.

5. G. Wenes, N. Yoshinaga and A.E.L. Dieperink, Nucl. Phys. **A443** (1985), 472 .

6. A.B. Balantekin, J.R. Bennett and N. Takigawa, Phys. Rev. **C44**, 145 (1991).

7. R. Bijker, R.D. Amado and D.A. Sparrow, Phys. Rev. **A33** (1986), 871 .

8. F. Iachello and D. Kusnezov, Phys. Lett. **B255** (1991), 493.

9. N.Ja. Vilenkin, 'Special functions and the theory of group representations' (American Mathematical Society, Providence, RI 1968), page 493.

10. A. Bohr, Kgl. Danske Videnskab. Selskab, Mat.-fys. Medd. **26**, No. 14 (1952).

11. L. Wilets and M. Jean, Phys. Rev. **102** (1956), 788 .

12. A. Arima and F. Iachello, Phys. Rev. Lett. **35** (1975), 1069 .

13. F. Iachello, Chem. Phys. Lett. **78** (1981), 581.

14. J. N. Ginocchio and M. W. Kirson, Phys. Rev. Lett. **44** (1980), 1744.
15. A.E.L. Dieperink, O. Scholten and F. Iachello, Phys. Rev. Lett. **44**(1980), 1747 .

THEORETICAL DIVISION, LOS ALAMOS NATIONAL LABORATORY, LOS ALAMOS, NM 87545
E-mail address: ginocchi@lampf.lanl.gov

Contemporary Mathematics
Volume **160**, 1994

QUASI-EXACT SOLVABILITY

ARTEMIO GONZÁLEZ-LÓPEZ, NIKY KAMRAN, AND PETER J. OLVER

ABSTRACT. This paper surveys recent work on quasi-exactly solvable Schrödinger operators and Lie algebras of differential operators.

1. INTRODUCTION.

Lie algebraic and Lie group theoretic methods have played a significant role in the development of quantum mechanics since its inception. In the classical applications, the Lie group appears as a symmetry group of the Hamiltonian operator, and the associated representation theory provides an algebraic means for computing the spectrum. Of particular importance are the exactly solvable problems, such as the harmonic oscillator or the hydrogen atom, whose point spectrum can be completely determined using purely algebraic methods. The fundamental concept of a "spectrum generating algebra" was introduced by Arima and Iachello, [4], [5], to study nuclear physics, and subsequently, by Iachello, Alhassid, Gürsey, Levine, Wu and their collaborators, was also successfully applied to molecular dynamics and spectroscopy, [19], [23], and scattering theory, [1], [2], [3]. The Schrödinger operators amenable to the algebraic approach assume a "Lie algebraic form", meaning that they belong to the universal enveloping algebra of the spectrum generating algebra. Lie algebraic operators reappeared in the discovery of Turbiner, Shifman, Ushveridze, and their collaborators, [27], [29], [30], [34], of a new class of physically significant spectral problems, which they named "quasi-exactly solvable", having the property that a (finite) part of the point spectrum can be determined using purely algebraic methods. This is an immediate consequence of the additional requirement that the hidden symmetry algebra preserve a finite-dimensional representation space consisting of smooth wave functions. In this case, the Hamiltonian restricts to a linear transformation on the representation space, and hence the associated eigenvalues can be computed by purely algebraic methods, meaning matrix

1991 *Mathematics Subject Classification.* Primary: 81C05, 81C40; Secondary: 17B15, 17B56.
Key words and phrases. Schrödinger operator, Lie algebra, differential operator, quasi-exactly solvable.
The first author was supported in part by DGICYT Grant PS 89–0011.
The second author was supported in part by an NSERC Grant.
The third author was supported in part by NSF Grant DMS 92–04192.
This paper is in final form and no version of it will be submitted for publication elsewhere

eigenvalue calculations. Finally, one must decide the "normalizability" problem of whether the resulting "algebraic" eigenfunctions are square integrable and therefore represent true bound states of the system. Connections with conformal field theory, [15], [26], [28], and the theory of orthogonal polynomials, [31], [32], [33], lend additional impetus for the study of such problems.

In this paper we will survey recent work done by the authors in the mathematical study of Lie algebraic and quasi-exactly solvable Schrödinger operators. Our principal focus has been in the classification of quasi-exactly solvable problems, especially those of physical importance. The one- dimensional case is relatively easy for two important reasons. First, there is just a single discrete family of finite-dimensional (quasi-exactly solvable) Lie algebras of first order differential operators on the line, which are well-known realizations of the $\mathfrak{sl}(2,\mathbb{R})$ algebra. Second, every second order differential operator on $\mathbb{R}$ can, up to an overall sign, be (locally) mapped to a Schrödinger operator. The determination of explicit normalizability conditions on the resulting operators and Lie algebraic coefficients relies on an unusual combination of elementary asymptotics and classical invariant theory. Therefore, in one dimension, we can determine a complete list of quasi-exactly solvable potentials and explicit normalizability conditions.

In higher dimensions, much less is known. Only a few special examples of quasi-exactly solvable problems in two dimensions have appeared in the literature to date, [29], all of which are constructed using semi-simple Lie algebras. Complete lists of finite-dimensional Lie algebras of differential operators are known in two (complex) dimensions; there are essentially 24 different classes, some depending on parameters and arbitrary functions. The quasi-exactly solvable condition imposes a remarkable quantization constraint on the cohomology parameters classifying these Lie algebras. This phenomenon of the "quantization of cohomology" has recently been been given an algebro-geometric interpretation, [9]. Any of the resulting quasi-exactly solvable Lie algebras of differential operators can be used to construct new examples of two-dimensional quasi-exactly solvable spectral problems. An additional complication is that, in higher dimensions, not every elliptic second order differential operator is equivalent to a Schrödinger operator (*i.e.*, minus Laplacian plus potential), so not every Lie algebraic operator can be assigned an immediate physical meaning. The resulting "closure conditions" are quite complicated to solve, and so the problem of completely classifying quasi-exactly solvable Schrödinger operators in two dimensions appears to be too difficult to solve in full generality. A variety of interesting examples are given in [14], and we present a few particular cases of interest here.

2. Quasi-Exactly Solvable Schrödinger Operators.

Let M denote an open subset of Euclidean space $\mathbb{R}^n$ with coordinates $x = (x^1, \ldots, x^n)$. The time-independent Schrödinger equation for a differential operator $\mathcal{H}$ is the eigenvalue problem

$$\mathcal{H}[\psi] = \lambda\psi. \tag{1}$$

In the quantum mechanical interpretation, a (self-adjoint) differential operator $\mathcal{H}$ plays the role of the quantum "Hamiltonian" of the system. A nonzero wave function $\psi(x)$ is called *normalizable* if it is square integrable, *i.e.*, lies in the Hilbert space $L^2(\mathbb{R}^n)$, and so represents a physical bound state of the quantum mechanical system, the corresponding eigenvalue determining the associated energy level. While it is of great interest to know the bound states and energy levels of a given operator, complete explicit lists of eigenvalues and eigenfunctions are known for only a handful of classical "exactly solvable" operators, such as the harmonic oscillator. For the vast majority of quantum mechanical problems, the spectrum can, at best, only be approximated by numerical computation. The quasi-exactly solvable systems occupy an important intermediate station, in that a finite part of the spectrum can be computed by purely algebraic means.

To describe the general form of a quasi-exactly solvable problem, we begin with a finite-dimensional Lie algebra $\mathfrak{g}$ spanned by r linearly independent first-order differential operators

$$T^a = \sum_{i=1}^{n} \xi^{ai}(x) \frac{\partial}{\partial x^i} + \eta^a(x), \quad a = 1, \ldots, r, \tag{2}$$

whose coefficients ξ^{ai}, η^a are smooth functions of x. The Lie algebra assumption requires that the commutator between two such operators can be written as a linear combination of the operators: $[T^a, T^b] = T^a T^b - T^b T^a = \sum_c C^c_{ab} T^c$, where the C^c_{ab} are the structure constants of the Lie algebra $\mathfrak{g}$. Note that each differential operator is a sum, $T^a = \mathbf{v}^a + \eta^a$, of a *vector field* $\mathbf{v}^a = \sum \xi^{ai} \partial / \partial x^i$ (which may be zero) and a *multiplication operator* η^a.

A differential operator is said to be *Lie algebraic* if it lies in the universal enveloping algebra $U(\mathfrak{g})$ of the Lie algebra $\mathfrak{g}$, meaning that it can be expressed as a polynomial in the operators T^a. In particular, a second order differential operator is Lie algebraic if it can be written as a quadratic combination

$$-\mathcal{H} = \sum_{a,b} c_{ab} T^a T^b + \sum_a c_a T^a + c_0, \tag{3}$$

for certain constants c_{ab}, c_a, c_0. (The minus sign in front of the Hamiltonian is taken for later convenience.) If some of the operators T^a generating the Lie algebra are pure multiplication operators, then one could allow higher degree combinations in (3); however, in the quasi-exactly solvable case to be considered here, such second order Lie algebraic operators can always be re-expressed in a quadratic form, (3), for some possibly larger Lie algebra $\mathfrak{g}$. Note that the commutator $[T^c, \mathcal{H}]$ of the Hamiltonian with any generator of $\mathfrak{g}$, while still of the same Lie algebraic form, is not in general a multiple of the Hamiltonian $\mathcal{H}$ (unless, for example, $\mathcal{H}$ happens to be a Casimir for $\mathfrak{g}$). Therefore, the "hidden symmetry algebra" $\mathfrak{g}$ is not a symmetry algebra in the traditional sense. Lie algebraic operators appeared in the early work of Iachello, Levine, Alhassid, Gürsey, and collaborators, in the algebraic approach to scattering theory, [1], [2], [23].

The condition of quasi-exact solvability imposes an additional constraint on the Lie algebra and hence on the type of operators which are allowed. A Lie algebra of first order differential operators $\mathfrak{g}$ will be called *quasi-exactly solvable* if it possesses a finite-dimensional representation space (or module) $\mathcal{N} \subset C^\infty$ consisting of smooth functions; this means that if $\psi \in \mathcal{N}$ and $T^a \in \mathfrak{g}$, then $T^a(\psi) \in \mathcal{N}$. A differential operator $\mathcal{H}$ is called *quasi-exactly solvable* if it lies in the universal enveloping algebra of a quasi-exactly solvable Lie algebra of differential operators. Clearly, the module $\mathcal{N}$ is an invariant space for the Hamiltonian $\mathcal{H}$, *i.e.*, $\mathcal{H}(\mathcal{N}) \subset \mathcal{N}$, and hence $\mathcal{H}$ restricts to a linear matrix operator on $\mathcal{N}$. We will call the eigenvalues and corresponding eigenfunctions for the restriction $\mathcal{H}|\mathcal{N}$ *algebraic* since they can be computed by algebraic methods for matrix eigenvalue problems. (This does not mean that these functions are necessarily "algebraic" in the traditional pure mathematical sense.) Note that the number of such "algebraic" eigenvalues and eigenfunctions equals the dimension of $\mathcal{N}$. So far we have not imposed any normalizability conditions on the algebraic eigenfunctions, but, if they are normalizable, then the corresponding algebraic eigenvalues give part of the point spectrum of the differential operator.

It is of great interest to know when a given differential operator is in Lie algebraic or quasi-exactly solvable form. There is not, as far as we know, any direct test on the operator in question that will answer this in general. Consequently, the best approach to this problem is to effect a complete classification of such operators under an appropriate notion of equivalence. In order to classify Lie algebras of differential operators, and hence Lie algebraic and quasi-exactly solvable Schrödinger operators, we need to precisely specify the allowable changes of variables.

Definition 1. Two differential operators are *equivalent* if they can be mapped into each other by a combination of change of independent variable,

$$\bar{x} = \varphi(x), \tag{4}$$

and "gauge transformation"

$$\overline{\mathcal{H}} = e^{\sigma(x)} \cdot \mathcal{H} \cdot e^{-\sigma(x)}. \tag{5}$$

The transformations (4), (5), have two key properties. First, they respect the commutator between differential operators, and therefore preserve their Lie algebra structure. Second, they preserve the spectral problem (1) associated with the differential operator $\mathcal{H}$, so that if $\psi(x)$ is an eigenfunction of $\mathcal{H}$ with eigenvalue λ, then the transformed (or "gauged") function

$$\overline{\psi}(\bar{x}) = e^{\sigma(x)}\psi(x), \qquad \text{where} \qquad \bar{x} = \varphi(x), \tag{6}$$

is the corresponding eigenfunction of $\overline{\mathcal{H}}$ having the same eigenvalue. Therefore this notion of equivalence is completely adapted to the problem of classifying quasi-exactly solvable Schrödinger operators. The gauge factor $\mu(x) = e^{\sigma(x)}$ in (5) is *not* necessarily unitary, *i.e.*, $\sigma(x)$ is not restricted to be purely imaginary, and hence does not necessarily preserve the normalizability properties of the associated eigenfunctions. Therefore, the problem of normalizability of the resulting algebraic wavefunctions must be addressed.

Definition 2. A quasi-exactly solvable Schrödinger operator is called *normalizable* if every algebraic eigenfunction is normalizable. An operator is called *partially normalizable* if some of the algebraic eigenfunctions are normalizable.

Let us summarize the basic steps that are required in order to obtain a complete classification of quasi- exactly solvable operators and their algebraic physical states.

1. Classify finite-dimensional Lie algebras of differential operators.
2. Determine which Lie algebras are quasi-exactly solvable.
3. Solve the equivalence problem for differential operators.
4. Determine normalizability conditions.
5. Solve the associated matrix eigenvalue problem.

The remainder of this survey will discuss what is now known about these problems, except the last which is merely an exercise in linear algebra.

3. Equivalence of Differential Operators.

Consider a second-order linear differential operator

$$-\mathcal{H} = \sum_{i,j=1}^{n} g^{ij}(x)\,\frac{\partial^2}{\partial x^i \partial x^j} + \sum_{i=1}^{n} h^i(x)\,\frac{\partial}{\partial x^i} + k(x), \tag{7}$$

defined on an open subset $M \subset \mathbb{R}^n$. We are interested in studying the problem of when two such operators are equivalent under the combination of change of variables and gauge transformations (4), (5). Of particular importance is the question of when $\mathcal{H}$ is equivalent to a Schrödinger operator, which we take to mean an operator $\mathcal{S} = -\Delta + V(x)$, where Δ denotes either the flat space Laplacian or, more generally, the Laplace-Beltrami operator over a curved manifold. (Operators on Riemannian manifolds with non-zero curvature can be viewed as constrained quantum mechanical systems, *e.g.*, a particle moving on a sphere, [6].) This definition of Schrödinger operator excludes the introduction of a magnetic field, which, however, can also be handled by these methods. There is an essential difference between one-dimensional and higher dimensional spaces in the solution to the equivalence problem for second order differential operators.

Theorem 3. *Let*

$$-\mathcal{H} = P(x)\,\frac{d^2}{dx^2} + \widehat{Q}(x)\,\frac{d}{dx} + \widehat{R}(x), \qquad x \in \mathbb{R},$$

be a second order differential operator such that $P(x) > 0$. *Then the change of variables*

$$\bar{x} = \varphi(x) = \int^x \frac{dy}{\sqrt{P(y)}}\,, \tag{8}$$

and gauge factor

$$\mu(x) = \left|P(x)\right|^{-1/4} \exp\left\{\int^x \frac{\widehat{Q}(y)}{2P(y)}\,dy\right\}, \tag{9}$$

will place the operator $\mathcal{H}$ *into Schrödinger form*

$$\mu(x)\cdot\mathcal{H}\cdot\frac{1}{\mu(x)} = -\frac{d^2}{d\bar{x}^2} + V(\bar{x}).$$

The potential is given by

$$V(\bar{x}) = -\frac{3P'^2 - 8\widehat{Q}P' + 4\widehat{Q}^2}{16P} - \widehat{R} + \frac{1}{2}\widehat{Q}' - \frac{1}{4}P'', \tag{10}$$

where the right hand side is evaluated at $x = \varphi^{-1}(\bar{x})$. Moreover, if $\psi(x)$ is a wave function in the original coordinates, then

$$\overline{\psi}(\bar{x}) = \mu(\varphi^{-1}(\bar{x}))\,\psi(\varphi^{-1}(\bar{x})) \tag{11}$$

will be the corresponding wave function in the physical (Schrödinger) coordinates.

In higher space dimensions, it is no longer true that every second order differential operator is locally equivalent to a Schrödinger operator of the form $-\Delta + V(x)$, where $-\Delta$ is the flat space Laplacian — explicit equivalence conditions were first found by É. Cotton, [7]. Since the symbol of a linear differential operator is invariant under coordinate transformations, we begin by assuming that the operator is elliptic, meaning that the symmetric matrix $\widehat{\mathbf{g}}(x) = \big(g^{ij}(x)\big)$ determined by the leading coefficients of $-\mathcal{H}$ is positive definite. Owing to the induced transformation rules under the change of variables (7), we interpret the inverse matrix $\mathbf{g}(x) = \widehat{\mathbf{g}}(x)^{-1} = \big(g_{ij}(x)\big)$ as defining a Riemannian metric

$$ds^2 = \sum_{i,j=1}^{n} g_{ij}(x)dx^i dx^j, \tag{12}$$

on the subset $M \subset \mathbb{R}^n$. We will follow the usual tensor convention of raising and lowering indices with respect to the Riemannian metric (12). We rewrite the differential operator (7) in a more natural coordinate-independent form

$$\mathcal{H} = -\sum_{i,j=1}^{n} g^{ij}(\nabla_i - A_i)(\nabla_j - A_j) + V, \tag{13}$$

where ∇_i denotes covariant differentiation using the associated Levi-Civita connection. Physically, $A(x) = (A_1(x), \ldots, A_n(x))$ can be thought of as a (generalized) magnetic vector potential; in view of its transformation properties, we define the associated *magnetic one-form*

$$\omega = \sum_{i=1}^{n} A_i(x)\, dx^i. \tag{14}$$

(Actually, to qualify as a physical vector potential, A must be purely imaginary and satisfy the stationary Maxwell equations, but we will not impose this additional physical constraint in our definition of the mathematical magnetic one-form (14).) The explicit formulas relating the covariant form (13) to the standard form (7) of the differential operator are

$$\begin{aligned} A^i &= \sum_{j=1}^{n} g^{ij} A_j = -\frac{h^i}{2} + \frac{1}{2\sqrt{g}} \sum_{j=1}^{n} \frac{\partial\big(\sqrt{g}\, g^{ij}\big)}{\partial x^j}, \\ V &= -k + \sum_{i=1}^{n} \left[A_i A^i - \frac{1}{\sqrt{g}} \frac{\partial}{\partial x^i} \big(\sqrt{g}\, A^i\big) \right], \end{aligned} \tag{15}$$

where $g(x) = \det(g_{ij}(x)) > 0$. Each second order elliptic operator then is uniquely specified by a metric, a magnetic one-form, and potential function $V(x)$. In particular, if the magnetic form vanishes, so $A = 0$, then $\mathcal{H}$ has the form of a Schrödinger operator $\mathcal{H} = -\Delta + V$, where Δ is the Laplace-Beltrami operator associated with the metric (12).

The application of a gauge transformation (5) does not affect the metric or the potential; however the magnetic one-form is modified by an exact one-form: $\omega \mapsto \omega + d\sigma$. Consequently, the "magnetic two-form" $\Omega = d\omega$, whose coefficients represent the associated magnetic field, *is* unaffected by gauge transformations.

Theorem 4. *Two elliptic second order differential operators $\mathcal{H}$ and $\overline{\mathcal{H}}$ are (locally) equivalent under a change of variables $\bar{x} = \varphi(x)$ and gauge transformation* (5) *if and only if their metrics, their magnetic two-forms, and their potentials are mapped to each other*

$$\varphi^*(d\bar{s}^2) = ds^2, \qquad \varphi^*(\overline{\Omega}) = \Omega, \qquad \varphi^*(\overline{V}) = V. \tag{16}$$

(Here φ^* denotes the standard pull-back action of φ on differential forms; in particular, $\varphi^*(\overline{V}) = \overline{V} \circ \varphi$.)

In particular, an elliptic second order differential operator is equivalent to a Schrödinger operator $-\Delta + V$ if and only if its magnetic one-form is closed: $d\omega = \Omega = 0$. Moreover, since the curvature tensor associated with the metric is invariant, the Laplace-Beltrami operator Δ will be equivalent to the flat space Laplacian if and only if the metric ds^2 is flat, *i.e.*, has vanishing Riemannian curvature tensor.

4. Lie Algebras of Differential Operators.

In this section, we summarize what is known about the classification problem for Lie algebras of first order differential operators. Any finite-dimensional Lie algebra $\mathfrak{g}$ of first order differential operators has a basis of the form

$$T^1 = \mathbf{v}^1 + \eta^1(x), \ldots, T^r = \mathbf{v}^r + \eta^r(x), \quad T^{r+1} = \zeta^1(x), \ldots, T^{r+s} = \zeta^s(x), \tag{17}$$

cf. (2). Here $\mathbf{v}^1, \ldots, \mathbf{v}^r$ are linearly independent vector fields spanning an r-dimensional Lie algebra $\mathfrak{h}$. The functions $\zeta^1(x), \ldots, \zeta^s(x)$ define multiplication operators, and span a finite-dimensional $\mathfrak{h}$-module $\mathcal{M} \subset C^\infty$ of smooth functions, hence $\mathbf{v}^i(\zeta^j) = \sum b_k^{ij}\zeta^k$ for constants b_k^{ij}. The functions $\eta^a(x)$ must satisfy additional constraints in order that the operators (17) span a Lie algebra; these conditions can be conveniently expressed using the basic theory of Lie algebra cohomology, [21]. Define the "one-cochain" on the vector field Lie algebra $\mathfrak{h}$ by the linear map $F\colon \mathfrak{h} \to C^\infty$ which satisfies $\langle F\,;\mathbf{v}^a\rangle = \eta^a$. Since we can add in any constant coefficient linear combination of the ζ^b's to the η^a's without changing the Lie algebra $\mathfrak{g}$, we should interpret the η^a's as lying in the quotient module $C^\infty/\mathcal{M}$, and hence regard F as a $C^\infty/\mathcal{M}$–valued cochain. It is straightforward to see that the collection of differential operators (17) spans a Lie algebra if and only if the cochain F satisfies

$$\mathbf{v}\langle F\,;\mathbf{w}\rangle - \mathbf{w}\langle F\,;\mathbf{v}\rangle - \langle F\,;[\mathbf{v},\mathbf{w}]\rangle \in \mathcal{M} \quad \text{forall} \quad \mathbf{v},\mathbf{w} \in \mathfrak{h}. \tag{18}$$

The left hand side of (18) is just the evaluation $\langle \delta_1 F ; \mathbf{v}, \mathbf{w} \rangle$ of the *coboundary* of the 1-cochain F, hence (18) expresses the fact that the cochain F must be a $C^\infty/\mathcal{M}$–valued *cocycle.* A 1-cocycle is itself a coboundary, written $F = \delta_0 \sigma$ for some $\sigma(x) \in C^\infty$ if and only if $\langle F ; \mathbf{v} \rangle = \mathbf{v}(\sigma)$ for all $\mathbf{v} \in \mathfrak{h}$. It can be shown that two cocycles will differ by a coboundary $\delta_0 \sigma$ if and only if the corresponding Lie algebras are equivalent under the gauge transformation (5). Therefore two cocycles lying in the same *cohomology class* in the cohomology space $H^1(\mathfrak{h}, C^\infty/\mathcal{M}) = \operatorname{Ker} \delta_1 / \operatorname{Im} \delta_0$, will give rise to equivalent Lie algebras of differential operators. In summary, then, we have the following fundamental characterization of Lie algebras of first order differential operators.

Theorem 5. *There is a one-to-one correspondence between equivalence classes of finite dimensional Lie algebras* $\mathfrak{g}$ *of first order differential operators on* M *and equivalence classes of triples* $[\mathfrak{h}, \mathcal{M}, [F]]$, *where*

1. $\mathfrak{h}$ *is a finite-dimensional Lie algebra of vector fields,*

2. $\mathcal{M} \subset C^\infty$ *is a finite-dimensional* $\mathfrak{h}$*-module of functions,*

3. $[F]$ *is a cohomology class in* $H^1(\mathfrak{h}, C^\infty/\mathcal{M})$.

Based on Theorem 5, there are three basic steps required to classify finite dimensional Lie algebras of first order differential operators. First, one needs to classify the finite dimensional Lie algebras of vector fields $\mathfrak{h}$ up to changes of variables; this was done by Lie in one and two dimensions under the assumption that the Lie algebra has no singularities — not every vector field in the Lie algebra vanishes at a common point. (Lie further claimed to have completed the classification in three dimensions, [24], but the complete results were never published.) Secondly, for each of these Lie algebras, one needs to classify all possible finite-dimensional $\mathfrak{h}$-modules $\mathcal{M}$ of C^∞ functions. Trivial modules, valid for any Lie algebra of vector fields are the zero module $\mathcal{M} = 0$, which consists of the zero function alone, and that containing just the constant functions, which we write $\mathcal{M} = \{1\}$. Finally, for each of the modules $\mathcal{M}$, one needs to determine the first cohomology space $H^1(\mathfrak{h}, C^\infty/\mathcal{M})$. As the tables indicate, the cohomology classes are parametrized by one or more continuous parameters or, in a few cases, smooth functions.

It is then a fairly straightforward matter to determine when a given Lie algebra satisfies the quasi-exactly solvable condition that it admit a non-zero finite-dimensional module $\mathcal{N} \subset C^\infty$. A simple lemma says that we can always, without loss of generality, take the Lie algebra $\mathfrak{g}$ to be represented by a triple $[\mathfrak{h}, \{1\}, [F]]$ with $\mathcal{M} = \{1\}$. (Indeed, $\mathfrak{g}$ admits a finite-dimensional module if and only if $\mathcal{M} = \{1\}$ or $\mathcal{M} = 0$, and, in the latter case, $\mathfrak{g}$ can always be enlarged to include constant functions without destroying its quasi-exact solvability.) Remarkably, *in all known cases*, the cohomology parameters are "quantized", the quasi-exact solvability requirement forcing them to assume at most a discrete set of distinct values. This intriguing phenomenon of "quantization of cohomology" has been geometrically explained in terms of line bundles on complex surfaces in [9].

In one dimension, meaning $M = \mathbb{R}$ or $\mathbb{C}$, the classification of non-singular quasi-exactly solvable Lie algebras is straightforward, and summarized in the following table for the real case. The complex classification is exactly the same, but with $\mathbb{C}$

replacing $\mathbb{R}$. (This, though, is particular to one-dimensional Lie algebras.)

Basis	Structure	Action	Module	Cohomology	Q.E.S.
∂_x	$\mathbb{R}$	$x+\beta$	$\operatorname{Ker}\mathcal{D}$	0	0
$\partial_x, x\partial_x$	$\mathfrak{h}_2$	$\alpha x+\beta$	$\mathcal{P}^{(n)}$	0 or $\mathbb{R}$	0
$\partial_x, x\partial_x, x^2\partial_x$	$\mathfrak{sl}(2,\mathbb{R})$	$\frac{\alpha x+\beta}{\gamma x+\delta}$	$\{1\}$	$\mathbb{R}$	$\mathbb{N}$

Here the first column provides a basis for the Lie algebra of vector fields $\mathfrak{h}$. The second column gives the structure as an abstract Lie algebra; we use $\mathfrak{h}_2 = \mathbb{R} \ltimes \mathbb{R}$ to denote the unique non-abelian two-dimensional Lie algebra, $\ltimes$ denoting semi-direct product. The third column indicates the associated group action. The fourth column lists the general $\mathfrak{h}$–module $\mathcal{M}$; in particular, $\operatorname{Ker}\mathcal{D} = \operatorname{Span}\{x^i e^{\lambda x} \mid 0 \leq i \leq n_\lambda\}$ denotes the solution space to a linear, constant coefficient ordinary differential equation $\mathcal{D}\psi = 0$, while $\mathcal{P}^{(n)}$ denotes the space of polynomials of degree at most n. As for the cohomology, in case 2 it is only nontrivial if $\mathcal{M} = 0$, in which case it is represented by the operators $\partial_x, x\partial_x + c$ for $c \in \mathbb{R}$; in Case 3, it is represented by $\partial_x, x\partial_x + c, x^2\partial_x + 2cx$, $c \in \mathbb{R}$. The final column discusses when a Lie algebra satisfies the requirement of quasi-exactly solvability, assuming $\mathcal{M} = \{1\}$, *i.e.*, the algebra $\mathfrak{g}$ contains the constant functions. In Case 3, the cohomology parameter is quantized, $c = -\frac{1}{2}n$, where $n \in \mathbb{N}$ is a non-negative integer (which plays the physical role of spin). The corresponding module is $\mathcal{N} = \mathcal{P}^{(n)}$, the space of polynomials of degree n. Therefore, we have proven the basic classification theorem for quasi-exactly solvable Lie algebras of differential operators in one dimension.

Theorem 6. *Every (non-singular) finite-dimensional quasi-exactly solvable Lie algebra of first order differential operators in one (real or complex) variable is, locally, equivalent to a subalgebra of one of the Lie algebras*

$$\widehat{\mathfrak{g}}_n = \operatorname{Span}\{\ \partial_x,\quad x\partial_x,\quad x^2\partial_x - n\,x,\quad 1\ \}, \tag{19}$$

where $n \in \mathbb{N}$. For $\widehat{\mathfrak{g}}_n$, the associated module $\mathcal{N} = \mathcal{P}^{(n)}$ consists of the polynomials of degree at most n.

Turning to the two-dimensional classification, a number of additional complications present themselves. First, as originally shown by Lie, there are many more equivalence classes of finite-dimensional Lie algebras of vector fields. Moreover, the classification results in $\mathbb{R}^2$ and $\mathbb{C}^2$ are no longer the same — here we just present the complex case since the real classification has yet to be completed. Another complication is that the modules $\mathcal{M}$ for the vector field Lie algebras are no longer necessarily spanned by monomials, a fact that makes the determination of the cohomology considerably more difficult. Tables 1 – 3 at the end of the paper summarize our classification results for finite-dimensional Lie algebras of differential operators in two complex variables, [10], [12]. Lie's classification of nonsingular finite-dimensional Lie algebras of vector fields on $\mathbb{C}^2$ is summarized in Table 1. The first column exhibits a basis of the algebra, and the second indicates its structure as an abstract Lie algebra. The last column indicates where the Lie algebra lies in Lie's "Gruppenregister", [25]. (We have, in a few cases, employed different coordinate systems than Lie.) Table 2 describes the different finite-dimensional modules for each of these Lie algebras. The first column tells whether the module is necessarily spanned by monomials, *i.e.*, single terms of the indicated form. (In cases 5 and 20, we have monomials unless $\alpha \in \mathbb{Q}^+$ or $r < \alpha \in \mathbb{Q}^+$ are positive rational numbers, respectively.) The second column indicates a typical term in a basis element for the module — the non-monomial basis elements will be linear combinations of terms of the indicated type; i, j always denote nonnegative integers. The third column either indicates ranges of indices which must be included, or, in the case of an arrow, indicates other indices which must be included if the given one is. For instance, in Case 19, if the monomial $x^i y^j e^{\mu x}$ belongs to the module, so must the monomials $x^{i-1} y^j e^{\mu x}$ and $x^{i+r} y^{j+1} e^{(\mu+\lambda)x}$ (provided $i > 0$ and/or $j > 0$) for each exponent λ appearing in the Lie algebra. Finally, $R_k^{n,m}(z)$ denotes the polynomial

$$R_k^{m,n}(z) = \frac{d^k}{dz^k}\left[(z-1)^{m+n}(z+1)^m\right],$$

which, for $n = 0$, is a multiple of the Gegenbauer (or ultraspherical) polynomial $C_{2m-k}^{k-m+(1/2)}(z)$, [8; vol. 2]. Table 3 describes the cohomology spaces $H^1(\mathfrak{h}, C^\infty/\mathcal{M})$ for each of the Lie algebras and corresponding modules. The first column indicates the dimension of the cohomology space, and the second column gives a representative cocycle of each nontrivial cohomology class. Only the vector fields $\mathbf{v}^a$ which are actually modified are indicated, *i.e.*, those for which $\eta^a = \langle F\,;\mathbf{v}^a\rangle \not\equiv 0$, *cf.* (17). In Case 4, $\operatorname{Div}\mathcal{M} = \{f_x + g_y \mid f, g \in \mathcal{M}\}$. Finally, Table 4 describes the quantization condition resulting from the quasi- exactly solvability assumption that, assuming $\mathcal{M} = \{1\}$, the Lie algebra admit a finite-dimensional module $\mathcal{N}$. If the cohomology is trivial, so $\mathfrak{g}$ is spanned by vector fields and the constant functions, then it automatically satisfies the quasi-exactly solvable condition, with the associated finite-dimensional modules being explicitly described in Table 2. The maximal algebras, namely Case 11, $\mathfrak{sl}(2) \oplus \mathfrak{sl}(2)$, Case 15, $\mathfrak{sl}(3)$, and Case 24, $\mathfrak{gl}(2) \ltimes \mathbb{R}^r$, play an important role in Turbiner's theory of differential equations in two dimensions with orthogonal polynomial solutions, [33].

5. Quasi-Exactly Solvable Operators on the Line.

Let us now specialize to problems in one dimension. In view of Theorem 6, we let $n \in \mathbb{N}$ be a nonnegative integer, and consider the Lie algebra $\mathfrak{g}_n$ spanned by the differential operators

$$T^- = T_n^- = \frac{d}{dz}, \qquad T^0 = T_n^0 = z\,\frac{d}{dz} - \frac{n}{2}, \qquad T^+ = T_n^+ = z^2\,\frac{d}{dz} - nz, \tag{20}$$

which satisfy the standard $\mathfrak{sl}(2)$ commutation relations. In this section, we shall use z instead of x for the "canonical coordinate", retaining x for the physical coordinate in which the operator takes Schrödinger form. The Lie algebra $\widehat{\mathfrak{g}}_n$ in Theorem 6 is merely a central extension (by the constant functions) of the subalgebra $\mathfrak{g}_n$, hence $\widehat{\mathfrak{g}}_n$ is isomorphic to the Lie algebra $\mathfrak{gl}(2)$ of all 2×2 matrices. Note that since $\widehat{\mathfrak{g}}_n$ and $\mathfrak{g}_n$ only differ by inclusion of constant functions, in our analysis of Lie algebraic differential operators we can, without loss of generality, concentrate on the Lie algebra $\mathfrak{g}_n$, since any Lie algebraic operator (3) for the full algebra $\widehat{\mathfrak{g}}_n$ is automatically a Lie algebraic operator for the subalgebra $\mathfrak{g}_n$.

Thus, the most general second order quasi-exactly solvable Hamiltonian in one space dimension can therefore be written in the form

$$\begin{aligned} -\mathcal{H} = c_{++}(T^+)^2 + c_{+0}[T^+T^0 + T^0T^+] + c_{00}(T^0)^2 + c_{+-}[T^+T^- + T^-T^+]+ \\ + c_{0-}[T^0T^- + T^-T^0] + c_{--}(T^-)^2 + c_+T^+ + c_0T^0 + c_-T^- + c_*. \end{aligned} \tag{21}$$

Notice that the constant matrix (c_{ab}) in (3) can, without loss of generality, be taken to be symmetric, *i.e.*, $c_{ab} = c_{ba}$. Substituting the explicit formulas (20) for T^+, T^0, T^- into (21), we find that every quasi-exactly solvable operator can be written in the canonical form

$$-\mathcal{H} = P(z)\frac{d^2}{dz^2} + \left\{Q(z) - \frac{n-1}{2}P'(z)\right\}\frac{d}{dz} + \left\{R - \frac{n}{2}Q'(z) + \frac{n(n-1)}{12}P''(z)\right\}, \tag{22}$$

where

$$\begin{aligned} P(z) &= c_{++}z^4 + 2c_{+0}z^3 + (2c_{+-} + c_{00})z^2 + 2c_{0-}z + c_{--}, \\ Q(z) &= c_+z^2 + c_0z + c_-, \\ R(z) &= \frac{n^2+2n}{12}c_{00} - \frac{n^2+2n}{3}c_{+-} + c_*, \end{aligned} \tag{23}$$

are (general) polynomials of respective degrees $4, 2, 0$. Since the module $\mathcal{N}$ is the space $\mathcal{P}^{(n)}$ of polynomials of degree at most n, the algebraic eigenfunctions (21) will, in the z- coordinates, just be polynomials $\chi_k(z) \in \mathcal{P}^{(n)}$. In terms of the standard basis $\chi_k(z) = z^k$, $k = 0, \ldots, n$, the restricted Hamiltonian $\mathcal{H} \mid \mathcal{P}^{(n)}$ takes the form of a pentadiagonal matrix, or, if $c_{++} = c_{--} = 0$, a tridiagonal matrix. Thus, for a normalizable one-dimensional quasi-exactly solvable operator, there are $n+1$ linearly independent algebraic eigenfunctions which, in the canonical z coordinates, are polynomials of degree at most n.

Specializing the solution to the equivalence problem given by Theorem 3 to the particular operator (22), we find that the change of variables (16) required to place the operator into physical (Schrödinger) form will, in general, be given by an elliptic integral

$$x = \varphi(z) = \int^z \frac{dy}{\sqrt{P(y)}}\,. \tag{24}$$

The corresponding gauge factor is

$$\mu(z) = \left|P(z)\right|^{-n/4} \exp\left\{\int^z \frac{Q(y)}{2P(y)}\,dy\right\}. \tag{25}$$

The potential is given by

$$V(x) = -\frac{n(n+2)\left(PP'' - \frac{3}{4}P'^2\right) + 3(n+1)(QP' - 2PQ') - 3Q^2}{12P} - R, \tag{26}$$

where the right hand side is evaluated at $z = \varphi^{-1}(x)$. In the physical coordinate, the associated algebraic wave functions will then take the form

$$\psi(x) = \mu(\varphi^{-1}(x))\,\chi(\varphi^{-1}(x)), \tag{27}$$

where $\chi(z)$ is a polynomial of degree at most n.

The canonical form (22) of a quasi-exactly solvable differential operator is not unique, since there is a "residual" symmetry group which preserves the Lie algebra $\mathfrak{g}_n$. Not surprisingly, this group is $\mathrm{GL}(2,\mathbb{R})$, which acts on the (projective) line by linear fractional (Möbius) transformations

$$z \longmapsto w = \frac{\alpha z+\beta}{\gamma z+\delta}, \qquad A = \begin{pmatrix} \alpha & \beta \\ \gamma & \delta \end{pmatrix}, \qquad \det A = \Delta = \alpha\delta - \beta\gamma \neq 0. \tag{28}$$

To describe the induced action of the transformations (28) on the quasi-exactly solvable operators (22), we first recall the basic construction of the finite-dimensional irreducible rational representations of the general linear group $\mathrm{GL}(2,\mathbb{R})$.

Definition 7. Let $n \geq 0$, i be integers. The irreducible multiplier representation $\rho_{n,i}$ of $\mathrm{GL}(2,\mathbb{R})$ is defined on the space $\mathcal{P}^{(n)}$ of polynomials of degree at most n by the transformation rule $\widehat{P} = \rho_{n,i}(P)$, where

$$\widehat{P}(z) = (\gamma z+\delta)^n(\alpha\delta-\beta\gamma)^i \, P\left(\frac{\alpha z+\beta}{\gamma z+\delta}\right), \qquad P \in \mathcal{P}^{(n)}. \tag{29}$$

The multiplier representation $\rho_{n,i}$ has infinitesimal generators given by the differential operators (5) combined with the operator of multiplication by i representing the diagonal subalgebra (center) of $\mathfrak{gl}(2,\mathbb{R})$. The action (28) induces an automorphism of the Lie algebra $\mathfrak{g}_n$, which is isomorphic to the representation $\rho_{2,-1}$, and, consequently, preserves the class of quasi-exactly solvable operators associated with the algebra $\mathfrak{g}_n$. Moreover, the corresponding gauge action

$$\widehat{\mathcal{H}}(z) = (\gamma z+\delta)^n \cdot \mathcal{H} \cdot (\gamma z+\delta)^{-n}, \tag{30}$$

will preserve the space of quasi-exactly solvable Hamiltonians (22). Identifying the operator $\mathcal{H}$ with the corresponding quartic, quadratic and constant polynomials (23), we find that the action (30) of $\mathrm{GL}(2,\mathbb{R})$ on the space of quasi-exactly solvable operators is isomorphic to the sum of three irreducible representations, $\rho_{4,-2} \oplus \rho_{2,-1} \oplus \rho_{0,0}$; the quartic $P(z)$ transforms according to $\rho_{4,-2}$, the quadratic $Q(z)$ according to $\rho_{2,-1}$, while R is constant. Finally, the associated module, which is just the space of polynomials $\mathcal{P}^{(n)}$, transforms according to the representation $\rho_{n,0}$.

Using the action of $\mathrm{GL}(2,\mathbb{R})$, we can place the gauged operator (22) into a simpler canonical form, based on the invariant theoretic classification of canonical forms for real quartic polynomials, [18].

Theorem 8. *Under the representation $\rho_{4,-2}$ of $\mathrm{GL}(2,\mathbb{R})$, every nonzero real quartic polynomial $P(z)$ is equivalent to one of the following canonical forms:*

$$\nu(1-z^2)(1-k^2z^2), \qquad \nu(1-z^2)(1-k^2+k^2z^2), \qquad \nu(1+z^2)(1+k^2z^2),$$

$$\nu(z^2-1), \quad \nu(z^2+1), \quad \nu z^2, \quad \nu(z^2+1)^2, \quad z, \quad 1,$$

where ν and $0 < k < 1$ are real constants.

The nine cases correspond to the positions of the complex roots of the quartic, which are, respectively, four simple real roots, two simple real roots and two simple complex conjugate roots, four simple complex roots, one double and two simple real roots, one double real and two simple complex roots, two double real roots, two double complex roots, one triple and one simple real roots, and one quadruple real root. Note that we are allowing ∞ to be a root, whose order is defined to be $4 - d$ where d is the degree of the polynomial. (This makes the concept of order of a root of a quartic polynomial invariant under the representation $\rho_{4,-2}$.)

Remark. If P has four simple roots, then the Schrödinger equation (1) for the operator (22), (23), has four regular singularities, and hence, by a *complex* linear fractional transformation, can be mapped to a form of Heun's equation

$$\frac{d^2y}{dz^2} + \left(\frac{\gamma}{z} + \frac{\varepsilon}{z-1} + \frac{\delta}{z-a}\right)\frac{dy}{dz} + \frac{\alpha\beta z + b}{z(z-1)(z-a)}\, y = 0, \tag{31}$$

with $\alpha + \beta - \gamma - \delta - \varepsilon + 1 = 0$, which amounts to placing P into the complex canonical form $P(z) = \nu z(z-1)(z-a)$ with the roots at $0, 1, a$ and ∞. Therefore, the algebraic eigenfunctions can be expressed in terms of Heun polynomials, or, in the case of multiple roots, confluent Heun polynomials, [8; vol. 3].

The solution to the normalizability problem begins with a detailed analysis of the elliptic integral (24). Consider an interval $z_0 < z < z_1$ where $P(z) > 0$ is positive and vanishes at the endpoints, $P(z_0) = P(z_1) = 0$. Simple asymptotic analysis of (24) immediately implies that if both z_0 and z_1 are simple roots, then (8) defines z as a periodic function of x; therefore, the potential $V(x)$ and the corresponding algebraic eigenfunctions are periodic functions of x, and cannot be normalizable. (However, if they have no singularities, they do contribute, albeit minutely, to the continuous spectrum of the operator.) If $P(z) > 0$ is everywhere positive and has no real roots including ∞, then the same conclusion of periodicity holds. Therefore, a necessary condition for the quasi-exactly solvable operator (22), (23), to be normalizable is that the quartic polynomial $P(z)$ have at least one multiple real root, which must lie at the end of an interval of positivity. Thus, the class of quasi-exactly solvable potentials naturally splits into two subclasses — the periodic potentials, which are never normalizable, and the non-periodic potentials, which are sometimes normalizable.

Tedious but direct calculations based on (24), (25), (27), produce the explicit change of variables, the potential, and eigenfunctions for the above normal forms in physical coordinates. Each of the classes of potentials is a linear combination of four elementary and/or elliptic functions, plus a constant which we absorb into the eigenvalue. The potentials naturally divide into two classes, which are listed in the following two Tables. In each case, the four coefficients are not arbitrary, but satisfy a single complicated algebraic equation and one or more inequalities; for simplicity, we only exhibit these in the non-periodic case. For the more general class of Lie algebraic potentials see [22].

First, the periodic quasi-exactly solvable potentials correspond to the cases when the roots of P are simple, of which there are five cases given in the following Table.

$$\begin{aligned} &1.\quad \nu(1-z^2)(1-k^2z^2),\\ &2.\quad \nu(1-z^2)(1-k^2+k^2z^2),\\ &3.\quad \nu(1+z^2)(1+(1-k^2)z^2),\\ &4.\quad \nu(1-z^2),\\ &5.\quad \nu(z^2+1)^2, \end{aligned} \tag{32}$$

where $\nu > 0$, $0 < k < 1$. The explicit formulas for the corresponding potentials follow. In the first three cases, the corresponding potentials are written in terms of the standard Jacobi elliptic functions of modulus k, [8; vol. 2]. Also, as remarked above, the coefficients A, B, C, D are not arbitrary, although the explicit constraints are too complicated to write here.

PERIODIC QUASI-EXACTLY SOLVABLE POTENTIALS

1. $\mathrm{dn}^{-2}\sqrt{\nu}\,x\,(A\,\mathrm{sn}\sqrt{\nu}\,x + B) + \mathrm{cn}^{-2}\sqrt{\nu}\,x\,(C\,\mathrm{sn}\sqrt{\nu}\,x + D)$
2. $\mathrm{dn}^{-2}\sqrt{\nu}\,x\,(A\,\mathrm{cn}\sqrt{\nu}\,x + B) + \mathrm{sn}^{-2}\sqrt{\nu}\,x\,(C\,\mathrm{cn}\sqrt{\nu}\,x + D)$
3. $A\,\mathrm{cn}\sqrt{\nu}\,x\,\mathrm{sn}\sqrt{\nu}\,x + B\,\mathrm{cn}^2\sqrt{\nu}\,x + C\,\mathrm{dn}^{-2}\sqrt{\nu}\,x\,(\mathrm{cn}\sqrt{\nu}\,x\,\mathrm{sn}\sqrt{\nu}\,x + D\,\mathrm{cn}^2\sqrt{\nu}\,x)$
4. $A\sin^2\sqrt{\nu}\,x + B\sin\sqrt{\nu}\,x + C\tan\sqrt{\nu}\,x\sec\sqrt{\nu}\,x + D\sec^2\sqrt{\nu}\,x$
5. $A\cos 4\sqrt{\nu}\,x + B\cos 2\sqrt{\nu}\,x + C\sin 2\sqrt{\nu}\,x + D\sin 4\sqrt{\nu}\,x$

The elliptic functions $\mathrm{sn}\,y, \mathrm{cn}\,y, \mathrm{dn}\,y$ are periodic of period $4K$ (or $2K$ in the case of $\mathrm{dn}\,y$), where

$$K = \int_0^1 \frac{dz}{\sqrt{(1-z^2)(1-k^2z^2)}},$$

is the complete elliptic integral of modulus k. Moreover, $\mathrm{dn}\,y$ is never zero, $\mathrm{cn}\,y$ vanishes at odd multiples of K, and $\mathrm{sn}\,y$ vanishes at even multiples of K; also $\mathrm{cn}\,4mK = 1 = \mathrm{sn}(4m+1)K$, $\mathrm{cn}(4m+2)K = -1 = \mathrm{sn}(4m+3)K$. Therefore, the potentials in cases 1, 2 and 4 have singularities unless $C = D = 0$. In Cases 3 and 5, the potential has no singularities, reflecting the fact that in these cases $P(z)$ has no real roots. Case 3 includes the Lamé equation, [8; vol. 3].

The non-periodic potentials correspond to the cases with one or two multiple roots, which are given in the following Table.

$$\begin{array}{lll} 1. & \nu(z^2+1), & (-\infty,\infty)\\ 2. & \nu(z^2-1), & [1,\infty) \quad \text{or} \quad (-\infty,-1]\\ 3. & \nu z^2, & (0,\infty) \quad \text{or} \quad (-\infty,0)\\ 4. & z, & [0,\infty)\\ 5. & 1. & (-\infty,\infty) \end{array} \tag{33}$$

Here $\nu > 0$, and the second column indicates the interval(s) where P is positive, with square brackets designating simple roots. In cases 2 and 3 there are two possible intervals of positivity, and hence potentially two different physical Hamiltonians, although we can readily switch from one to the other by a discrete reflection $z \mapsto -z$.

The explicit formulas for the corresponding potentials follow.

Non-periodic Quasi-Exactly Solvable Potentials

1. $A\sinh^2\sqrt{\nu}\,x + B\sinh\sqrt{\nu}\,x + C\tanh\sqrt{\nu}\,x\,\mathrm{sech}\,\sqrt{\nu}\,x + D\,\mathrm{sech}^2\sqrt{\nu}\,x$
2. $A\cosh^2\sqrt{\nu}\,x + B\cosh\sqrt{\nu}\,x + C\coth\sqrt{\nu}\,x\,\mathrm{csch}\,\sqrt{\nu}\,x + D\,\mathrm{csch}^2\sqrt{\nu}\,x$
3. $Ae^{2\sqrt{\nu}\,x} + Be^{\sqrt{\nu}\,x} + Ce^{-\sqrt{\nu}\,x} + De^{-2\sqrt{\nu}\,x}$
4. $Ax^6 + Bx^4 + Cx^2 + \dfrac{D}{x^2}$
5. $Ax^4 + Bx^3 + Cx^2 + Dx$

In cases 2 and 4, the potential has a singularity at $x = 0$ unless $C + D = 0$ (Case 2) or $D = 0$ (Case 4). The nonsingular potentials in Case 4 are the anharmonic oscillator potentials discussed in detail in [30], [28]. The algebraic constraints satisfied by the coefficients are given in the following table.

Non-Periodic Constraints

1. $[B \pm 2(n+1)\sqrt{\nu A}]^4 + A(4D-\nu)[B \pm 2(n+1)\sqrt{\nu A}]^2 - 4A^2C^2 = 0, \qquad A \geq 0.$
2. $[B \pm 2(n+1)\sqrt{\nu A}]^4 - A(4D+\nu)[B \pm 2(n+1)\sqrt{\nu A}]^2 + 4A^2C^2 = 0, \qquad A \geq 0.$
3. $\pm C\sqrt{A} \pm B\sqrt{D} = 2(n+1)\sqrt{\nu AD}, \qquad A, D \geq 0.$
4. $16A^3[(4n+5)(4n+3) - 4D] \pm 32(n+1)A^{3/2}(B^2-4AC) + (B^2-4AC)^2 = 0,$
   $A \geq 0, \qquad D \geq -\frac{1}{4}.$
5. $8A^2D + B(B^2-4AC) = \pm 16(n+1)A^{5/2}, \qquad A \geq 0.$

According to Turbiner, [30], a potential is *exactly solvable* if it does not explicitly depend on the discrete "spin" parameter n, since, in this case, one can find representation spaces of arbitrarily large dimension and thereby (if normalizable) produce infinitely many eigenvalues by algebraic methods. Note that since the gauge transformation (8), (9), can explicitly depend on n, exact solvability cannot be detected in the canonical coordinates, but depends on the final physical form of the operator. The exactly solvable nonperiodic potentials are characterized by the condition $A = B = 0$, or, in case 3, $C = D = 0$. In Case 2, there is an additional inequality, $|C| \leq D + \frac{1}{4}\nu$, to be satisfied. The exactly solvable potentials include the (restricted) Pöschl–Teller potentials (Case 1), the Morse potentials (Case 3), the radial harmonic oscillator (provided $D = l(l+1)$, $l \in \mathbb{N}$) (Case 4), and the harmonic oscillator (Case 5). Interestingly, all of the non-periodic exactly solvable Schrödinger operators appear in Infeld and Hull's factorization method,

[20], coinciding with the set of second order operators admitting factorizations of types A–D. The underlying reason for this coincidence is not known. Also, we do not know whether any generalization of the factorization method will incorporate the periodic exactly solvable operators, or, conversely, how to generalize the quasi-exactly solvable method to incorporate the type E and F factorizations.

Analysis of the explicit formulas for the eigenfunctions based on Theorem 3 yields a complete set of conditions for the normalizability of the non-periodic potentials, which are written in terms of the coefficients of the quadratic polynomial $Q(z)$ as given in (23). Note that this requires that $P(z)$ be in canonical form. Thus, although the following table provides a complete list of normalizable potentials, it does not give the most general set of normalizable Lie algebraic coefficients, but rather, describes a single representative set of coefficients for each $\mathrm{GL}(2,\mathbb{R})$ orbit of normalizable quasi-exactly solvable operators. See [13] for partial normalizability conditions.

NORMALIZABILITY CONDITIONS

1. $c_{++} = c_{+0} = c_{0-} = 0, \quad c_{--} = 2c_{+-} + c_{00}, \quad c_+ = 0, \quad c_0 < -nc_{--}.$
2. $c_{++} = c_{+0} = c_{0-} = 0, \quad c_{--} = -2c_{+-} - c_{00},$
   $\bigl(c_+ + c_0 + c_- = -nc_{--} \quad \text{or} \quad c_+ + c_0 + c_- = -(n+2)c_{--}\bigr),$
   $\bigl(c_+ < 0 \quad \text{or} \quad c_+ = 0, \quad c_0 < nc_{--}\bigr).$
3. $c_{++} = c_{+0} = c_{0-} = c_{--} = 0,$
   $\bigl(c_+ < 0, \quad \text{or} \quad c_+ = 0, \quad c_0 < -2nc_{+-} - nc_{00}\bigr),$
   $\bigl(c_- > 0, \quad \text{or} \quad c_- = 0, \quad c_0 > 2nc_{+-} + nc_{00}\bigr).$
4. $c_{++} = c_{+0} = 2c_{+-} + c_{00} = c_{--} = 0, \qquad c_{0-} = \tfrac{1}{2},$
   $\bigl(c_- = \tfrac{1}{2}n \quad \text{or} \quad c_- = \tfrac{1}{2}n + 1\bigr),$
   $\bigl(c_+ < 0, \quad \text{or} \quad c_+ = 0, \quad c_0 < 0\bigr).$
5. $c_{++} = c_{+0} = 2c_{+-} + c_{00} = c_{0-} = 0, \quad c_{--} = 1, \quad c_+ = 0, \quad c_0 < 0.$

It is possible to deduce explicit, general normalizability conditions on the Lie algebraic coefficients c_{ab}, by using the fact that such conditions must be invariant under the action of the group $\mathrm{GL}(2,\mathbb{R})$. Therefore, normalizability conditions can be written in terms of the classical joint invariants and covariants of the pair of polynomials P, Q. The simplest of these necessary conditions stems from our earlier observation that, for normalizability, the quartic polynomial P must have a multiple root. This condition can be expressed in an invariant manner by the vanishing of the discriminant of P, which results in a sixth degree algebraic constraint, having the explicit form

$$\bigl[12c_{++}c_{--} - 12c_{+0}c_{0-} + (2c_{+-} + c_{00})^2\bigr]^3 = \\ = \left(\det \begin{vmatrix} 6c_{++} & 3c_{+0} & 2c_{+-} + c_{00} \\ 3c_{+0} & 2c_{+-} + c_{00} & 3c_{0-} \\ 2c_{+-} + c_{00} & 3c_{0-} & 6c_{--} \end{vmatrix}\right)^2. \tag{34}$$

Additional normalizability conditions require further joint covariants, which can be explicitly written in terms of the classical transvectants.

Definition 9. Let $F(z) \in \mathcal{P}^{(m)}$ and $G(z) \in \mathcal{P}^{(n)}$ be polynomials, and suppose that $r \leq \min\{m, n\}$. The r-th *transvectant* of F and G is the polynomial

$$(F, G)^{(r)} = \sum_{k=0}^{r} (-1)^k \binom{r}{k} \frac{(m-r+k)!}{(m-r)!} \frac{(n-k)!}{(n-r)!} F^{(r-k)}(z)\, G^{(k)}(z). \tag{35}$$

Theorem 10. *Suppose $F \in \mathcal{P}^{(m)}$ and $G \in \mathcal{P}^{(n)}$ are polynomials transforming under* $\mathrm{GL}(2,\mathbb{R})$ *according to the representations $\rho_{m,i}$, $\rho_{n,j}$, respectively. Then the r-th transvectant $(F, G)^{(r)} \in \mathcal{P}^{(m+n-2r)}$ is a polynomial of degree at most $m+n-2r$ and transforms according to the representation $\rho_{m+n-2r,i+j+r}$.*

The fact that $(F, G)^{(r)}$ has degree at most $m+n-2r$ is *not* obvious from the formula (20). The First Fundamental Theorem of classical invariant theory, [16], [18], states that every joint covariant of a system of polynomials is a suitably homogeneous polynomial in the succesive transvectants of the system. The transvectants are not all independent, and, except for a few low order cases (which, fortunately, include the case of a quartic and a quadratic), the determination of a complete list of algebraically independent covariants (a Hilbert Basis) is a difficult, unsolved problem.

Theorem 11. *Let P be a quartic polynomial and Q a quadratic polynomial. Then a complete system of irreducible covariants for the pair P, Q is provided by the polynomials themselves, the discriminant $\Delta = (Q, Q)^{(2)}$ of the quadratic, the covariants $H = (P, P)^{(2)}$, $T = (H, P)^{(1)}$, and invariants $i = (P, P)^{(4)}$, $j = (H, P)^{(4)}$ of the quartic, and the following 11 joint covariants:*

$$(P, Q)^{(1)}, \quad (P, Q)^{(2)}, \quad (P, Q^2)^{(3)}, \quad (P, Q^2)^{(4)}, \quad (H, Q)^{(1)}, \quad (H, Q)^{(2)}$$
$$(H, Q^2)^{(3)}, \quad (H, Q^2)^{(4)}, \quad (T, Q)^{(2)} \quad (T, Q^2)^{(4)}, \quad (T, Q^3)^{(6)}.$$

For example, the discriminant of a quartic is given by the combination

$$\frac{1}{9216} \left\{ \frac{1}{96} \left[(P, P)^{(4)}\right]^3 - \left[(H, P)^{(4)}\right]^2 \right\}.$$

Using the covariants in Theorem 11, we can now state the explicit necessary and sufficient conditions for normalizability of a quasi-exactly solvable operator on the line. In each case, the first line gives the invariant conditions for a quartic P to be equivalent to one of our five canonical forms with a multiple root; these conditions automatically imply the discriminantal condition (34). The subcases a) and b) are different alternatives; also, in Cases 2 and 4, n^* means either n or $n+2$. Finally, we write $F \succ G$ if $F(z) \geq G(z)$ for all z, and $F \not\equiv G$. Each of these conditions, when written out, gives a (very) complicated, but explicit condition on the Lie algebraic coefficients of our quasi-exactly solvable operator.

Theorem 12. *A one-dimensional quasi-exactly solvable operator* (21) *is normalizable if and only if its Lie algebraic coefficients satisfy one of the following sets of constraints:*

I. $i^3 = j^2 \neq 0, \qquad iH - jP \succ 0.$
$(iH - jP,\, Q^2)^{(4)} = 0, \qquad 3(T,\, Q)^{(2)} \succ 10n(iH - jP).$

II. $i^3 = j^2 \neq 0, \qquad iH - jP \prec 0.$

a) $(iH - jP,\, Q^2)^{(4)} \neq 0,$

$$12i^2(T,\, Q^3)^{(6)} + 1800n^* i^2 (iH - jP,\, Q^2)^{(4)} = \\ = 25\sqrt{(3jP - 3iH,\, Q^2)^{(4)}}\left[(2jH + i^2P,\, Q^2)^{(4)} - 144ij(Q,\, Q)^{(2)}\right].$$

b) $(iH - jP,\, Q^2)^{(4)} = 0,$

$$25\left[(P,\, Q^2)^{(3)}\right]^2 (iH - jP) = 864(Q,\, Q)^{(2)}\left[3(T,\, Q)^{(2)} + 10n^*(iH - jP)\right]^2,$$

$3(T,\, Q)^{(2)} \prec 10n(iH - jP).$

III. $i^3 = j^2 \neq 0, \qquad T = 0.$

a) $(H,\, Q^2)^{(4)} - 48i(Q,\, Q)^{(2)} > 0.$

b) $(H,\, Q^2)^{(4)} = 48i(Q,\, Q)^{(2)},\ (H,\, Q)^{(2)} - 8iQ \not\equiv 0, \quad \dfrac{(P,\, Q^2)^{(3)}}{(H,\, Q)^{(2)} - 8iQ} < 12n.$

IV. $i = j = 0, \qquad H \not\equiv 0.$

a) $(T,\, Q^3)^{(6)} > 0,\ 2400(Q,\, Q)^{(2)}(H,\, Q^2)^{(4)} - 25[(P,\, Q^2)^{(4)}]^2 = 64n^*(T,\, Q^3)^{(6)},$

b) $(T,\, Q^3)^{(6)} = 0, \qquad (P,\, Q)^{(2)} \succ 0, \qquad 3P(H,\, Q)^{(1)} - 6H(P,\, Q)^{(1)} = 8n^*H^2.$

V. $i = j = T = H = 0,$
$(P,\, Q^2)^{(4)} = 0, \qquad \text{and} \qquad (P,\, Q)^{(1)} \succ 0.$

6. Two Dimensional Problems.

There are a number of additional difficulties in the two dimensional problem which do not appear in the scalar case. First, there are several different classes of quasi-exactly solvable Lie algebras available. Even more important is the fact that, according to Theorem 3, there are nontrivial "closure conditions" which must be satisfied in order that the magnetic one-form associated with a given Hamiltonian operator be closed and hence the operator be equivalent, under a gauge transformation (5), to a Schrödinger operator. Unfortunately, in all but trivial cases, the closure conditions associated with a quasi-exactly solvable Hamiltonian (3) corresponding to the generators of one of the quasi-exactly solvable Lie algebras on our list are *nonlinear algebraic equations* in the coefficients c_{ab}, c_a, c_0, and it appears to be impossible to determine their general solution. Nevertheless, there are useful simplifications of the general closure conditions which can be effectively used to generate large classes of planar quasi-exactly solvable and exactly solvable Schrödinger operators, both for flat space as well as curved metrics.

Suppose that the Lie algebra $\mathfrak{g}$ is spanned by linearly independent first-order differential operators as in (2). Substituting these into the general Lie algebraic form (3), we find that the operator assumes the form (7) with

$$g^{ij} = \sum_{a,b=1}^{r} c_{ab}\xi^{ai}\xi^{bj},$$
$$h^i = \sum_{a,b=1}^{r} \left[c_{ab}\left(\xi^{aj}\frac{\partial \xi^{bi}}{\partial x^j} + 2\eta^a\xi^{bi}\right) + c_a\xi^{ai}\right], \tag{36}$$
$$k = \sum_{a,b=1}^{r} \left[c_{ab}\left(\xi^{aj}\frac{\partial \eta^{b}}{\partial x^j} + \eta^a\eta^{b}\right) + c_a\eta^{a}\right].$$

The magnetic form ω, (14), and potential V for the covariant form (13) of the Hamiltonian are then computed using formulas (15). The *closure conditions* $d\omega = 0$ are equivalent to the solvability of the system of partial differential equations

$$\sum_{a,b=1}^{r} c_{ab}\xi^{ai}\sum_{j=1}^{n}\left(\xi^{bj}\frac{\partial \tau}{\partial x^j} + \frac{\partial \xi^{bj}}{\partial x^j}\right) = \sum_{a=1}^{r}\xi^{ai}\left[2\sum_{b=1}^{r}c_{ab}\eta^b + c_a\right], \qquad i = 1, \dots, n, \tag{37}$$

for a scalar function $\tau(x)$, given by $\tau = 2\sigma + \frac{1}{2}\log g$ in terms of the gauge factor e^σ required to place the operator in Schrödinger form. The closure conditions (37) are extremely complicated to solve in full generality, but a useful subclass of solutions can be obtained from the *simplified closure conditions*

$$\sum_{i=1}^{n}\left(\xi^{ai}\frac{\partial \tau}{\partial x^i} + \frac{\partial \xi^{ai}}{\partial x^i}\right) - 2\eta^a = k^a, \qquad a = 1, \dots, r. \tag{38}$$

where $k^1, \dots, k^r$ are constants. Any solution $\tau(x)$ of equations (38) will generate an infinity of solutions to the full closure conditions (37), with c_{ab} arbitrary, and $c_a = \sum_b c_{ab}k^b$. The case $k^a = 0$ and $\mathfrak{g}$ semi-simple was investigated in [26]. Although the simplified closure conditions can be explicitly solved for such Lie algebras, with the exception of $\mathfrak{so}(3)$, their solutions are found to generate quasi-exactly solvable Schrödinger operators that are *not* normalizable, and hence of limited use. Note that even when the simplified closure conditions do not have any acceptable solutions, the *full* closure conditions (37) may be compatible and may give rise to normalizable operators.

Consider, in the first place, the Lie algebra $\mathfrak{g} \simeq \mathfrak{sl}(2) \oplus \mathfrak{sl}(2)$ of type 11 spanned by the first-order differential operators

$$T^1 = \partial_x, \quad T^2 = \partial_y, \quad T^3 = x\partial_x, \quad T^4 = y\partial_y, \quad T^5 = x^2\partial_x - nx, \quad T^6 = y^2\partial_y - my,$$

where $n, m \in \mathbb{N}$. The particular choice

$$(c_{ab}) = \begin{pmatrix} 2 & 1 & 0 & 0 & 0 & 1 \\ 1 & 2 & 0 & 0 & 1 & 0 \\ 0 & 0 & 3 & 0 & 0 & 0 \\ 0 & 0 & 0 & 3 & 0 & 0 \\ 0 & 1 & 0 & 0 & 1 & 1 \\ 1 & 0 & 0 & 0 & 1 & 1 \end{pmatrix},$$

$$(c_a) = \Big(0, 0, -(1+4n), -(1+4m), 0, 0\Big), \qquad c_0 = \tfrac{3}{4} + m^2 + n^2,$$

of Lie algebraic coefficients leads to a quasi-exactly solvable Hamiltonian with Riemannian metric

$$g^{11} = (1+x^2)(2+x^2), \quad g^{12} = (1+x^2)(1+y^2), \quad g^{22} = (1+y^2)(2+y^2),$$

which has complicated curvature, and potential

$$\begin{aligned} 4V = -y^2 - \frac{(1+2n)(3+2n)}{1+x^2} - \frac{(1+2m)(3+2m)}{1+y^2} \\ - \frac{17+12y^2-y^4+2xy(6+5y^2)}{3+x^2+y^2} + \frac{5(3+2xy)(1+y^2)(2+y^2)}{(3+x^2+y^2)^2}. \end{aligned}$$

A second interesting solution is

$$(c_{ab}) = \begin{pmatrix} 1 & 1 & 0 & 0 & 0 & 1 \\ 1 & 4 & 0 & 0 & 1 & 0 \\ 0 & 0 & 2 & 0 & 0 & 0 \\ 0 & 0 & 0 & 8 & 0 & 0 \\ 0 & 1 & 0 & 0 & 1 & 1 \\ 1 & 0 & 0 & 0 & 1 & 4 \end{pmatrix},$$

$$(c_a) = \left(0, 0, -2n, -8m, 0, 0\right), \qquad c_0 = -n - 4m.$$

The Riemannian metric is

$$g^{11} = (1+x^2)^2, \qquad g^{12} = (1+x^2)(1+y^2), \qquad g^{22} = 4(1+y^2)^2,$$

which has zero curvature. The potential $V = 0$ also vanishes. The coordinate transformation

$$\bar{x} = \frac{1}{2\sqrt{3}}\left(4 \arctan x - \arctan y\right), \qquad \bar{y} = \frac{1}{2} \arctan y, \tag{39}$$

maps the whole plane onto a bounded rectangle, so this example describes the physical situation of a free particle confined to a bounded rectangle. The algebraic eigenfunctions have the form

$$\psi(\bar{x}, \bar{y}) = \frac{P(x, y)}{(1+x^2)^{\frac{n}{2}}\,(1+y^2)^{\frac{m}{2}}},$$

where P is a polynomial of degree less than or equal to n in x and less than or equal to m in y, and we re-express x, y using (39).

The Lie algebra of type 15 provides a realization of $\mathfrak{sl}(3, \mathbb{R})$ in terms of first-order planar differential operators, given by

$$T^1 = \partial_x, \quad T^2 = \partial_y, \quad T^3 = x\partial_x, \quad T^4 = y\partial_x, \quad T^5 = x\partial_y, \quad T^6 = y\partial_y,$$

$$T^7 = x^2\partial_x + xy\partial_y - nx, \qquad T^8 = xy\partial_x + y^2\partial_y - ny,$$

with $n \in \mathbb{N}$ admits the finite-dimensional module consisting of polynomials of total degree in x and y less than or equal to n. The quasi-exactly solvable Hamiltonian

$$\begin{aligned}\mathcal{H} = \left(T^1\right)^2 + \left(T^2\right)^2 + 2\left(T^7\right)^2 + 2\left(T^8\right)^2 + \\ + T^1T^7 + T^7T^1 + T^2T^8 + T^8T^2 + (3+2n)\left(T^3 + T^6\right)\end{aligned}$$

has contravariant metric coefficients

$$g^{11} = x^2(1+\rho) + 1, \qquad g^{12} = xy(1+\rho), \qquad g^{22} = y^2(1+\rho) + 1,$$

where $\rho = 1+2(x^2+y^2)$, whose Gaussian curvature $\kappa = -2\rho$ is negative everywhere. The potential is given by

$$4V = -3\rho - (7+16n+8n^2) + \frac{14+24n+8n^2+(22+24n+8n^2)\rho}{\rho^2+1}.$$

If we look for solutions of the Schrödinger equation depending only on the "radial" coordinate ρ, we end up with an effectively one-dimensional Schrödinger operator

$$-\widehat{\mathcal{H}} = 4(\rho-1)(\rho^2+1)\,\frac{d^2}{d\rho^2} + \left[(6-4n)\rho^2+(8n+4)\rho-4n-2\right]\frac{d}{d\rho} + \left[(n^2-n)\rho-n^2-n\right].$$

which does appear among the list of purely one-dimensional quasi-exactly solvable Hamiltonians, albeit with a different cohomology parameter. The question of whether the class of one-dimensional quasi-exactly solvable Schrödinger operators can be significantly enlarged via looking at reductions of two-dimensional quasi-exactly solvable Hamiltonians remains unanswered.

Next let $\mathfrak{g}$ be the noncompact Lie algebra of type 24 for $r=1$, spanned by

$$T^1 = \partial_x, \quad T^2 = \partial_y, \quad T^3 = x\partial_x, \quad T^4 = x\partial_y, \quad T^5 = y\partial_y, \tag{40}$$

and

$$T^6 = x^2\partial_x + xy\partial_y - nx, \qquad \text{where} \qquad n \in \mathbb{N}.$$

This Lie algebra admits the finite-dimensional module $\mathcal{N}$ spanned by the monomials x^iy^j with $i+j \le n$. The Lie algebraic coefficients

$$(c_{ab}) = \begin{pmatrix} 1 & 0 & 0 & 0 & 0 & 0 \\ 0 & 1 & 0 & 0 & 0 & 0 \\ 0 & 0 & 2 & 0 & 1 & 0 \\ 0 & 0 & 0 & 1 & 0 & 0 \\ 0 & 0 & 1 & 0 & 1 & 0 \\ 0 & 0 & 0 & 0 & 0 & 1 \end{pmatrix},$$

$$(c_a) = \Big(0, 0, -2n, 0, -1-2n, 0\Big), \qquad c_0 = n^2+n+1,$$

give a quasi-exactly solvable operator with flat Riemannian metric

$$g^{11} = (1+x^2)^2, \qquad g^{12} = xy(1+x^2), \qquad g^{22} = (1+x^2)(1+y^2).$$

Flat coordinates are given by

$$\bar{x} = \arctan x, \qquad \bar{y} = \operatorname{arcsinh} \frac{y}{\sqrt{1+x^2}},$$

which maps the plane to an open infinite strip $(-\pi/2, \pi/2) \times \mathbb{R}$. The potential in these coordinates is

$$V(\bar{x}, \bar{y}) = -\frac{(n+1)(n+2)}{2} \operatorname{sech}^2 \bar{y}.$$

which is simply a Pöschl–Teller potential in $\bar{y}$. The algebraic eigenfunctions take the form

$$\psi(\bar{x}, \bar{y}) = \frac{\cos^n \bar{x}}{\cosh^{n+1} \bar{y}} P(\tan \bar{x}, \sec \bar{x} \sinh \bar{y}),$$

where $P(x, y)$ is a polynomial of total degree at most n in (x, y). Notice that this potential, the preceding zero potential, and, indeed, all other flat potentials that we have found satisfy a conjecture of Turbiner: any quasi-exactly solvable Hamiltonian on a flat manifold in more than one dimension is necessarily separable. However, we do not know whether this conjecture holds in general.

Finally, let $\mathfrak{g}$ be a general Lie algebra of type 24, spanned by the first-order differential operators (40), $T^6 = x^2\partial_x + rxy\partial_y - nx$, and $T^{6+i} = x^{i+1}\partial_y$, $i = 1, \ldots, r-1$. The module $\mathcal{N}$ is spanned by the monomials $x^i y^j$ with $i + rj \leq n$, $j \leq l$. For $m \in \mathbb{N}$, $A, B > 0$, the Schrödinger operator with metric

$$g^{11} = Ax^2 + B, \quad g^{12} = (1+m)Axy, \quad g^{22} = (Ax^2+B)^m + A(1+m)^2y^2,$$

and potential

$$V = -\frac{\lambda AB(1+m)^2(Ax^2+B)^m}{(Ax^2+B)^{1+m} + AB(1+m)^2y^2}, \qquad m \leq r \neq 2(m+1),$$

is normalizable and quasi-exactly solvable with respect to $\mathfrak{g}$, provided that the parameter λ is large enough. The metric in this case has constant negative Gaussian curvature $\kappa = -A$. Furthermore, since the potential V does not depend on the cohomology parameter n, the above Hamiltonian is *exactly solvable.* Moreover, the potential is also independent of r; hence we have constructed a single exactly solvable Hamiltonian which is associated to an infinite number of inequivalent Lie algebras of arbitrarily large dimension.

TABLE 1

FINITE-DIMENSIONAL LIE ALGEBRAS OF VECTOR FIELDS IN $\mathbb{C}^2$

	Generators	Structure	Label
1.	$\{\partial_x\}$	$\mathbb{C}$	E
2.	$\{\partial_x, x\partial_x\}$	$\mathfrak{h}_2$	$C1$
3.	$\{\partial_x, x\partial_x, x^2\partial_x\}$	$\mathfrak{sl}(2)$	$C4$
4.	$\{\partial_x, \partial_y\}$	$\mathbb{C}^2$	$D1$
5.	$\{\partial_x, \partial_y, x\partial_x + \alpha y\partial_y\}$ $\alpha \neq 0$	$\mathbb{C} \ltimes \mathbb{C}^2$	$C8, D3$
6.	$\{\partial_x, \partial_y, x\partial_x, y\partial_y\}$	$\mathfrak{h}_2 \oplus \mathfrak{h}_2$	$C3$
7.	$\{\partial_x, \partial_y, x\partial_x - y\partial_y, y\partial_x, x\partial_y\}$	$\mathfrak{sl}(2) \ltimes \mathbb{C}^2$	$A3$
8.	$\{\partial_x, \partial_y, x\partial_x, y\partial_x, x\partial_y, y\partial_y\}$	$\mathfrak{gl}(2) \ltimes \mathbb{C}^2$	$A2$
9.	$\{\partial_x, \partial_y, x\partial_x, x^2\partial_x\}$	$\mathfrak{gl}(2)$	$C5$
10.	$\{\partial_x, \partial_y, x\partial_x, y\partial_y, x^2\partial_x\}$	$\mathfrak{sl}(2) \oplus \mathfrak{h}_2$	$C6$
11.	$\{\partial_x, \partial_y, x\partial_x, y\partial_y, x^2\partial_x, y^2\partial_y\}$	$\mathfrak{sl}(2) \oplus \mathfrak{sl}(2)$	$C7$
12.	$\{\partial_x + \partial_y, x\partial_x + y\partial_y, x^2\partial_x + y^2\partial_y\}$	$\mathfrak{sl}(2)$	$C9$
13.	$\{\partial_x, 2x\partial_x - y\partial_y, x^2\partial_x - xy\partial_y\}$	$\mathfrak{sl}(2)$	$B\delta1$
14.	$\{\partial_x, x\partial_x, y\partial_y, x^2\partial_x - xy\partial_y\}$	$\mathfrak{gl}(2)$	$B\delta2$
15.	$\{\partial_x, \partial_y, x\partial_x, y\partial_x, x\partial_y, y\partial_y,$ $x^2\partial_x + xy\partial_y, xy\partial_x + y^2\partial_y\}$	$\mathfrak{sl}(3)$	$A1$
16.	$\{\xi_1(x)\partial_y, \dots, \xi_r(x)\partial_y\}$	$\mathbb{C}^r$	$B\alpha1$
17.	$\{\xi_1(x)\partial_y, \dots, \xi_r(x)\partial_y, y\partial_y\}$	$\mathbb{C} \ltimes \mathbb{C}^r$	$B\alpha2$
18.	$\{\partial_x, x^i e^{\lambda x}\partial_y \mid 0 \leq i \leq r_\lambda\}$	$\mathbb{C} \ltimes \mathbb{C}^r$	$B\beta1, D2$
19.	$\{\partial_x, y\partial_y, x^i e^{\lambda x}\partial_y \mid 0 \leq i \leq r_\lambda\}$	$\mathbb{C}^2 \ltimes \mathbb{C}^r$	$B\beta2, C2$
20.	$\{\partial_x, \partial_y, x\partial_x + \alpha y\partial_y, x\partial_y, \dots, x^r\partial_y\}$	$\mathfrak{h}_2 \ltimes \mathbb{C}^{r+1}$	$B\gamma1, 2$
21.	$\{\partial_x, \partial_y, x\partial_y, \dots, x^{r-1}\partial_y, x\partial_x + (ry + x^r)\partial_y\}$	$\mathbb{C} \ltimes (\mathbb{C} \ltimes \mathbb{C}^r)$	$B\gamma3$
22.	$\{\partial_x, \partial_y, x\partial_x, x\partial_y, y\partial_y, x^2\partial_y, \dots, x^r\partial_y\}$	$(\mathfrak{h}_2 \oplus \mathbb{C}) \ltimes \mathbb{C}^{r+1}$	$B\gamma4$
23.	$\{\partial_x, \partial_y, 2x\partial_x + ry\partial_y, x\partial_y, x^2\partial_x + rxy\partial_y,$ $x^2\partial_y, \dots, x^r\partial_y\}$	$\mathfrak{sl}(2) \ltimes \mathbb{C}^{r+1}$	$B\delta3$
24.	$\{\partial_x, \partial_y, x\partial_x, x\partial_y, y\partial_y, x^2\partial_x + rxy\partial_y,$ $x^2\partial_y, \dots, x^r\partial_y\}$	$\mathfrak{gl}(2) \ltimes \mathbb{C}^{r+1}$	$B\delta4$

In Cases 16 and 17 we assume $r > 1$. In Cases 18 and 19, $r = \sum r_\lambda$, and the exponents λ belong to some finite set Λ. In cases 20–24, we assume $r \geq 1$.

TABLE 2

FINITE-DIMENSIONAL MODULES FOR LIE ALGEBRAS OF VECTOR FIELDS

	Monomials?	Generators	Rules
1.	No	$x^i e^{\lambda x} g(y)$	$(i, \lambda, g) \longrightarrow (i-1, \lambda, g)$,
2.	Yes	$x^i g(y)$	$(i, g) \longrightarrow (i-1, g)$
3.	Yes	$g(y)$	
4.	No	$x^i y^j e^{\lambda x + \mu y}$	$(i, j, \lambda, \mu) \longrightarrow (i-1, j, \lambda, \mu), (i, j-1, \lambda, \mu)$
5.	$\alpha \in \mathbb{Q}^+$	$x^i y^j$	$(i, j) \longrightarrow (i-1, j), (i, j-1)$
6.	Yes	$x^i y^j$	$(i, j) \longrightarrow (i-1, j), (i, j-1)$
7.	Yes	$x^i y^j$	$0 \le i + j \le n$
8.	Yes	$x^i y^j$	$0 \le i + j \le n$
9.	Yes	$y^j e^{\mu y}$	$0 \le j \le n_\mu$
10.	Yes	y^j	$0 \le j \le n$
11.	Yes	1	
12.	No	$(x-y)^{m-k} R_k^{m,0}\left(\dfrac{x+y}{x-y}\right)$	$0 \le k \le 2m$
13.	Yes	$x^i y^n$	$0 \le i \le n,\ n \in S$
14.	Yes	$x^i y^n$	$0 \le i \le n,\ n \in S$
15.	Yes	1	
16.	No	$y^j g(x)$	$(j, g) \longrightarrow (j-1, g \cdot \xi_k)$
17.	Yes	$y^j g(x)$	$(j, g) \longrightarrow (j-1, g \cdot \xi_k)$
18.	No	$x^i y^j e^{\mu x}$	$(i, j, \mu) \longrightarrow (i-1, j, \mu), (i + r_\lambda, j-1, \mu+\lambda)$
19.	Yes	$x^i y^j e^{\mu x}$	$(i, j, \mu) \longrightarrow (i-1, j, \mu), (i + r_\lambda, j-1, \mu+\lambda)$
20.	$r < \alpha \in \mathbb{Q}^+$	$x^i y^j$	$(i, j) \longrightarrow (i-1, j), (i+r, j-1)$
21.	Yes	$x^i y^j$	$(i, j) \longrightarrow (i-1, j), (i+r, j-1)$
22.	Yes	$x^i y^j$	$(i, j) \longrightarrow (i-1, j), (i+r, j-1)$
23.	Yes	1	
24.	Yes	1	

TABLE 3

COHOMOLOGIES FOR LIE ALGEBRAS OF VECTOR FIELDS

	Dimension	Representatives
1.	0	
2.	∞	$x\partial_x + h(y), h \notin \mathcal{M}$
3.	∞	$x\partial_x + h(y), \quad x^2\partial_x + 2xh(y)$
4.	$\dim(\mathcal{M}/\operatorname{Div}\mathcal{M}) < \infty$	$\partial_y + h(x,y),\ h_x \in \mathcal{M}, h \neq \psi_y$ with $\psi_x \in \mathcal{M}$
5.	$0\ (\alpha \notin \mathbb{Q}^-), 1$ or $0\ (\alpha \in \mathbb{Q}^-)$	$x\partial_x + \alpha y\partial_y + c_1x^iy^j$ or $\partial_y + c_1x^iy^j$
6.	$0\ (\mathcal{M} \neq 0),\ 2\ (\mathcal{M} = 0)$	$x\partial_x + c_1, \quad y\partial_y + c_2$
7.	$0\ (\mathcal{M} \neq \{1\}),\ 1\ (\mathcal{M} = \{1\})$	$\partial_y + 2c_1x, \quad y\partial_x + c_1y^2, \quad x\partial_y + c_1x^2$
8.	$0\ (\mathcal{M} \neq 0),\ 1\ (\mathcal{M} = 0)$	$x\partial_x + c_1, \quad y\partial_y + c_1$
9.	1	$x\partial_x + c_1, x^2\partial_x + 2c_1x$
10.	$1\ (\mathcal{M} \neq 0),\ 2\ (\mathcal{M} = 0)$	$x\partial_x + c_1, \quad x^2\partial_x + 2c_1x, \quad y\partial_y + c_2$
11.	2	$x\partial_x + c_1,\ x^2\partial_x + 2c_1x,\ y\partial_y + c_2, y^2\partial_y + 2c_2y$
12.	1	$x^2\partial_x + y^2\partial_y + c_1(x-y)$
13.	1	$x^2\partial_x - xy\partial_y + c_1y^{-2}$
14.	$0\ (1 \in \mathcal{M}),\ 1\ (1 \notin \mathcal{M})$	$x\partial_x + c_1, \quad y\partial_y + 2c_1$
15.	1	$x\partial_x + c_1, x^2\partial_x + xy\partial_y + 3c_1x, y\partial_y + c_1, xy\partial_x + y^2\partial_y + 3c_1y$
16.	$\infty^r + k, \quad k < \infty$	$\xi_i(x)\partial_y + f_i(x)y^j$
17.	∞	$y\partial_y + f(x)$
18.	$< \infty$	$x^ke^{\lambda x}\partial_y + c_{i,j}^{\lambda,k}x^{i+k}y^je^{\lambda x}$
19.	1	$y\partial_y + c_1x^n$
20.	$0\ (\alpha \notin \mathbb{Q}),\ 1$ or $0\ (\alpha \in \mathbb{Q})$	$x\partial_x + \alpha y\partial_y + c_1x^iy^j$, or $\xi_k\partial_y + c_kx^{i+k}y^j,\ k \geq l \geq 0$
21.	$0\ (\mathcal{M} \neq 0),\ 1\ (\mathcal{M} = 0)$	$x\partial_x + (ry + x^ry)\partial_y + c_1$
22.	$0\ (\mathcal{M} \neq 0),\ 2\ (\mathcal{M} = 0)$	$x\partial_x + c_1, \quad y\partial_y + c_2$
23.	$1\ (r > 2)$	$2x\partial_x + ry\partial_y + c_1, \quad x^2\partial_x + rxy\partial_y + c_1x$
	$2\ (r = 2)$	$x\partial_x + y\partial_y + c_1, x\partial_y + c_2, x^2\partial_x + 2xy\partial_y + 2c_1x + 2c_2y, x^2\partial_y + 2c_2x$
24.	$1\ (\mathcal{M} \neq 0),\ 2\ (\mathcal{M} = 0)$	$x\partial_x + c_1, y\partial_y + c_2, x^2\partial_x + rxy\partial_y + (2c_1 + rc_2)x$

TABLE 4

QUASI-EXACTLY SOLVABLE LIE ALGEBRAS OF DIFFERENTIAL OPERATORS

	Quantization condition	Module
1.	0	
2.	0	
3.	$h = -\frac{n}{2}, \quad n \geq 0,$	$\{x^i g(y) \mid i \leq n, g \in S\}$
4.	0	
5.	0	
6.	0	
7.	0	
8.	0	
9.	$c_1 = -\frac{n}{2}, \quad n \geq 0,$	$\{x^i y^j e^{\mu y} \mid i \leq n, j \leq m_\mu\}$
10.	$c_1 = -\frac{n}{2}, \quad n \geq 0,$	$\{x^i y^j \mid i \leq n, j \leq m\}$
11.	$c_1 = -\frac{n}{2},\ c_2 = -\frac{m}{2}, \quad n, m \geq 0,$	$\{x^i y^j \mid i \leq n, j \leq m\}$
12.	$c_1 = \frac{n}{2},$ †	$\left\{(x-y)^{m+\frac{n}{2}-k} R_k^{m,n}\left(\frac{x+y}{x-y}\right) \,\middle|\, 0 \leq k \leq 2m+n, m \in S\right\}$
13.	0	
14.	0	
15.	$c_1 = -\frac{n}{3}, \quad n \geq 0,$	$\{x^i y^j \mid i + j \leq n\}$
16.	0	
17.	0	
18.	0	
19.	0	
20.	0	
21.	0	
22.	0	
23.	$c_1 = -n, \quad n \geq 0,$	$\{x^i y^j \mid i + rj \leq n, j \leq l\}$
24.	$c_1 = -\frac{n}{2},\ c_2 = 0, \quad n \geq 0,$	$\{x^i y^j \mid i + rj \leq n, j \leq l\}$

† In case 12, there is no positivity restriction on n, and $S = \{m \mid m \geq \max(0, -n)\}$ is a finite set of integers.

REFERENCES

1. Y. Alhassid, J. Engel and J. Wu, *Algebraic approach to the scattering matrix*, Phys. Rev. Lett. **53** (1984), 17–20.
2. Y. Alhassid, F. Gürsey and F. Iachello, *Group theory approach to scattering*, Ann. Phys. **148** (1983), 346–380.
3. Y. Alhassid, F. Gürsey and F. Iachello, *Group theory approach to scattering. II. The euclidean connection*, Ann. Phys. **167** (1986), 181–200.
4. A. Arima and F. Iachello, *Interacting boson model of collective nuclear states. I – The vibrational limit*, Ann. Phys. **99** (1976), 253–317.
5. A. Arima and F. Iachello, *Interacting boson model of collective nuclear states. I – The rotational limit*, Ann. Phys. **111** (1978), 201–238.
6. N.L. Balazs and A. Voros, *Chaos on the pseudosphere*, Physics Reports **143** (1986), 109–240.
7. É. Cotton, *Sur les invariants différentiels de quelques équations linéaires aux dérivées partielles du second ordre*, Ann. École Norm. **17** (1900), 211–244.
8. A. Erdélyi and H. Bateman, *Higher Transcendental Functions*, McGraw–Hill, New York, 1953.
9. A. González–López, J. Hurtubise, N. Kamran and P.J. Olver, *Quantification de la cohomologie des algèbres de Lie de champs de vecteurs et fibrés en droites sur des surfaces complexes compactes*, Comptes Rendus Acad. Sci. (Paris) **316** (1993), 1307–1312.
10. A. González–López, N. Kamran and P.J. Olver, *Lie algebras of first order differential operators in two complex variables*, Differential Geometry, Global Analysis, and Topology, A. Nicas and W.F. Shadwick, eds., Canadian Math. Soc. Conference Proceedings, vol. 12, Amer. Math. Soc., Providence, R.I., 1991, pp. 51–84.
11. A. González–López, N. Kamran and P.J. Olver, *Quasi–exactly solvable Lie algebras of first order differential operators in two complex variables*, J. Phys. A **24** (1991), 3995–4008.
12. A. González–López, N. Kamran and P.J. Olver, *Lie algebras of differential operators in two complex variables*, American J. Math. **114** (1992), 1163–1185.
13. A. González–López, N. Kamran and P.J. Olver, *Normalizability of one–dimensional quasi–exactly solvable Schrödinger operators*, Commun. Math. Phys. **153** (1993), 117–146.
14. A. González–López, N. Kamran and P.J. Olver, *New quasi–exactly solvable Hamiltonians in two dimensions*, preprint, *Commun. Math. Phys.*, to appear.
15. A. Gorsky, *Relationship between exactly solvable and quasi-exactly solvable versions of quantum mechanics with conformal block equations in 2D theories*, JETP Lett. **54** (1991), 289–292.
16. J.H. Grace and A. Young, *The Algebra of Invariants*, Cambridge Univ. Press, Cambridge, 1903.
17. I.S. Gradshteyn and I.W. Ryzhik, *Table of Integrals, Series and Products*, Academic Press, New York, 1965.
18. G.B. Gurevich, *Foundations of the Theory of Algebraic Invariants*, P. Noordhoff Ltd., Groningen, Holland, 1964.
19. F. Iachello and R. Levine, *Algebraic approach to rotation–vibration spectra. I – Diatomic molecules*, J. Chem. Phys. **77** (1982), 3046–3055.
20. L. Infeld and T.E.. Hull, *The factorization method*, Rev. Mod. Phys. **23** (1951), 21–68.
21. N. Jacobson, *Lie Algebras*, Interscience Publ. Inc., New York, 1962.
22. N. Kamran and P.J. Olver, *Lie algebras of differential operators and Lie-algebraic potentials*, J. Math. Anal. Appl. **145** (1990), 342–356.
23. R.D. Levine, *Lie algebraic approach to molecular structure and dynamics*, Mathematical Frontiers in Computational Chemical Physics, D.G. Truhlar, ed., IMA Volumes in Mathematics and its Applications, Vol. 15, Springer-Verlag, New York, 1988, pp. 245–261.
24. S. Lie, *Theorie der Transformationsgruppen*, vol. 3, B.G. Teubner, Leipzig, 1893.
25. S. Lie, *Gruppenregister*, Gesammelte Abhandlungen, vol. 5, B.G. Teubner, Leipzig, 1924, pp. 767–773.

26. A.Y. Morozov, A.M. Perelomov, A.A. Rosly, M.A. Shifman and A.V. Turbiner, *Quasi-exactly solvable quantal problems: one-dimensional analogue of rational conformal field theories*, Int. J. Mod. Phys. **5** (1990), 803–832.
27. M.A. Shifman, *New findings in quantum mechanics (partial algebraization of the spectral problem)*, Int. J. Mod.. Phys. A **126** (1989), 2897–2952.
28. M.A. Shifman, *Quasi-exactly solvable spectral problems and conformal field theory*, preprint, *Contemp. Math.*, this volume..
29. M.A. Shifman and A.V. Turbiner, *Quantal problems with partial algebraization of the spectrum*, Commun. Math. Phys. **126** (1989), 347–365.
30. A.V. Turbiner, *Quasi-exactly solvable problems and* $\mathfrak{sl}(2)$ *algebra*, Commun. Math. Phys. **118** (1988), 467–474.
31. A.V. Turbiner, *Lie algebras and polynomials in one variable*, J. Phys. A **25** (1992), L1087–L1093.
32. A.V. Turbiner, *On polynomial solutions of differential equations*, J. Math. Phys. **33** (1992), 3989–3993.
33. A.V. Turbiner, *Lie algebraic approach to the theory of polynomial solutions. III. Differential equations in two real variables and general outlook*, preprint, ETH-TH/92-34, 1992.
34. A. G. Ushveridze, *Quasi-exactly solvable models in quantum mechanics*, Sov. J. Part. Nucl. **20** (1989), 504–528.

Departamento de Física Teórica II, Universidad Complutense, 28040 Madrid, Spain
E-mail address: aglopez@fis.ucm.es

Department of Mathematics, McGill University, Montréal, Québec, Canada H3A 2K6
E-mail address: nkamran@gauss.math.mcgill.ca

School of Mathematics, University of Minnesota, Minneapolis, Minnesota 55455
E-mail address: olver@ima.umn.edu

Contemporary Mathematics
Volume **160**, 1994

QUANTIZATION AND DEFORMATION OF LIE ALGEBRAS

PALLE E. T. JORGENSEN

ABSTRACT. We consider the role of finite-dimensional Lie algebras as symmetries of a class of simple (i.e., no two-sided ideals) C^*-algebras which arise by a quantum deformation of classical mechanics. The symmetries under consideration are defined relative to the Connes non-commutative differential geometric formalisms. Among the Lie algebras which occur, the Heisenberg Lie algebra is shown to be generic. We then show how deformations of the Heisenberg generators lead to representations of the infinite C^*-algebra defined on a Hilbert space (also called the Cuntz algebra). Finally, we consider the *stability question* for the deformations, where stability refers to the C^*-algebraic isomorphism class.

1. INTRODUCTION

Consider canonical variables x_i, p_i $(1 \leq i \leq n)$ from classical mechanics (see [6,24]) with Poisson bracket, defined on the C^∞ functions in the $2n$ variables,

$$\{f,g\} = \sum_i \left(\frac{\partial f}{\partial x_i}\frac{\partial g}{\partial p_i} - \frac{\partial f}{\partial p_i}\frac{\partial g}{\partial x_i}\right) \tag{1.1}$$

There is an indexed family of C^*-algebras related to the Poisson bracket (1.1) by a *deformation* of the symplectic matrix which defines $\{\cdot,\cdot\}$. They have been studied earlier by many authors, see, e.g., [1,5,10,11,23,26,35], and recently Rieffel [31] made precise the realization of the C^*-algebras as a *Poisson deformation.* Here we will only hint at this by sketching the $n = 1$ case: To simplify, compactify the two real variables to $\mathbb{T}^2$ (the two-torus), and expand C^∞ functions on $\mathbb{T}^2$ in their Fourier series

$$\sum\sum a_{n,m}e^{i(ns+mt)} \tag{1.2}$$

1991 *Mathematics Subject Classification.* 46L40, 46L55, 47A67, 81R50, 17B15, 22D35, 22E70.
Key words and phrases. Quantization, Deformation, Lie Algebras.

This paper is in final form and no version of it will be submitted for publication elsewhere.

This paper represents joint research involving the following co-authors: O. Bratteli, G. A. Elliott, F. M. Goodman (papers [3,4,5,19]); L. M. Schmitt, and R. F. Werner (the paper [21] which is submitted). Jorgensen's joint work has been supported by the US NSF, NATO, a University of Iowa Faculty Scholar Award, and by the UI Oakdale Research Center with summation over $\mathbb{Z}^2$. The pointwise product on $\mathbb{T}^2$ turns into Cauchy product on the associated Fourier series, i.e.,

$$c_{n,m} = \sum_k \sum_\ell a_{k,\ell} b_{n-k,m-\ell} \, .$$

If it is redefined with a phase factor depending on a parameter $\theta \in \mathbb{R}$ by

$$\sum_k \sum_\ell a_{k,\ell} b_{n-k,m-\ell} e^{i\ell(n-k)2\pi\theta} \, , \tag{1.3}$$

then we denote the corresponding operation on $C^\infty(\mathbb{T}^2)$

$$(f *_\theta g)(s,t) = \sum \sum c(\theta)_{n,m} e^{i(ns+mt)}$$

where $c_{..}(\theta)$ is given by the Cauchy-formula (1.3) with phase θ. It was shown by Rieffel [31] that then

$$\lim \frac{1}{\theta}(f *_\theta g - g *_\theta f) = \{f, g\} \, , \tag{1.4}$$

and further that this limit is uniform (in a certain sense) making each $*_\theta$ into a C^*-completion of $C^\infty(\mathbb{T}^2)$, and providing a *continuous field of* C^*-algebras (see also [1,15,31]). When θ is fixed (we need only the interval $\theta \in [0,1)$) the C^*-algebras, denoted $\mathfrak{A}_\theta$, may be described alternatively as the (unital) C^*-algebra on two unitary generators u, v, and the relation

$$vu = e^{i2\pi\theta} uv \, . \tag{1.5}$$

The C^*-algebra is called the rotation-algebra since it is given as a *foliation* of $\mathbb{T}^2$ with angle θ, see [29], and it is a simple C^*-algebra when θ is irrational, and the field $\theta \to \mathfrak{A}(\theta)$ is a *field of C^*-isomorphism classes* : there is a complete invariant $(\mathbb{Z}+\theta\mathbb{Z}) \cap [0,1]$, and it follows that the isomorphism classes are distinct when θ, θ' are irrational $\theta \neq \theta'$. The case $\theta = 0$ corresponds just to the abelian algebra of functions on $\mathbb{T}^2$. (We refer to [14,25,26,30] for details on the C^*- algebraic invariants.)

It is clear from (1.5) that the unitary generators u, v may be scaled $u \longmapsto z_1 u$, $v \longmapsto z_2 v$ by $(z_1, z_2) \in \mathbb{T}^2$, and we get the same algebra $\mathfrak{A}(\theta)$. It follows that the group $\mathbb{T}^2$ acts as a group of automorphisms on $\mathfrak{A}(\vartheta)$, for each θ. The smooth functions (1.2) are characterized by the Schwartz- condition on the Fourier coefficients: The series $a_{n,m}$ is said to be smooth if, for all $k \in \mathbb{N}$, there is a constant C (depending only on k and $a_{..}$) such that

$$|a_{n,m}| \leq \frac{C}{(1+n^2+m^2)^k} \, , \tag{1.6}$$

and we will use this condition to define a (dense) subalgebra $\mathfrak{A}^\infty \subset \mathfrak{A}_\theta$ for each θ. It is then immediate that the $\mathbb{T}^2$-automorphisms are *smooth*, i.e., map $\mathfrak{A}^\infty$ into itself. Since the Poisson deformation (1.3) is made for $C^\infty(\mathbb{T}^2)$, we shall focus on the study of the *smooth* groups of automorphisms. (It is also technically easier to find the smooth automorphisms on $\mathfrak{A}_\theta$, see the discussion in Section 2 below.)

2. Smooth Lie Group Actions

In this section we shall report on joint work [3,4,5,20] where we find *all* the *smooth* Lie group actions on $\mathfrak{A}_\theta$, θ fixed, in a certain generic subset of $[0,1)$. It further turns out (Theorem 2.2) that there is a *duality* between the *multiplicative* Heisenberg-relation (1.5) for $\mathfrak{A}_\theta$ itself, and the Lie algebraic Heisenberg relation

$$xp - px = \hbar i I \tag{2.1}$$

($\hbar$ Planck's constant). We already noted that both (1.5) and (2.1) are quantizations of the variables x, p from classical mechanics. (See [37] on quantization.)

Let G be a Lie group, $\mathfrak{g}$ the corresponding Lie algebra, and $\exp : \mathfrak{g} \to G$ the exponential mapping. Let $\alpha : G \to Aut(\mathfrak{A})$ be a group of automorphisms on a C^*-algebra (See [6] for the general theory). Then we have the *derived representation* $\dot{\alpha}$ of $\mathfrak{g}$, acting by *derivations* on the G-smooth elements in $\mathfrak{A}$, see [4,20] and (1.6); for $X \in \mathfrak{g}$, $a \in \mathfrak{A}^\infty$,

$$\dot{\alpha}_X(a) = \lim_{t\to 0} t^{-1}(\alpha_{\exp tX}(a) - a) \ ;$$

and

$$\begin{aligned} \dot{\alpha}_X(ab) &= \dot{\alpha}_X(a)b + a\dot{\alpha}_X(b) \ , \\ \dot{\alpha}_{[X,Y]} &= \dot{\alpha}_X\dot{\alpha}_Y - \dot{\alpha}_Y\dot{\alpha}_X \end{aligned} \tag{2.2}$$

hold for $a, b \in \mathfrak{A}^\infty$, X, $Y \in \mathfrak{g}$. Let $\rho : \mathbb{T}^2 \longrightarrow Aut(\mathfrak{A}_\theta)$ be the torus-action on the rotation algebra $\mathfrak{A}_\theta$. Then we say that a given Lie action (α, G) on $\mathfrak{A}_\theta$ is *smooth* if $(g, z) \longmapsto \rho_z(\alpha_g(a))$ is smooth on $\mathbb{T}^2 \times G$ for all $a \in \mathfrak{A}_\theta^\infty$. (See (1.6) for the definition of the smooth elements in $\mathfrak{A}_\theta$.)

The representation $(\rho, \mathbb{T}^2)$ has a derived Lie action with two generators δ_1, δ_2 (both satisfying Leibniz' rule (2.2)) and given by

$$\begin{cases} \delta_1(u) = iu \ , & \delta_1(v) = 0 \\ \delta_2(u) = 0 \ , & \delta_2(v) = iv \ , \end{cases} \tag{2.3}$$

and we say that $\mathfrak{A}_\theta$ is a two-dimensional *non-commutative differentiable manifold*, see [10]. Linear combinations $\delta = c_1\delta_1 + c_2\delta_2$ are called *tangent vectors*, $c_1, c_2 \in \mathbb{C}$. The derivations δ corresponding to $c_i \in \mathbb{R}$, are precisely the derived ones $\dot{\rho}_X$ when X is in the Lie algebra ($\simeq \mathbb{R}^2$) of the scaling action $(\rho, \mathbb{T}^2)$.

A derivation δ is said to be *inner* if there is some element h say in the C^*-algebra $\mathfrak{A}$ such that $\delta(a) = ha - ah =: ad_h(a)$, $\forall a \in \mathfrak{A}$. Suppose a derivation $\delta : \mathfrak{A}_\theta^\infty \to \mathfrak{A}_\theta$ is given, and there is a sequence (h_n) such that $\|\delta(a) - ad_{h_n}(a)\| \to 0$ for $\forall a \in \mathfrak{A}_\theta^\infty$. Then we say that δ is *approximately inner*. We have

Theorem 2.1. *[5] Suppose $|1 - e^{i2\pi n\theta}|^{-1}$ is of polynomial growth in n, then every derivation $\delta : \mathfrak{A}_\theta^\infty \to \mathfrak{A}_\theta$ is a unique sum $\delta = \delta_0 + ad_h$ where δ_0 is in the tangent space, and $h \in \mathfrak{A}_\theta$. If θ is general (irrational), then the same conclusion holds, but the second term in the sum is approximately inner. If $\theta \in (0,1)$ is rational, then the δ_0 term may be taken of the form $c_1\delta_1 + c_2\delta_2$ with $c_i (i = 1, 2)$ in the center of $\mathfrak{A}_\theta$.*

We shall not discuss the proof here, but refer to [5]. The result solved a conjecture [32,33] of Sakai, and it is also the basis for analysis of the smooth Lie actions.

Theorem 2.2. *[3,4] Let θ satisfy the Diophantine condition, restricting the sequence $n \longmapsto |1 - e^{i2\pi n\theta}|^{-1}$ to polynomial growth. Then the smooth Lie actions on $\mathfrak{A}_\theta$ are determined by the Levi decomposition $\mathfrak{S} + \mathfrak{R}$ with $\mathfrak{S}$ semisimple and $\mathfrak{R}$ the solvable Lie radical; and*

(i) *$\mathfrak{S}$ is compact and consists of inner derivations.*
(ii) *$\mathfrak{S}$ and $\mathfrak{R}$ are commuting Lie ideals, $\mathfrak{S} \cap \mathfrak{R} = 0$.*
(iii) *The Lie subalgebra $\mathfrak{R}_b$ of bounded (inner) elements in $\mathfrak{R}$ is abelian, and $[\mathfrak{R}, \mathfrak{R}] \subset \mathfrak{R}_b$*
(iv) *The adjoint representation of $\mathfrak{R}$ on $\mathfrak{R}_b^{\mathbb{C}}$ is diagonalizable with purely imaginary weights.*
(v) *$\mathfrak{R}/\mathfrak{R}_b$ is isomorphic to a subalgebra of the tangent space.*
(vi) *Either $\mathfrak{R} \to \mathfrak{R}/\mathfrak{R}_b$ splits in the category of Lie algebras, or else $\mathfrak{R}$ contains a copy of the Heisenberg algebra (2.1).*
(vii) *In the latter case, the Heisenberg Lie algebra is generated by ξ_i, $\zeta (i = 1, 2)$ where $[\xi_1, \xi_2] = \zeta$ (the central element $\neq 0$), and the ξ_i's may be scaled such that $\xi_i = \delta_i + ad_{h_i} (i = 1, 2)$, $\zeta = ad_k$, $k = \delta_1(h_2) - \delta_2(h_1) + [h_1, h_2]$ where $h_i \in \mathfrak{A}_\theta^\infty$, $[h_1, h_2] = h_1h_2 - h_2h_1$, and the δ_i's are given by (2.3).*

3. HEISENBERG C^*-ALGEBRAS AND q DEFORMATIONS

In this section, we report on very recent joint work [21] with L. Schmitt and R. Werner. In describing the background, we shall generalize the Heisenberg relation (2.1) in two ways: First we shall allow the variables x, p to run over a complex Hilbert space $\mathcal{H}$ say; and secondly we shall put the scaling into the commutator, rather than the traditional place for Planck's constant $\hbar$ (see (2.1).) Let $\mathcal{H}^*$ be the conjugate Hilbert space with some fixed isomorphism (conjugate linear) $f \longmapsto f^* : \mathcal{H} \longmapsto \mathcal{H}^*$, and finally, let $\langle \cdot, \cdot \rangle$ be the inner product in $\mathcal{H}$. (It becomes *bilinear* on $\mathcal{H}^* \times \mathcal{H}$.)

In this section, we shall give the C^*-algebraic quantization of the q-Heisenberg relation

$$fg^* - qg^*f = \langle f, g \rangle 1 \tag{3.1}$$

where we are taking $\mathcal{H} \times \mathcal{H}^*$ as a set of generators, (3.1) as a relation on the generators, 1 as the unit in the universal unital algebra $\mathcal{A}(q)$ on (3.1), and finally a $*$-involution on this universal algebra which is determined on the generators by $f \longmapsto f^* : \mathcal{H} \to \mathcal{H}^*$ mentioned above.

We shall consider the interval $-1 < q < 1$ for the variable as $q^2 = 1$ leads to singular behavior, both for the *algebra* and the *analysis*.

Then the universal algebra on (3.1) may be taken, as a *vector space*, to be a copy of the tensor algebra $T(\mathcal{H} \times \mathcal{H}^*)$ over $\mathcal{H} \times \mathcal{H}^*$. As a vector space V, this tensor algebra is spanned by product-elements $f_1^* \cdots f_n^* g_1 \cdots g_m$, $1 \leq n, m < \infty$, and the unit 1. (We have chosen Wick order with star to the left.) This means that V acquires a product for each q, denoted $\cdot_q$, and a deformation expansion

$$A \cdot_q B = \sum_{i=0}^{\infty} q^i f_i(A, B) \tag{3.2}$$

for $A, B \in V$, where the sum in (3.2) is finite when A, B are fixed, and $f_i : V \times V \longrightarrow V$ is bilinear for $i = 0, 1, 2, \cdots$. But the case $q = 0$, $f_0(A, B) = AB$ is a familiar product which was studied earlier by Cuntz [12], and the $q = 0$ relation (3.1) generates this famous C^*-algebra, also called $\mathcal{O}(\mathcal{H})$. It is *unital* and simple. It is known that there is a C^*-isomorphism class for each value of the dimension $\dim \mathcal{H} = 2, 3, \cdots, \infty$. (The case $\dim \mathcal{H} = 1$ is special, see [7].)

Our formula (3.2) is a deformation in the sense of Gerstenhaber [18], i.e., the f_i's satisfy cohomological conditions, where the first one is

$$f_1(A, B)C + f_1(AB, C) = Af_1(B, C) + f_1(A, BC) \tag{3.3}$$

for all $A, B, C \in V$.

Contrasting the two deformations (1.5) and (3.1), we say that (1.5) is *multiplicative* and (3.1) *additive*. (This is because $(u, v) := uvu^{-1}v^{-1}$ is a multiplicative commutator, as opposed to $[u, v] := uv - vu$.)

When $q = 0$, the product $f_0(A, B) = AB$ in (3.1) and (3.3) is the associative product of the Cuntz-algebra $\mathcal{O}(\mathcal{H})$. To see this, note that, in an orthonormal basis e_i for $\mathcal{H}$, (3.1) becomes

$$e_i e_j^* - q e_j^* e_i = \delta_{ij} 1 \tag{3.4}$$

with δ_{ij} denoting the Kronecker delta, and 1 the unit is the universal algebra $\mathcal{A}(q)$. In a representation, 1 is mapped into the unit operator I on the representation Hilbert space $\mathcal{K}$. Formula (3.4) then becomes an operator-identity, now with the right-hand side $\delta_{ij} I$, and the terms e_i become operators on $\mathcal{K}$, and e_i^* the adjoint operators, i.e.,

$$\langle e_i \phi, \psi \rangle_{\mathcal{K}} = \langle \phi, e_i^* \psi \rangle_{\mathcal{K}} \tag{3.5}$$

for all vectors ϕ, ψ in the common domain of the operators. Our first result is the *existence* of such a *bounded* universal representation of $\mathcal{A}(q)$. The restriction $q < 1$ is important for that, since (3.1), with $q = 1$, does not have bounded representations. Recall that the two extreme cases $q = -1$, and $q = 1$, are related to the canonical anticommutation relations (CAR), respectively the canonical commutation relations (CCR), see [6], and the C^*-algebras are known in both cases; but in the CCR–case, the operators e_i are not bounded, not even when $\dim \mathcal{H} < \infty$.

Theorem 3.1. *[21] There is a unital C^*-algebra $\mathcal{E}^q(\mathcal{H})$ which is a faithful representation of (3.1) and (3.5), and therefore of the universal $*$-algebra $\mathcal{A}(q)$ on a given Hilbert space $\mathcal{H}$. When $\mathfrak{A}$ is any C^*-algebra with unit, and $a : \mathcal{H} \to \mathfrak{A}$ is a mapping such that*

$$a(e_i)a(e_j)^* - q a(e_j)^* a(e_i) = \delta_{ij} 1 \,, \tag{3.6}$$

then there is a unique unital $*$-homomorphism

$$\rho : \mathcal{E}^q(\mathcal{H}) \to \mathfrak{A}$$

satisfying $\rho(e_i) = a(e_i)$, and $\rho(e_i^*) = a(e_i)^*$, i.e., the adjoint operator for all i in the index-set for the given $\mathcal{H}$ basis e_i. Moreover, $\mathcal{E}^q(\mathcal{H})$ is functorial and continuous in the sense that $U \longmapsto \mathcal{E}^q(U)$ is continuous on the unitary group of $\mathcal{H}$ in the sense that strong convergence $U_\alpha \to U$ implies $\mathcal{E}^q(U_\alpha)(x) \to \mathcal{E}^q(U)(x)$ for all x in the C^*-algebra $\mathcal{E}^q(\mathcal{H})$, and this convergence is in norm.

The proof is based on the following (also from our [21]):

Lemma 3.2. *Let some mapping $a : \mathcal{H} \to \mathfrak{A}$ into a C^*-algebra satisfy the q-relations (3.6), assuming just that $\mathcal{H}$ carries a sesquilinear form $\langle \cdot, \cdot \rangle$. Then it follows that $\langle \cdot, \cdot \rangle$ is positive semidefinite, and $\|a(f)\|_{\mathfrak{A}} = c(q)\langle f, f\rangle^{\frac{1}{2}}$, where*

$$c(q) = \begin{cases} (1-q)^{-\frac{1}{2}} & \text{for } 0 \leq q < 1 \\ 1 & \text{for } -1 < q \leq 0 \,. \end{cases}$$

The following two results are also used and they highlight the significance of $\dim \mathcal{H}$.

Lemma 3.3. *[21]. Let a be a bounded operator, and let $q, c \in \mathbb{R}$, $-1 < q < 1$. Then, if $aa^* - qa^*a = cI$, and $a \neq 0$, then it follows that $c > 0$, and (i) or (ii) must hold:*

(i) $aa^* = a^*a = c(1-q)^{-1}I$

(ii) $sp(aa^*) = sp(a^*a) =$ *closure* $\{c(1-q^n)(1-q)^{-1}\}_{n=0}^{\infty}$

including $n = 0$ for a^*a.

This last lemma is crucial, since we show in [21] that, when $\dim \mathcal{H} \geqq 2$, then possibility (i) may be excluded for $a(f)$, $f \in \mathcal{H} \backslash \{0\}$; and it follows from this that 0 must be in the spectrum of $a(f)^*a(f)$.

Now the stability result for the C^*-isomorphism class:

Theorem 3.4. *[21] Let $\dim \mathcal{H} \geq 2$, $|q| < \sqrt{2} - 1$; then the two C^*-algebras $\mathcal{E}^q(\mathcal{H})$ and $\mathcal{E}^0(\mathcal{H})$ are isomorphic, and when $\dim \mathcal{H} < \infty$, there is a short exact sequence of C^*-algebras*

$$0 \longrightarrow \mathcal{C} \longrightarrow \mathcal{E}^q(\mathcal{H}) \longrightarrow \mathcal{O}(\mathcal{H}) \to 0 \tag{3.7}$$

where $\mathcal{C}$ denotes the C^*-algebra of all compact operators on a separable Hilbert space, and $\mathcal{O}(\mathcal{H})$ the simple C^*-algebra of Cuntz. Moreover, the isomorphism $\mathcal{E}^q(\mathcal{H}) \longrightarrow \mathcal{E}^0(\mathcal{H})$ may be defined on generators by $a_i \longmapsto v_i$, $1 \leq i \leq \dim \mathcal{H}$, with the a_i's satisfying (3.6), $v_i v_j^* = \delta_{ij} 1$,

$$a_i = v_i T^{\frac{1}{2}} \tag{3.8}$$

where $T \in \mathcal{E}^0(\mathcal{H})$ may be found s.t.

$$v_i T v_j^* = \delta_{ij} + qT^{\frac{1}{2}} v_j^* v_i T^{\frac{1}{2}} \,, \tag{3.9}$$

(by use of a Banach-contraction argument.)

In conclusion, we mention that the results from [21] are stronger than the presentation in Theorems 3.1 and 3.4. The same results (3.7), (3.8) and (3.9), are in fact established in [21] for a class of relations which includes the q relation (3.4) above, as well as

$$a_i a_j^* - q a_i^* a_j = \delta_{ij} 1 \,. \tag{3.10}$$

Notice the difference in the position of the indices in the term containing the parameter. But, for (3.10), it is required that $d = \dim \mathcal{H}$ be finite as is illustrated by the following:

Proposition 3.5. *[21] Let $\{a_i\}_{i=1}^d$ be a bounded operator representation of (3.10). Then $f(x) = 1 + qd + q^2x$ is spectrum generating for the operator $T = \sum_i a_i^* a_i$, i.e., $x \in sp(T) \Leftrightarrow f(x) \in sp(T)$, provided that both $f(x) \neq 1$ and $f(x) \neq 0$.*

More generally we consider functions Γ_{ij} of $a = (a_1, \dots, a_d)$ as a non- commutative C^*-functional calculus (see [34]), and show stability for the relation

$$a_i a_j^* = \Gamma_{ij}(a_1, \dots, a_d) \ , \ 1 \le i, j \le d \ . \tag{3.11}$$

Theorem 3.6. *[21] The conditions for the C^*-stability of the (3.11) isomorphism class are as follows (see [21] for more details):*

(i) *$\Gamma(a)^* = \Gamma(a)$ for all C^*-algebras $\mathfrak{A}$ and all $a = (a_1, \dots, a_d)$, $a_i \in \mathfrak{A}$.*

(ii) *There are constants $\mu, \in$ and λ, such that $\mu, \epsilon > 0$, $\lambda < 2\sqrt{\epsilon}$, and*

$$||\sum_i a_i^* a_i|| \le \mu$$

for all a satisfying (3.11).

(iii) *For such $a, b \in \mathfrak{A}^d$ with $||\sum_i a_i^* a_i|| \le \mu$, and $||\sum_i b_i^* b_i|| \le \mu$, we have following the three estimates*

(1) $||\Gamma(a)|| \le \mu$,

(2) $\Gamma(a) \ge \epsilon 1$ (as an estimate on hermitian elements in $\mathfrak{A}$,) and

(3) $||\Gamma(a) - \Gamma(b)|| \le \lambda ||a - b||$.

Then there is a unique C^*-algebra $\mathcal{E}(\Gamma)$ satisfying (3.7), $T \in \mathcal{E}^0(d)$ such that $T^{\frac{1}{2}} \sum_i v_i^* v_i = T^{\frac{1}{2}}$, and (3.8) holds for the $\mathcal{E}^0(d)$ generators v_i. Again our C^*-algebra $\mathcal{E}(\Gamma)$ is universal and functorial.

We conjecture in Theorem 3.4 (but not 3.5–3.6) that the stability interval may in fact be extended from $|q| < \sqrt{2} - 1$, to all $-1 < q < 1$. We also expect such an extension to be difficult.

4. Conclusions and Notes

The C^*-algebras from Section 2 were studied first in mathematical physics [35] by Slawny, and reapeared again in [23] in connection with the theory of duality for C^*-algebras and locally compact abelian groups. The rotation algebras also supplied examples (see, e.g., [8,10]) for the non-commutative differential geometry of Connes, and then again (for the Yang-Mills theory) from mathematical physics [11]. Both the C^*-algebras in Sections 2 and 3 serve as generic examples for K-theory, since the K-groups can be computed explicitly (see [14,25,26,29]), and other C^*-algebras arising in applications may be compared with the generic models, using methods from algebraic topology. The q- relations in Section 3 are related to (but different from) other q-relations studied first by Pusz and Woronowicz (see [27,28,39]), and (3.6) were studied by Bozejko-Speicher, Speicher (see [2,36]), and Biedenharn [1a] and Greenberger [17] in connection with quantum deformations and non-commutative probability theory.

In the present paper, we have contrasted the deformation approach to the two classes of C^*-algebras, based on the *additive* commutator, and the *multiplicative* commutator, to illustrate a dichotomy for the stability question for C^*-isomorphism classes (–instability for the multiplicative case, and stability for the additive case).

REFERENCES

1. J. Anderson and W. Paschke, *The rotation algebra*, Houston Math. J..

1a. L. C. Biedenharn, *The quantum group $SU_q(2)$ and a q-analogue of the Boson operators*, J. Phys. A. **22** (1989), L873–L878.

2. M. Bozejko and R. Speicher, *An example of a generalized Brownian motion*, Comm. Math. Phys. **137** (1991), 519–531.

3. O. Bratteli, G. A. Elliott, F. M. Goodman and P. E. T. Jorgensen, *On Lie algebras of operators*, J. Funct. Anal. **85** (1989), 341–359.

4. ______, *Smooth Lie group actions on the irrational rotation algebra*, Nonlinearity **2** (1989), 271–286.

5. O. Bratteli, G. A. Elliott, and P. E. T. Jorgensen, *Decomposition of unbounded derivations into invariant and approximately inner parts*, J. reine angw. Math. (Crelle's J.) **346** (1984), 166–193.

6. O. Bratteli and D. W. Robinson, *Operator Algebras and Quantum Statistical Mechanics*, (v. I, 1979 and 1987; v. II, 1981), Springer-Verlag, New York.

6a. B. Brenken, *Representations and automorphisms of the irrational rotation algebra.*

7. L. A. Coburn, *The C^*-algebra generated by an isometry I*, Bull. Amer. Math. Soc. **73** (1967), 722–726.

8. A. Connes, *C^*-algèbres et géométric différentielle*, C. R. Acad. Sci. (Paris) **290** (1980), 599–604.

9. ______, *Essay on physics and non-commutative geometry*, The Interface of Mathematics and Particle Physics (D. G. Quillen et al., eds.), Oxford U. Press, Oxford UK, 1990.

10. ______, *Non-commutative differential geometry*, Publ. Math. IHES **62** (1985), 257–360.

11. A. Connes and M. A. Rieffel, *Yang-Mills for non-commutative two-tori*, Contemp. Math. **61** (1987), 237–266.

12. J. Cuntz, *Simple C^*-algebras generated by isometries*, Comm. Math. Phys. **57** (1977), 173–185.

13. J. Cuntz, G. A. Elliott, F. M. Goodman and P. E. T. Jorgensen, *On the classification of non-commutative two-tori II*, C. R. Math. Rep. Acad. Sci. Canada **7** (1985), 189–194.

14. G. A. Elliott, *The K-groups of the C^*-algebra generated by a projective representation of a torsion-free discrete abelian group*, Operator Algebras and Group Representations (Gr. Arsene et al., eds.), Pitman, London, 1984.

15. ______, *Gaps in the spectrum of an almost periodic Schrödinger operator*, C. R. Math. Rep. Acad. Sci. Canada **4** (1982), 255–259.

16. D. E. Evans, *The C^*-algebra of topological Markov chains*, Lecture Notes, Kyoto University RIMS, Kyoto, 1983.

17. O. W. Greenberger, *Particles with small violations of Fermi or Bose statistics*, Phys. Rev. D **43** (1991), 4111–4120.

18. M. Hazewinkel and M. Gerstenhaber (eds.), *Deformation theory of algebras of structures and applications*, Kluwer Publ., Dordrecht-Boston, 1988.

18a. M. Gerstenhaber, *On the deformation rings and algebras*, Ann. Math. (2) **79** (1964), 59–103.

19. P. E. T. Jorgensen, *Approximately inner derivations, decompositions, and vector fields of simple C^*-algebras*, Birkhauser, Boston, 1990.

20. P. E. T. Jorgensen and R. T. Moore, *Operator commutation relations*, Kluwer Publ., Dordrecht-Boston, 1984.

21. P. E. T. Jorgensen, L. M. Schmitt and R. F. Werner, *q-Canonical commutation relations and stability of the Cuntz algebra*, Preprint, (U. Osnabrück and U. Iowa), 1992.

22. Yu. I. Manin, *Topics in non-commutative geometry*, Princeton U. Press, Princeton, 1991.

23. D. Olesen, G. K. Pedersen and M. Takesaki, *Ergodic actions of compact abelian groups*, J. Operator Theory **3** (1980), 237–269.

24. P. J. Olver, *Applications of Lie groups to differential equations*, Springer-Verlag, New York, 1986.
25. M. Pimsner and S. Popa, *The Ext-groups of some C^*-algebras considered by J. Cuntz*, Rev. Roumaine Math. Pure Appl. **23** (1978), 1069–1076.
26. M. Pimsner and D. Voiculescu, *Exact sequences for K-groups and Ext-groups of certain cross-product C^*-algebras*, J. Operator Theory **4** (1980), 93–118.
27. W. Pusz, *Twisted canonical anticommutation relations*, Reports Math. Phys. **27** (1989), 349–360.
28. W. Pusz and S. L. Woronowicz, *Twisted second quantization*, Reports Math. Phys. **27** (1989), 231–257.
29. M. A. Rieffel, *C^*-algebras associated with irrational rotations*, Pacific J. Math. **93** (1981), 425–429.
30. M. A. Rieffel, *Vector bundles over higher-dimensional non-commutative tori*, LNM 1132, SpringerVerlag, New York, 1985.
31. ———, *Deformation quantization of Heisenberg manifolds*, Comm. Math. Phys. **122** (1989), 531–562.
32. S. Sakai, *Developments in the theory of unbounded derivations in C^*-algebras*, Proc. Symp. Pure Math. (v. II) **38** (1982), 309–331.
33. ———, *Operator algebras in dynamical systems*, Cambridge U. Press, New York, 1991.
34. L. M. Schmitt and R. F. Werner, *Non-commutative functional calculus*, Preprint, U. Osnabrück, 1992.
35. J. Slawny, *On factor-representations and the C^*-algebra of the canonical commutation relations*, Comm. Math. Phys. **24** (1971), 151–170.
36. R. Speicher, *A non-commutative central limit theorem*, Preprint, Heidelberg SFB.
37. V. S. Varadarajan, *Geometry of quantum theory*, (2nd ed.), Springer-Verlag, New York, 1985.
38. D. Voiculescu, *Symmetries of some reduced free product C^*-algebras*, LNM 1132, Springer-Verlag, New York, 1985.
39. S. L. Woronowicz, *Twisted $SU(2)$-group. An example of a non-commutative differential calculus*, vol. 23, Publ. RIMS, Kyoto Univ., 1987.

DEPARTMENT OF MATHEMATICS, THE UNIVERSITY OF IOWA, IOWA CITY, IA 52240

Contemporary Mathematics
Volume **160**, 1994

Algebraic theory

FRANCESCO IACHELLO

ABSTRACT. The logic scheme of algebraic theory is briefly reviewed. Examples of application of this theory are presented.

1. Algebraic theory

Algebraic theory is a formulation of quantum mechanics in terms of algebraic structures (Lie algebras, graded Lie algebras, ...) according to the logic scheme depicted in Figure 1. A quantum mechanical system is mapped onto an algebraic structure. This structure is then used to compute observable quantities which are in turn compared with experiment.

2. Ingredients of algebraic theory

The main ingredient of algebraic theory is the spectrum generating algebra (SGA). This is the set of elements, $X_i \in \mathcal{G}$, onto which all physical operators are expanded. For example, the Hamiltonian operator, H, is written as

---

1991 Mathematical Subject Classification. Primary 17B, 22E.
Supported in part under U.S. DOE Grant DE-FG02-31ER40608.
This paper is in final form and no version of it will be submitted for publication elsewhere.

Quantum Mechanical System

↓

Algebraic Structure — { Lie Algebras; Graded Lie algebras; Infinite dimensional algebras; q-deformed algebras; ... }

↓

Observables — { Spectra; Transitions; ... }

↓

Experiment

Fig. 1. Logic scheme of algebraic theory

$$H = f(X_i) \quad , \qquad X_i \in \mathcal{G} \qquad . \tag{1}$$

Usually the expansion is a polynomial in the X_i's,

$$H = E_0 + \sum_i \epsilon_i X_i + \sum_{ij} v_{ij} X_i X_j + \dots \quad , \tag{2}$$

that is H is in the envelopping algebra of $\mathcal{G}$. A special case of (2) is that in which H contains only invariant (Casimir) operators of $\mathcal{G}$ and its subalgebras $\mathcal{G} \supset \mathcal{G}' \supset \mathcal{G}'' \supset \dots$,

$$H = f(G_\alpha) \quad , \tag{3}$$

usually written as a polynomial in the G_α,

$$H = E_0 + \sum_\alpha \epsilon_\alpha G_\alpha + \dots \quad . \tag{4}$$

Here G_α denotes generically a Casimir invariant. Since these are linear combinations of the elements of the algebra $\mathcal{G}$ and their powers, the expansion (4) is a special case of (2). This special situation is called a dynamic symmetry (DS) and it is particularly useful in comparing with experiment, since whenever it occurs the eigenvalues of H can be obtained in closed form.

The choice of the SGA is dictated by the problem under consideration and in particular by the number of its degrees of freedom. As a result of a systematic investigation of several problems in physics [1,2,3], it has been suggested that, in general, *the Lie algebra u(n+1) be taken as spectrum generating algebra of non-relativistic quantum mechanics in n dimensions*. This statement applies to bound state problems with no spin. For problems with half-integer spin, graded Lie algebras u(n+1,m) are needed, and for unbound state problems (scattering) the algebra u(n+1) is changed into its non-compact extension u(n,1). Finally, for other problems, such as those having to do with internal spaces, other algebras $\mathcal{G}$ may be used, including all the classical Lie algebras o(n), sp(n) and the exceptional algebras.

3. Spectrum generating Lie algebras

Most applications of algebraic theory (like those to non-relativistic quantum mechanics mentioned above) make use of Lie algebras. The properties of the SGA needed to study problems in physics are:

(i) The commutation relations of the elements of $\mathcal{G}$

$$[X_i, X_j] = \sum_k c_{ij}^k X_k \quad , \quad X_i \in \mathcal{G} \quad , \tag{5}$$

plus the Jacobi identity

$$[X_i, [X_j, X_k]] + [X_j, [X_k, X_i]] + [X_k, [X_i, X_j]] = 0 \quad . \tag{6}$$

(ii) A list of all the possible subalgebras of $\mathcal{G}$

$$[Y_i, Y_j] = \sum_k c_{ij}^k Y_k \quad , \quad X_i \supset Y_i \quad , \tag{7}$$

$$X_i \in \mathcal{G} \quad , \quad Y_i \in \mathcal{G}' \quad , \quad \mathcal{G} \supset \mathcal{G}' \quad .$$

Here Y_i is an element of the algebra $\mathcal{G}'$.

(iii) The representation theory of $\mathcal{G}$, usually done in terms of Young tableau

(8) $$[\lambda_1,\lambda_2,\ldots,\lambda_n] \quad , \quad \lambda_1 \geq \lambda_2 \geq \ldots \geq \lambda_n \quad .$$

(iv) The branching rules

(9) $$\left| \begin{array}{ccc} \mathcal{G} & \supset & \mathcal{G}' \\ \downarrow & & \downarrow \\ [\lambda_1,\lambda_2,\ldots,\lambda_n] & & [\lambda_1',\lambda_2',\ldots,\lambda_{n'}'] \end{array} \right\rangle \quad ,$$

which enumerate the irreducible representations of $\mathcal{G}'$ contained in a given irreducible representation of $\mathcal{G}$.
(v) The knowledge of the invariant (Casimir) operators, G_α, which satisfy

(10) $$[G_\alpha, X_i] = 0 \quad , \quad \text{for any } X_i \in \mathcal{G} \quad .$$

These invariant operators are particularly important for physical applications because they are associated with conserved quantities.
(vi) The eigenvalues of the invariant operators in a given irreducible representation of $\mathcal{G}$

(11) $$\langle \lambda_1',\lambda_2',\ldots | G_\alpha | \lambda_1,\lambda_2,\ldots \rangle = c_\alpha(\lambda_1,\lambda_2\ldots)\, \delta_{\lambda_1\lambda_1'}\, \delta_{\lambda_2\lambda_2'} \quad .$$

4. Example 1

Algebraic theory has been used in a variety of problems in physics.One of the most extensive applications to date is in the study of the structure of atomic nuclei[1]. For the purpose of discussion here, atomic nuclei can be imagined as liquid drops with radii given by

(12) $$R = R_0(1+\sum_\mu \alpha_\mu Y_{2\mu}(\theta,\phi)) \quad ,$$

as shown in Fig. 2a. The surface (12) is characterized by

five coordinates $\alpha_\mu(\mu=0,\pm1,\pm2)$. The surface vibrates and rotates. The number of degrees of freedom is here n=5. According to algebraic theory, the algebraic structure of this problem is $\mathcal{G}=u(6)$. Both for practical reasons and in order to provide a physical interpretation of the algebraic structure involved, it is necessary to realize the algebra u(6) in some space. It has been found convenient to do so in terms of creation and annihilation operators. This realization (called interacting boson model[1]), makes use of six boson creation and annihilation operators, $b^\dagger_\alpha, b_\alpha (\alpha=1,\ldots,6)$ satisfying the commutation relations

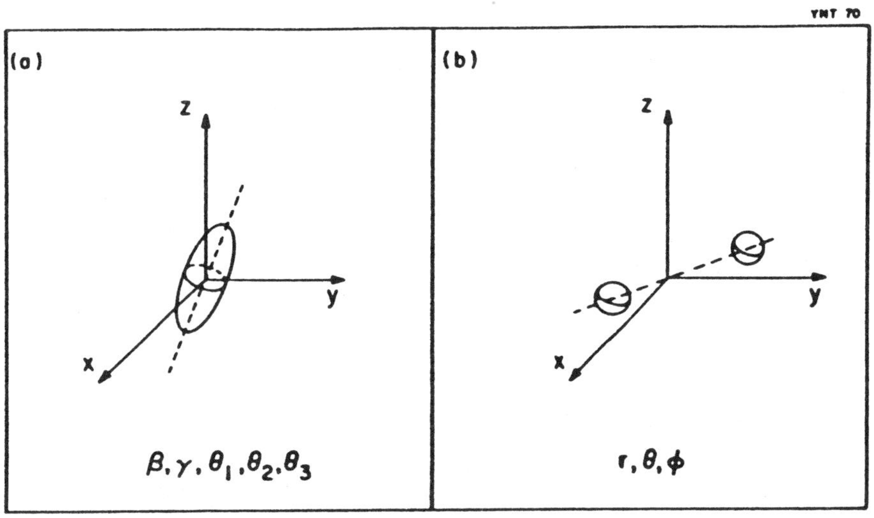

Fig.2. Schematic representation of the structure of atomic nuclei (a) and diatomic molecules (b). In part (a) the five variables $\beta,\gamma,\theta_1,\theta_2,\theta_3$ (Bohr variables) are used instead of the variables α_μ $(\mu=0,\pm1,\pm2)$ of Eq.(12). In part (b) the three variables r,θ,ϕ (spherical variables) are used instead of the variables r_μ $(\mu=0,\pm1)$ of Eq.(27).

(13) $$\left[b_\alpha, b^\dagger_{\alpha'}\right] = \delta_{\alpha\alpha'} \quad , \quad [b_\alpha, b_{\alpha'}] = \left[b^\dagger_\alpha, b^\dagger_{\alpha'}\right] = 0 \quad .$$

The elements of u(6) can be written as bilinear products of creation and annihilation operators

(14) $$\mathcal{G} \doteq \quad G_{\alpha\beta} = b^{\dagger}_{\alpha} b_{\beta} \quad , \quad (\alpha,\beta=1,\ldots,6).$$

They satisfy commutation relations

(15) $$\left[G_{\alpha\beta}, G_{\gamma\delta}\right] = G_{\alpha\delta}\delta_{\beta\gamma} - G_{\gamma\beta}\delta_{\delta\alpha} \quad .$$

The basis states on which the elements $G_{\alpha\beta}$ act are constructed by applying creation operators on a vacuum state, $|0\rangle$, the state with no particles,

(16) $$\mathcal{B}: \quad \frac{1}{\sqrt{N!}} \ldots b^{\dagger}_{\alpha'} b^{\dagger}_{\alpha} |0\rangle \quad .$$

Since the basis $\mathcal{B}$ is constructed with boson operators, it corresponds to a totally symmetric representation of u(6), characterized by the Young tableau

(17) $$[N] = [N,0,0,0,0,0] = \overbrace{\square\,\square\,\ldots\,\square}^{N\text{ - times}} \quad .$$

The interacting boson model is an expansion of the physical operators describing properties of atomic nuclei in terms of the elements $G_{\alpha\beta}$ of u(6). For example, the model Hamiltonian is

(18) $$H = E_0 + \sum_{\alpha\beta} \epsilon_{\alpha\beta} G_{\alpha\beta} + \frac{1}{2} \sum_{\alpha\beta} u_{\alpha\beta\gamma\delta} G_{\alpha\beta} G_{\gamma\delta} + \ldots$$

while the transition operators are

(19) $$T_{\alpha\beta} = t_0 + t_{\alpha\beta} G_{\alpha\beta} + \ldots$$

where $\epsilon_{\alpha\beta}$, $u_{\alpha\beta\gamma\delta}$, $t_{\alpha\beta}$, ... are parameters describing a given nucleus. The physical observables (energies and transition rates) are obtained by diagonalizing H in the basis (17) and by computing the matrix elements of $T_{\alpha\beta}$

between the corresponding eigenstates of H. These computed quantities are then compared with experiment.

5. Dynamic symmetries 1

The main advantage of algebraic theory, is that all observable quantities can be computed in a straightforward way by making use of properties of the algebra (in this case the Lie algebra u(6)). Another advantage is that it is possible to find, in a relatively simple way, all those problems, within the given model space, which can be solved in closed form (exactly solvable problems). These correspond to dynamic symmetries of the algebraic structure $\mathcal{G}$, i.e. situations in which H can be written in terms of invariant operators of $\mathcal{G}$ and its subalgebras $\mathcal{G} \supset \mathcal{G}' \supset \mathcal{G}'' \ldots$. All exactly solvable problems correspond to all possible subalgebras of $\mathcal{G}$. In the present case, there are three chains that one can form

$$(20) \qquad u(6) \begin{cases} u(5) \supset so(5) \supset so(3) \supset so(2) & ,\ (I) \\ su(3) \supset so(3) \supset so(2) & ,\ (II) \\ so(6) \supset so(5) \supset so(3) \supset so(2) & ,\ (III) \end{cases}$$

under the (physical) assumption that the rotation algebra so(3) be contained as a subalgebra. In order to construct the corresponding solutions, one must first solve the branching problem for u(6). This problem has been completely solved for the three chains (20). For example, for chain III, the solution of the branching problem yields[1] the classification scheme

$$(21) \qquad \left| \begin{matrix} u(6) & \supset & so(6) & \supset & so(5) & \supset & so(3) & \supset & so(2) \\ \downarrow & & \downarrow & & \downarrow & & \downarrow & & \downarrow \\ [N] & & (\sigma,0,0) & & (\tau,0)\nu_\Delta & & L & & M \end{matrix} \right\rangle ,$$

with

$$(22) \qquad \begin{aligned} &\sigma = N, N-2, \ldots, 1 \text{ or } 0 \quad ; \quad N = \text{odd or even} , \\ &\nu_\Delta = \text{missing label} = 0, 1, \ldots \quad ; \\ &\tau = 3\nu_\Delta + \lambda \quad ; \end{aligned}$$

$$L = \lambda,\ \lambda+1, \ldots, 2\lambda-2, 2\lambda \quad ;$$
$$M = L, L-1, \ldots, -L+1, -L \quad .$$

In (21) $[N], (\sigma,0,0), \ldots$ denote the representations of u(6), so(6),.... Finding the branching rules and constructing the irreducible representations is usually the most difficult part in applications of algebraic theory. This is because the chains of physical interest are often not the canonical chains $u(n) \supset u(n-1) \supset \ldots$, as it is clear from (20). There are therefore quite often missing labels, like in the case of the reduction $so(5) \supset so(3)$ where there is, for the symmetric representations $(\tau,0)$ of so(5), one missing label ν_Δ.

Having solved the branching problem, one can then proceed to the study of all possible dynamic symmetries. For example, a dynamic symmetry corresponding to chain III, occurs whenever the Hamiltonian H has the form

$$(23) \qquad H^{(III)} = E_0 + AC_2(so6) + BC_2(so5) + CC_2(so3) \quad ,$$

where $C_2(so6)$, $C_2(so5)$, $C_2(so3)$ denote the quadratic Casimir operators of so(6), so(5) and so(3) respectively. Most physical applications assume that the Hamiltonian is at most quadratic in the elements of the algebra X_i, and thus that at most quadratic Casimir operators appear in H. Linear and quadratic operators of u(6) are not included explicitly in (23) since they can be reabsorbed in E_0. (Their eigenvalues are a constant within a given representation of u(6)). The invariant operator of so(2) is also not included in (23) since it causes a splitting of the angular momentum degeneracy which occurs only if the nucleus is placed in an external electric or magnetic field.

The eigenvalues of the Hamiltonian operator, $H^{(III)}$ in the representation (21) can be simply evaluated and are given by

$$(24) \quad E^{(III)}(N,\sigma,\tau,\nu_\Delta,L,M) = E_0 + A\sigma(\sigma+4) + B\tau(\tau+3) + CL(L+1) \quad .$$

Repeating the procedure for the other two cases, one obtains the full set of solutions of the u(6) dynamic symmetry problem (exactly solvable problems). The eigenvalues are:

$$E(N,n_d,v,n_\Delta,L,M)=E_0+\epsilon n_d+\alpha n_d(n_d+4)+\beta v(v+3)+\gamma L(L+1), \quad (I)$$

$$(25) \quad E(N,\lambda,\mu,K,L,M)=E_0+\kappa(\lambda^2+\mu^2+\lambda\mu+3\lambda+3\mu)+\kappa' L(L+1), \quad (II)$$

$$E(N,\sigma,\tau,\nu_\Delta,L,M) = E_0+A\sigma(\sigma+4)+B\tau(\tau+3)+CL(L+1), \quad (III).$$

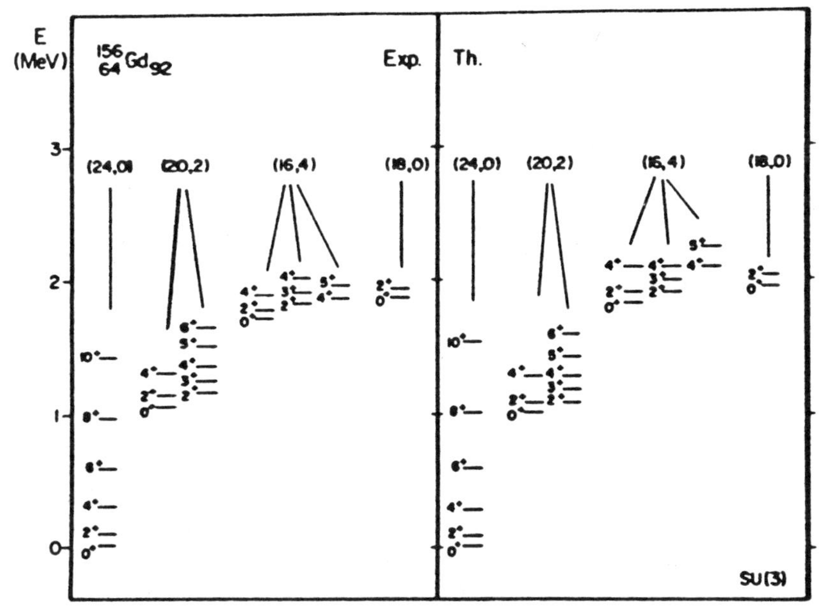

Fig.3. The energy levels of the nucleus ^{156}Gd are shown as an example of dynamic symmetries in physics. On the left the experimental spectrum, on the right the spectrum obtained by making use of Eq.(25), II. The extent to which the two parts of the figure agree is an indication of the goodness of the symmetry.

These formulas provide the classification scheme for dynamic symmetries of atomic nuclei with an even number of particles. Several examples of nuclei that can be described by one of the three formulas of Eq.(25) have been found experimentally. One such case is shown in Fig.3.

Dynamic symmetries correspond to special values of the coefficients $\epsilon_{a\beta}, u_{\alpha\beta\gamma\delta}$ in Eq. (18), such that H can be written in terms of Casimir operators of $\mathcal{G}\supset\mathcal{G}'\supset\mathcal{G}''\supset\ldots$. When the conditions leading to a dynamic symmetry are not met, one must return to Eq.(18) and diagonalize H numerically. Dynamic symmetries are also useful here in providing a basis in which the diagonalization of the Hamiltonian can be done.

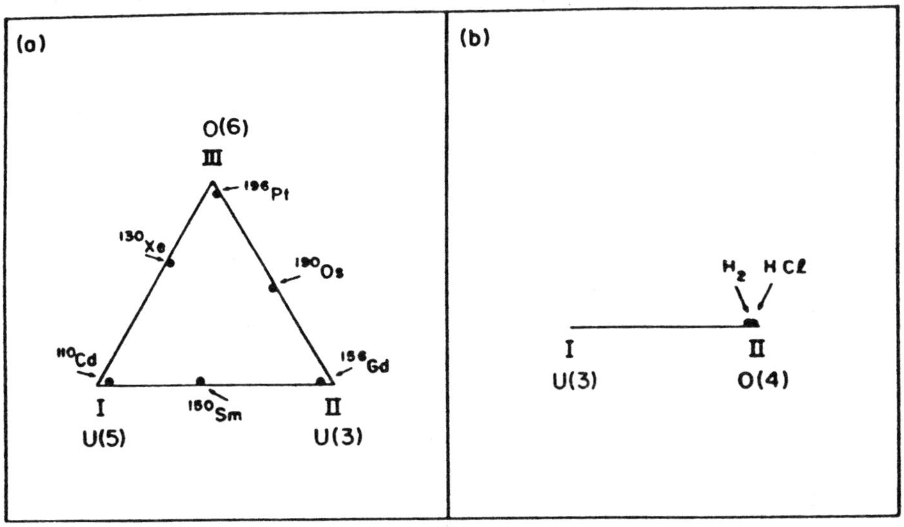

Fig.4. Schematic representation of the space of solutions of the interacting boson model (a) and of the vibron model (b). Some representative nuclei are shown in part (a) and some representative diatomic molecules in part (b).

In addition to providing a way to classify the exactly solvable problems in n dimensions through the study of its dynamic symmetries, algebraic theory also allows one to construct the complete space of solutions of a given problem. As the parameters $\epsilon_{\alpha\beta}, u_{\alpha\beta\gamma\delta}, \ldots,$ in Eq.(18) vary, one covers the entire space of solutions of the eigenvalue problem for H. It has become customary to show the space of solutions of quantum mechanical problems in n=5 dimensions by a diagram, called Casten's triangle. In this diagram, Fig.4, one places the three dynamic symmetries at the vertices of the triangle. Intermediate situations are shown as points inside the triangle, as discussed in Chapt. 2 of Ref.[4]. The points are identified by dimensionless ratios of parameters, called control parameters.

Another use of algebraic theory is in the evaluation of matrix elements of physical operators. A set of operators of particular interest are the transition operators, which induce transitions from one quantum level to another. Since these operators are also in the enveloping algebra of $\mathcal{G}$ (in fact in most applications are in the algebra $\mathcal{G}$), as shown in Eq.(19), the evaluation of the matrix elements of the transition operators, T, is straightforward, since it is of the type

$$\text{(26)} \qquad \langle \text{Irrep of } \mathcal{G} | \text{Element of } \mathcal{G} | \text{Irrep of } \mathcal{G} \rangle \quad ,$$

where Irrep of $\mathcal{G}$ denotes an irreducible representation of $\mathcal{G}$. This evaluation can be done once for all by using the properties of the algebras $\mathcal{G}$. Extensive results for $\mathcal{G}$=u(6) are given in Ref.[4].

6. Example 2

Another extensive application of algebraic theory is in the study of the structure of molecules[2,3]. In the case of diatomic molecules, there are three coordinates $r_\mu(\mu=0,\pm1)$, characterizing the position of one atom with respect to the other, as shown in Fig. 2b. This situation can be represented as a surface

$$\text{(27)} \qquad R = R_0(1 + \sum_\mu r_\mu Y_{1\mu}(\theta,\phi)) \quad .$$

The number of degrees of freedom is here n=3. According to algebraic theory, the algebraic structure of this problem is $\mathcal{G}$=u(4). It is again convenient to realize the algebra of u(4) by means of boson creation and annihilation operators, $b^\dagger_\alpha, b_\alpha(\alpha=1,\ldots,4)$, called vibron model[2]. The study of this case can be done precisely in the same way as discussed in the previous sections.

7. Dynamic symmetries 2

The only difference between problems in five

dimensions and problems in three dimensions is that there are now only two chains of subalgebras that contain the rotation algebra, so(3),

$$
(28) \qquad u(4) \begin{cases} u(3) \supset so(3) \supset so(2) \quad , & (I) \\ so(4) \supset so(3) \supset so(2) \quad . & (II) \end{cases}
$$

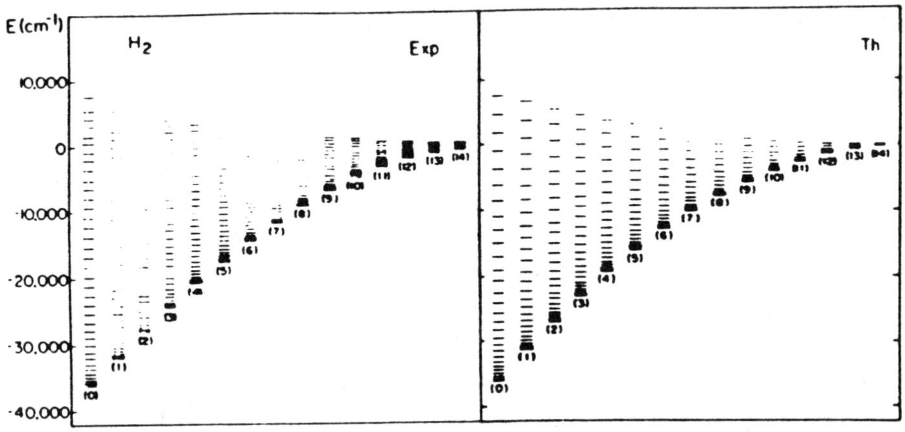

Fig.5. The energy levels of the molecule H_2 in its electronic ground state are shown as an example of dynamic symmetries in molecules. On the left the experimental spectrum, on the right the spectrum obtained by making use of Eq.(29), II. The extent to which the two parts of the figure agree is an indication of the goodness of the symmetry.

Correspondingly, there are only two solutions of the dynamic symmetry problem (exactly solvable problems). The eigenvalues are:

$$
(29) \qquad \begin{aligned} E(N,n_\pi,L,M) &= E_0+\epsilon n_\pi+\alpha n_\pi(n_\pi+2)+\beta L(L+1) \quad , & (I) \\ E(N,\omega,L,M) &= E_0+A\omega(\omega+2)+BL(L+1) \quad , & (II). \end{aligned}
$$

These formulas provide the classification scheme for dynamic symmetries of diatomic molecules. Several examples of molecules that can be described by one of the two formula of Eq.(29) have been found experimentally. One such case is shown in Fig.5.

8. Example 3

As a last example of algebraic theory, consider the case of quantum mechanical problems with only one coordinate. The algebraic structure of these problems is $\mathcal{G}=u(2)$. The dynamic symmetries are

$$u(2) \begin{matrix} \nearrow & u(1) \quad , \\ \searrow & so(2) \quad . \end{matrix} \tag{30}$$

This case is different from the previous ones because:
(i) there is no angular momentum in one dimension and thus no constraint in the subalgebra chains;
(ii) the subalgebras u(1) and so(2) are trivial abelian algebras and are isomorphic

$$u(1) \approx so(2) \quad . \tag{31}$$

As a result, there are many more exactly solvable quantum mechanical problems in one dimension than in three or five dimensions. The energy eigenvalues of all these problems are given by

$$E(N,m) = E_0 + Am^2 \quad . \tag{32}$$

Since they have the same eigenvalue structure, they are called isospectral problems.

9. Geometry

Although all properties of a physical system can be computed by making use of purely algebraic methods (this is indeed the advantage of algebraic theory), it is also interesting to study the geometric properties of algebraic models. This is because algebraic equations can then be converted into differential equations for geometric variables and one can make a connection with the formulation of quantum mechanics in terms of differential equations (Schrödinger-like equations).

Given a Lie algebra $\mathcal{G}$, there is a straightforward procedure by means of which one can attach geometric spaces to it. The procedure consists in decomposing the algebra $\mathcal{G}$ into two parts,

(33) $$\mathcal{G} = \mathcal{H} \oplus \mathcal{P} \quad ,$$

where $\mathcal{H}$ is a subalgebra of $\mathcal{G}$

(34) $$\mathcal{G} \supset \mathcal{H}$$

and thus closed with respect to commutation, while $\mathcal{P}$, the remainder, is not closed with respect to commutation. Here $\mathcal{H}$ is called the stability algebra and $\mathcal{G}/\mathcal{H}$ the factor algebra. Geometric variables can then be associated with each element of $\mathcal{P}$. Their number is just the number of elements in $\mathcal{P}$.

Since the algebra $\mathcal{G}$ may have several possible subalgebras $\mathcal{H}$, there are in principle several possible geometric spaces. In particular applications, some possibilities are eliminated by some physical constraint. For example, if one insists that $\mathcal{H}$ contains the rotation algebra, so(3), one obtains constraints on the possible subalgebras chains of u(6), as in Eq.(20). Even with these constraints there may be several possibilities left. One usually defines as geometric space associated with $\mathcal{G}$ that which is connected with the maximum stability subalgebra $\mathcal{G}\supset\mathcal{H}$ where $\mathcal{H}$ is the largest possible algebra compatible with the physical conditions on $\mathcal{G}$. The procedure then uniquely determines the geometric variables associated with $\mathcal{G}$.

For non-relativistic quantum mechanics in n dimensions and no intrinsic spin, described algebraically by u(n+1), the spaces of interest are u(n+1)/u(n)$\oplus$u(1). These spaces (also called coset spaces) are Riemannian globally symmetric spaces[5,6]. In the cases discussed in the previous sections, in which the elements of the algebra $\mathcal{G}$ are written as in Eq.(14), one has

$$\mathcal{G}: \qquad G_{\alpha\beta} = b^\dagger_\alpha b_\beta \;, \qquad (\alpha,\beta=1,\ldots,n+1)$$

$$(35) \qquad \mathcal{H}: \qquad b^\dagger_{n+1} b_{n+1} \;,\; b^\dagger_\alpha b_\beta \;, \qquad (\alpha,\beta=1,\ldots,n)$$

$$\mathcal{P}: \qquad b^\dagger_{n+1} b_\alpha \;,\; b_\alpha{}^\dagger b_{n+1} \;. \qquad (\alpha=1,\ldots,n)$$

The coset spaces are labelled by the complex quantities $\eta_\alpha(\alpha=1,\ldots,n)$, associated with the elements of $\mathcal{P}$. An explicit construction of coset spaces can be done by introducing coherent (or intrinsic) states, defined as

$$(36) \qquad |N,\eta_\alpha> = \exp\left[\sum_\alpha \eta_\alpha\, b^\dagger_\alpha b_{n+1} + \sum_\alpha \eta^*_\alpha\, b^\dagger_{n+1} b_\alpha\right] |\Lambda_{ext}>$$

where

$$(37) \qquad |\Lambda_{ext}> = \frac{1}{\sqrt{N!}}\,(b^\dagger_{n+1})^N|0> \quad .$$

Instead of the complex quantities η_α, when making a connection with Schrödinger-like equations, one usually introduces coordinates, q_α, and momenta, p_α, through

$$(38) \qquad \eta_\alpha=(q_\alpha+ip_\alpha)/\sqrt{2} \quad , \quad \eta^*_\alpha = (q_\alpha-ip_\alpha)/\sqrt{2} \quad .$$

In the Schrödinger equation, the momenta p_α become differential operators $p_\alpha = i\,\partial/\partial q_\alpha$.

10. Geometry 1

As an example of explicit construction of the coset space, consider the case discussed in Sect. 4 and divide the six boson creation operators into 5+1 operators $b^\dagger_\alpha(\alpha=1,\ldots,6) = d^\dagger_\mu(\mu=0,\pm1,+2)$, $s^\dagger$. (The index $\mu=0,\pm1,+2$ is used here, instead of a running index 1,...,5 in order to identify the coordinates and momenta with the physical coordinates and momenta of Fig.2a.). Similarly, $b_\alpha = d_\mu, s$. The coherent state is

$$|N,\eta_\mu\rangle = \exp\left[\sum_\mu \eta_\mu d^\dagger_\mu s + \sum_\mu \eta^*_\mu s^\dagger d_\mu\right] |\Lambda_{ext}\rangle ,$$

(39)

$$|\Lambda_{ext}\rangle = \frac{1}{\sqrt{N!}} s^{\dagger N} |0\rangle .$$

By expanding the exponential and keeping only the term with a fixed number N (i.e. a given representation of u(6)), one can rewrite the coherent state as

$$|N,\eta_\mu\rangle = (s^\dagger + \sum_\mu \eta_\mu d^\dagger_\mu)^N |0\rangle . \tag{40}$$

The η_μ are complex numbers which live in the coset space $u(6)/u(5)\oplus u(1)$. One can introduce coordinates, α_μ, and momenta, ρ_μ, in this space through

$$\eta_\mu = (\alpha_\mu + i\rho_\mu)/\sqrt{2} \quad , \quad \eta^*_\mu = (\alpha_\mu - i\rho_\mu)/\sqrt{2} . \tag{41}$$

The coordinates α_μ of this five dimensional space are called Bohr variables[7] (see also Eq.(12) and Fig.2a). By making use of the techniques discussed in Chapt.3 of Ref.[4] and in Chapt. 9.IV of Ref.[8], one can then convert the eigenvalue equation

$$H\psi = E\psi , \tag{42}$$

solved previously with algebraic methods into a differential equation

$$\mathcal{D}\psi(\alpha_\mu) = E\psi(\alpha_\mu) . \tag{43}$$

This equation is in n=5 variables. The differential

operator $\mathcal{D}$ depends on the form of the algebraic Hamiltonian. If this Hamiltonian is at most quadratic in the elements of the algebra, $\mathcal{G}$, the differential operator is at most of fourth order in the derivatives $\partial/\partial\alpha_\mu$. The differential equation (43) is a generalized version of the Bohr equation (the Bohr equation is the Schrödinger equation in five dimension). Explicit expressions for the differential operator $\mathcal{D}$ are given in Chapt. 3.10 of Ref.[4]. One can see explicitly in this reference that the differential operator is of fourth order. Dynamic symmetries of the algebraic Hamiltonian correspond then to differential equations which are exactly solvables. This is because the algebraic Hamiltonians with dynamic symmetry are written only in terms of Casimir operators. The Casimir operators are Laplace-Beltrani operators in the coset space α_μ. (See Ref.[8] for details).

Riemannian symmetric spaces provide a way, as discussed above, to construct exactly solvable problem in any number of dimensions. However, sometimes one is interested in differential equations with further restrictions. Two conditions are often imposed:
(i) that the differential equation (43) corresponds to algebraic Hamiltonians which are scalars under transformations induced by the subalgebras so(3) of u(n+1), i.e. states have good angular momentum;
(ii) that the differential operator $\mathcal{D}$ be at most quadratic in the derivatives, and of the form

$$(44) \qquad \mathcal{D} = -\frac{\hbar^2}{2B'}\sum_\mu \left|\frac{\partial^2}{\partial\alpha_\mu^2}\right| + V(|\alpha_\mu|) \quad .$$

Under these conditions, there appear to be only two exactly solvable problems in $n\geq 2$ dimensions, that corresponding to

$$(45) \qquad V(|\alpha_\mu|) = \frac{1}{2}C'\sum_\mu |\alpha_\mu|^2 \quad ,$$

called the harmonic oscillator problem, and that corresponding to

$$V(|\alpha_\mu|) = \frac{C}{\sqrt{\sum_\mu |\alpha_\mu|^2}} \quad , \tag{46}$$

called the Kepler problem.

11. Geometry 2

The same procedure can be used to discuss the example 2 of Sect. 6. One divides the four boson operators into 3+1 operators, usually called $\pi^\dagger_\mu (\mu=0,\pm 1)$, $\sigma^\dagger$ and π_μ, σ. The coset space here is the usual three dimensional space. Denoting by r_μ and p_μ the coordinates and momenta one has

$$\begin{aligned} &|N,\eta_\mu\rangle = (\sigma^\dagger + \sum_\mu \eta_\mu \pi^\dagger_\mu)^N |0\rangle \quad , \\ &\eta_\mu = (r_\mu + ip_\mu)/\sqrt{2} \quad , \quad \eta^*_\mu = (r_\mu - ip_\mu)/\sqrt{2} \quad . \end{aligned} \tag{47}$$

By using the techniques discussed above, one can then convert the algebraic Hamiltonian into a differential operator

$$\mathcal{D}\psi(r_\mu) = E\psi(r_\mu) \quad . \tag{48}$$

This is a generalized Schrödinger equation containing fourth order derivatives.

If one insists on having $\mathcal{D}$ of second order in the derivatives and having rotational invariance,

$$\mathcal{D} = -\frac{\hbar^2}{2m} \sum_\mu \left|\frac{\partial^2}{\partial r_\mu^2}\right| + V(|r_\mu|) \quad , \tag{49}$$

only two problems appear to be exactly solvable,

corresponding again to the harmonic oscillator and Kepler problems.

12. Geometry 3

The statements made in the previous two sections apply only to problems in $n \geq 2$ dimensions. For $n=1$ the situation is very different, since there is no rotational invariance to impose. Thus in $n=1$ dimension the class of exactly solvable problems is much larger than that given above. A large number of examples was given in Ref.[8] and the complete classification is given in Ref.[9] and the accompanying paper by Gonzales-Lopez, Olver and Kamran[10]. The coherent state for these one-dimensional problems is

$$|N,\eta\rangle = (\sigma^{\dagger}+\eta\tau^{\dagger})^{N}|0\rangle \tag{50}$$

with

$$\eta = (x+ip)/\sqrt{2} \quad , \quad \eta^{*} = (x-ip)/\sqrt{2} \quad , \tag{51}$$

where x is the coordinate and p the associated momentum, and the differential (Schrödinger) equation is

$$\begin{aligned} \mathcal{D}\psi(x) &= E\psi(x) \quad , \\ \mathcal{D} &= -\frac{\hbar^{2}}{2m}\frac{d^{2}}{dx^{2}} + V(x) \quad . \end{aligned} \tag{52}$$

13. Conclusions

Algebraic theory is a powerful tool that allows one to analyze problems in physics and chemistry. In this article, the logic scheme of this theory has been given, and examples discussed. The discussion has been limited to non-relativistic quantum mechanical problems in n dimensions and no intrinsic spin. The Lie algebras u(n+1), which describe these problems, can be realized in terms of boson operators, as in Eq.(14). However, several problems in non-relativistic quantum mechanics involve particles with spin. To treat these problems, one needs the introduction of algebraic structures which are more complex than Lie algebras, in particular, for problems with half-integer spin, one needs the introduction of graded Lie algebras[11]. There has been also in recent years an extensive use of these more complex structures. These algebraic structures have not been described here, but the interested reader may consult Refs.[12] and [13] for an exposition.

REFERENCES

1. A. Arima and F. Iachello, "Interacting Boson Model of Collective Nuclear States.I-The Vibrational Limit", Ann. Phys.(N.Y.) 99 (1976), 253-317;
A. Arima and F. Iachello, "Interacting Boson Model of Collective Nuclear States. II-The Rotational Limit", Ann. Phys. (N.Y.) 111 (1978), 201-238;
A. Arima and F. Iachello, "Interacting Boson Model of Collective Nuclear States. IV-The O(6) Limit", Ann. Phys.(N.Y.) 123 (1979), 468-492.
2. F. Iachello, "Algebraic Methods for Molecular Rotation-Vibration Spectra", Chem. Phys. Lett. 78 (1981), 581-585.
3. F. Iachello and R.D. Levine, "Algebraic Approach to Rotation-Vibration Spectra. I-Diatomic Molecules", J. Chem. Phys. 77 (1982), 3046-3055.
O.S. van Roosmalen, A.E.L. Dieperink, and F. Iachello, "A Dynamic Algebra for Rotation-Vibration Spectra of Complex Molecules",Chem. Phys. Lett. 85 (1982), 32-36.
O.S. van Roosmalen, F. Iachello, R.D. Levine, and A.E.L. Dieperink, "Algebraic Approach to Molecular Rotation-Vibration Spectra. II-Triatomic Molecules", J. Chem. Phys. 79 (1983), 2515-2536.
4. F. Iachello and A. Arima, "The Interacting Boson Model",Cambridge University Press, Cambridge, 1987.
5. R. Gilmore, "Lie Groups, Lie Algebras and Some of Their Applications", Wiley, New York, 1974.
6. S. Helgason, "Differential Operators on Homogeneous Spaces", Acta Math. 102 (1959), 239-299.
7. A. Bohr and B.R. Mottelson, "Nuclear Structure", Benjamin, New York (1976).

8. J. Wu, "Group Theory Approach to Scattering", Ph.D. Thesis, Yale University (1985), unpublished.
9. N. Kamran and P.J. Olver, "Lie Algebras of Differential Operators and Lie Algebraic Potentials", J. Math. Anal. Appl. 145 (1990), 342-356.
10. A. Gonzales-Lopez, N. Kamran, and P.J. Olver, These Proceedings.
11. A.B. Balantekin, I. Bars and F. Iachello, "U(6/4) Supersymmetry in Nuclei", Nucl. Phys. A370 (1981), 284-316.
12. F. Iachello and P. Van Isacker, "The Interacting Boson-Fermion Model", Cambridge University Press, Cambridge, 1991.
13. A. Frank, R. Lemus, and F. Iachello, "Algebraic Approach to Molecular Electronic Spectrum-I-Energy Levels", J. Chem. Phys. 91 (1989), 29-41.

Current address: Center for Theoretical Physics, Sloane Laboratory, Yale University, New Haven, Connecticut 06511, USA

Contemporary Mathematics
Volume **160**, 1994

The Time-Dependent Schrödinger Equation in Multidimensional Integrable Evolution Equations

D.J. Kaup

ABSTRACT The one-dimensional, time dependent Schröd- inger equation plays a key role in two multidimensional integrable evolution equations. In the so-called Davey-Stewartson I equation, it appears as the time-evolution operator for the reflection and transmission coefficients of the scattering data. As a consequence of this, the transmission coefficient is no longer a constant of the motion. An asymptotic expansion in the spectral parameters reveals that the transmission coefficient contains only one conservation law and then an infinity of dynamical relations. How this result is related to an underlying group structure is not known. The second equation is the KPI equation where the time dependent Schrödinger equation is <u>the</u> scattering problem for the IST of KPI. The solution of the inverse scattering problem requires

1. Introduction

The connection between integrable evolution equations and Lie groups has had a long and fruitful history. Estabrook and Wahlquist[1] observed that an infinite algebraic prolongation structure existed for the KdV equation, which was also reflected in the algebraic structure of the Lax pair. In fact, one could use this algebraic structure to construct the Lax pair[2]. The Sato Theory[3] of the τ-function also depends critically

---

1991 Mathematics Subject Classification. Primary 35Q40 Secondary 35Q58.

This work has been supported in part by the Air Force Office of Scientific Research through URI Grant No. AFOSR-89-0510.

This paper is in final form and no version of it will be submitted for publication elsewhere.

unbounded wave functions which are not allowed in the quantum mechanical Hilbert space. The consequences and implications of this are discussed.

on this rich algebraic structure as well as the infinite symmetries and conservation laws that these integrable equations possess.

For one-dimensional systems, this algebraic structure in the underlying spectral problem has been quite well detailed in general and is considered to be well understood. However for multidimensional integrable problems, although some work has been done[4,5], there are aspects of these scattering problems that are not so well understood. What I wish to do here is to detail and point out two aspects of one multidimensional scattering problem which are not fully understood, and hopefully to encourage some other work on these aspects. Both of these aspects deal with one of the key subjects of this session: the time-dependent Schrödinger equation in one-dimension with a time-dependent potential.

This scattering problem plays a key role in two of the major multidimensional integrable equations. These two equations are called, in the inverse scattering literature, KPI and DSI. For the KPI, it is the key scattering problem[6–8]: it is that scattering problem with which the Inverse Scattering Transform (IST) is constructed. In DSI, it plays a lesser role. Here the IST is constructed around a sector of the IST for the 3D3WRI[9,10], while the time-dependent Schrödinger equation is used to determine the time evolution of the scattering data[11]. Thus in DSI it plays a role not as a scattering problem, but instead the key role in determining the time evolution of any symmetries and conservation laws. In a one-dimensional integrable system, there is almost always an infinite set of conservation laws and conserved quantities. These conserved quantities can invariably be obtained easily from an asymptotic expansion of a transmission coefficient in its spectral parameter. Such is not true for DSI. Expansion of its transmission coefficients yield only <u>one</u> conservation law and an <u>infinity</u> of "dynamical relations". This is in stark contrast to one dimensional integrable systems and indeed is not well understood.

In the next section, we will describe this feature in more detail. In Section 3, we will consider very interesting situation in regard to KPI. Here the time-dependent Schrödinger equation is <u>the</u> scattering problem for the IST. In the solution of the inverse scattering problem, it was observed that scattering data from unnormalizable wave functions must be used. This violates fundamental assumptions of quantum mechanics and could have interesting consequences. Finally we summarize our main points in the last section after we have presented more of the details.

2. The Time-Dependent Schrödinger Equation in DSI

The details of the following are to be published elsewhere[11], but we shall outline the general approach here and give some basic background information. The so-called DSI equations are

$$-i\partial_t q = 1/2(\partial_x^2 + \partial_y^2)q + q(\partial_y \phi + rq) \tag{1a}$$

$$(\partial_x^2 - \partial_y^2)\phi - 2\partial_y(rq) = 0 \tag{1b}$$

where $r = \pm q^*$. The method of solution for this system was first detailed by Fokas and Ablowitz[13]. This method is based on the Lax pair whose spatial part is[9,10]

$$(\partial_x + \partial_y)v_1 = qv_2 \tag{2a}$$

$$(\partial_y - \partial_x)v_2 = rv_1 \tag{2b}$$

where v_j (j=1,2) are the components of a rank 2 column vector. The time part of this Lax pair is

$$i\partial_t v_1 = -(\partial_y^2 - a_1 + \frac{1}{2}rq)v_1 + q\partial_y v_2 - \frac{1}{2}(\partial_x q - \partial_y q)v_2 \tag{3a}$$

$$i\partial_t v_2 = (\partial_y^2 - a_2 + \frac{1}{2}rq)v_2 - r\partial_y v_1 - \frac{1}{2}(\partial_x r + \partial_y r)v_2 \tag{3b}$$

where

$$a_1 = \frac{1}{2}(\partial_x - \partial_y)\phi \tag{4a}$$

$$a_2 = -\frac{1}{2}(\partial_x + \partial_y)\phi \tag{4b}$$

The direct scattering problems for (2) has two transmission coefficients. It is the expansion of these transmission coefficients in which we are interested. First, let's review the one-dimensional case. In one-dimensional systems such as the Korteweg-de Vires (KdV) equation or the nonlinear Schrödinger (NLS) equation, the transmission coefficient, $a(k)$, is a constant of the motion. Since it is a function of k (the spectral parameter), one can therefore expand it in an asymptotic series about $k = \infty$, in which case every coefficient of $k^{-n}(n = 1, 2, 3, ...)$ will also be a constant of the motion. Thus

$$(5a) \qquad a(k) \to 1 + C_1/k + C_2/k^2 + ...$$

where C_n is the infinite set of constants of the motion [see for example Ref. 14]. The same general procedure also works for the 3D3WRI[9] although the conserved quantities are in general nonlocal. Of course there are the known exceptions to this which arise whenever one has singular dispersion relations[15] such as self- induced transparancy[16,17] and cavitons[18]. However one still has the expansion of $a(k)$ as in (5) where each C_n is a functional of the field and its spatial derivatives only. The only change is that each C_n varies in time due to boundary effects. There is still the infinity of differential conservation laws

$$(5b) \qquad \partial_t \rho_n + \partial_x \sigma_n = 0, \quad C_n = \int_{-\infty}^{\infty} \rho_n dx$$

with the exception arising only because the σ_n's do not in general vanish at the boundaries.

However, this pattern is broken in the case of the DSI system. After the first conservation law is obtained, the same general procedure inexplicably returns only trivially conserved quantities, although an initial cursorary (and incomplete) inspection of the result would lead one to believe that nontrivial conserved quantities did exist. What we find in the asymptotic expansion of the transmission coefficient of the DSI system is instead an infinite set of dynamical relations, not conservation laws. The analogous quantities in (5) now contain time-derivatives of the fields as well as spatial derivatives. If one then uses the field equations to eliminate the time-derivatives of all field quantities, one is left with only a triviality. Thus an infinite set of conservation laws are not found in these transmission coefficients, but what is found in an infinite set of dynamical relations. (Recently Fokas and Santini[5] have used a two-point recursion operator to obtain an infinite set of constants-of-the-motion for this equation as well as KPI and KPII. However explicit forms for the conserved quantities are still to be given.)

In multidimensional integrable systems, the transmission coefficient is in general a function of two spectral parameters and is a true constant only when both spectral parameters are equal. When they are not equal, it has a nontrivial time evolution in general. For the DSI system, this time evolution is guided by the quantum mechanical time-dependent Schrödinger equation with a nontrivial time-dependent potential. In order to obtain the hierarchy of dynamical relations, it is necessary to expand in both spectral parameters, and calculate the appropriate time dependence. In order to do the latter, it is necessary to determine the time dependence of the quantum-mechanical wave function (the solution of the time-dependent Schrödinger equation) for large spectral parameters. This has been done and with this result[11], we are able to give the complete solution of the asymptotic expansion (in both spectral parameters) of the transmission coefficient.

We then demonstrate that the coefficients in this expansion contain two parts. The first part is nonzero in general but is simply a convolution of the initial data and is devoid of any information about the dynamics. The second part must vanish and this set is the infinite hierarchy of dynamical relations. We shall now briefly outline the results.

In the following, we shall use the characteristic coordinates

$$\chi_1 = \frac{1}{2}(y - x),\ \chi_2 = \frac{1}{2}(x + y) \tag{6a}$$

for which

$$\partial_x + \partial_y = \partial_2,\ \partial_y - \partial_x = \partial_1 \tag{6b}$$

and eq (2) becomes

$$\partial_2 v_1 = q v_2 \tag{7a}$$

$$\partial_1 v_2 = r v_1 \tag{7b}$$

The scattering problem (7) yields two transmission coefficients, each a funtion of two spectral parameters. If we take the double Fournier transform, one can show that the time evolution of these transmission coefficients are govern by

$$i\partial_t a_1^1(s,s') = -[H_{1f}(s) - H_{10}(s')]a_1^1(s,s') \tag{8a}$$

$$i\partial_t a_2^2(s,s') = [H_{20}(s) - H_{2f}(s')]a_2^2(s,s') \tag{8b}$$

where the operator H is

$$H = -\frac{1}{4}\partial^2 + A \tag{9}$$

and the potential, A, is either A_{10}, A_{20}, A_{1f} or A_{2f} depending on the subscripts on H in (8). The final values are related to the initial values by

$$A_{1f} = A_{10} + \frac{1}{2}\partial_1 \int_{-\infty}^{\infty} rq d\chi_2 \tag{10a}$$

$$A_{2f} = A_{20} + \frac{1}{2}\partial_2 \int_{-\infty}^{\infty} rq d\chi_1 \tag{10b}$$

Note that the first subscript on the potentials give the characteristic coordinate that A is a function of. These A's are simply to the initial and final boundary values of the a_1 and a_2 in (4). For simplicity, we shall only consider the case where $A_{10} = 0 = A_{20}$. It is quite simple[11] to construct constants of the motion from (8) and (9). Take any set of Ψ's which satisfy

$$i\partial_t \Psi_{10} = H_{10}\Psi_{10} \tag{11a}$$

$$i\partial_t \Psi_{1f} = H_{1f}\Psi_{1f} \tag{11b}$$

(11c) $$i\partial_t \Psi_{20} = H_{20}\Psi_{20}$$

(11d) $$i\partial_t \Psi_{2f} = H_{2f}\Psi_{2f}$$

then it follows that

(12a) $$R_1^1 = \int_{-\infty}^{\infty} ds \int_{-\infty}^{\infty} ds' \Psi_{1f}(s) a_1^1(s,s') \Psi_{10}^*(s')$$

(12b) $$R_2^2 = \int_{-\infty}^{\infty} ds \int_{-\infty}^{\infty} ds' \Psi_{20}^*(s) a_2^2(s,s') \Psi_{2f}(s')$$

are each constants of the motion. These quantities are actually each a double continuum of constants of the motion since the functional form of the Ψ's is arbitrary. In other words, for any choice of $\Psi(s)$ at $t = 0$, R_1^1 and R_2^2 will always satisfy

(13) $$\partial_t R_1^1 = \partial_t R_2^2 = 0$$

provided the Ψ's satisfy the appropriate form of (10). Thus once we find an asymptotic expansion for the Ψ's and also for the a_n^n's, then the coefficients of this expansion would be constants of the motion.

The natural choice is to expand in plane waves. So at $t = 0$ we choose

(14) $$\Psi(s) = e^{iks} e^{-\frac{i}{4}k^2 t}$$

for all the Ψ's. Each Ψ will evolve in time according to its own Hamiltonian. Since (10) and (11) is simply the one-dimensional quantum-mechanical Schrödinger equation with a time dependent potential, there is a lot we know about the solutions of this equation. Since the set (14) is orthonormal and complete, as it evolves under (10), it will remain orthonormal and complete. Thus this set would give us the complete set of constants of the motion contained in R_1^1 and R_2^2. Furthermore, for $k \to \pm\infty$, asymptotic expansions in k will exist for Ψ.

The expansion of the transmission coefficients is straight-forward[11], but the expansion of the Ψ's involve some subtle points that we should mention. At first, one would suspect that we could choose

$$\Psi(s) = e^{iks} e^{-i\frac{k^2}{4}t} \sum_{j=0}^{\infty} \frac{\mu_j(s,t)}{(ik)^r} \tag{15}$$

However this expansion gives

$$\mu_1(s,t) = 2 \int^s A(s',t) ds' \tag{16}$$

which clearly violates the initial condition (14) since $A(s, t = 0) \neq 0$ in general. Thus there is an initial secular growth which has to be considered. It is from this secular growth that our hierarchy of dynamic relations will emerge.

One can resolve this problem of a secular growth by considering the quantum mechanical picture of a single, high-speed particle being incident on the scattering potential, $A(s,t)$. The $\Psi(s,t)$ is the probability amplitude of locating the particle at time t at position s. This Ψ will be localized about the classical path, which is $s = s_0 + \frac{1}{2}kt$. For large k, the particle is moving very rapidly through the potential and feels its effect only briefly, but on the time scale of L/k where L is the typical width of A. Consequently, we can expect a fast time scale (kt) to be present in the solution for Ψ. To include this fast time scale into Ψ, we simply have to expand as

$$\Psi(s,t) = e^{iks} e^{-i\frac{k^2}{4}t} \sum_{j=0}^{\infty} \frac{M_j(z,\tau,t)}{(ik)^j} \tag{17}$$

where

$$z = s - \frac{1}{2}\tau \tag{18a}$$

$$\tau = kt \tag{18b}$$

$$M_0 = 1 \tag{18c}$$

In (36), z is the comoving coordinate of the particle and τ is the fast time scale.

Inserting (17) into (9) and (11) gives the recursion relations

$$\partial_\tau M_1 = A \tag{19a}$$

$$\partial_\tau M_2 + DM_1 = AM_1 \tag{19b}$$

and for j $\geq$ 1

$$\partial_\tau M_{j+1} + DM_j = AM_j \tag{19c}$$

where the operator D is

$$D = i\partial_t + \frac{1}{4}\partial_z^2 \tag{20a}$$

Note that due to (18)

$$A(s,t) = A(z + \frac{1}{2}\tau, t) \tag{20b}$$

Thus the solution of (19a) is

$$M_1 = m_1(z,t) + \mathcal{M}_1(z,\tau,t) \tag{21}$$

where m_1 is independent of the fast time and is the asymptotic value of M_1 for large τ. $\mathcal{M}_1$ is the transient part and may be taken to be

$$\mathcal{M}_1 = -\int_\tau^\infty d\tau' A(z + \frac{1}{2}\tau', t) \tag{22}$$

so that it vanishes as $\tau \to \infty$.

The equation for m_1 follows in next order. From (19b) and (21)

$$\partial_\tau M_2 = -Dm_1 + [AM_1 - D\mathcal{M}_1] \tag{23}$$

the quantity in brackets vanishes as $\tau \to \infty$. Then if M_2 is to be nonsecular, we must choose

$$Dm_1 = 0 \tag{24}$$

and the solution of (23) is

$$M_2 = m_2(z,t) + \mathcal{M}_2(z,\tau,t) \tag{25}$$

where

$$\mathcal{M}_2 = -\int_\tau^\infty d\tau[AM_1 - D\mathcal{M}_1] \tag{26}$$

Obviously this can be continued to all orders, given for $j \geq 2$

$$M_j = m_j(z,t) + \mathcal{M}_j(z,\tau,t) \tag{27}$$

where

$$\mathcal{M}_j = -\int_\tau^\infty d\tau[AM_{j-1} - D\mathcal{M}_{j-1}] \tag{28}$$

$$Dm_j = 0 \tag{29}$$

It only remains to give the initial data for m_j. From (14) and (17), it follows that $M_j(z,0,0)$ must vanish for $j \geq 1$. Thus

$$m_j(z,0) = -\mathcal{M}_j(z,0,0) =$$

$$\int_0^\infty d\tau[A(z+\frac{1}{2}\tau,0)M_{j-1}(z,\tau,0) - \lim_{t\to 0} D\mathcal{M}_{j-1}(z,\tau,t)] \tag{30}$$

In general, (30) will not vanish as $z \to -\infty$ although it does for $z \to +\infty$. The quantum-mechanical interpretation of this is as follows. The particle is incident from the left, and one would not expect to find that the particle has moved faster than $\frac{1}{2}k$. Thus for large $z(= s - \frac{1}{2}kt)$ one would not expect to find any particles. However the potential could temporarily trap the particle causing the particle to be reflected back-and-forth before re-emerging. Thus in the wake of the classical path ($z < 0$), one would easily expect a nonzero probability for finding the particle.

Note that $\mathcal{M}_j$, defined by (26) depends on $\partial_t A$. And the higher orders will depend on even higher derivatives of A. Thus the asymptotic expansion for Ψ decomposes into two basic parts; m_j which depends only on initial data at $t = 0$ (and $\tau = 0$), and $\mathcal{M}_j$ which contains the dynamics and time derivatives of the potential.

We now expand R_1^1. One obtains[11]

$$R_1^1 = 2\pi\delta(k-\ell) + \sum_{j=1}^{\infty} \frac{1}{(ik)^j} \int_{-\infty}^{\infty} ds e^{i\frac{\ell^2-k^2}{4}t} e^{i(k-\ell)s} C_j(s, z, \tau, t) \tag{31}$$

where the first three C_j's are

$$C_1 = m_1(z, t) \tag{32a}$$

$$C_2 = m_2(z,t) + i[\frac{1}{2}P_1 - \partial_t \int_\tau^\infty d\tau' \Gamma_1(z + \frac{1}{2}\tau', t)] \tag{32b}$$

$$C_3 = m_3 - E_1 + \frac{1}{4}\partial_1^2 \Gamma_1 + \Gamma_1 \partial_1 \Gamma_1 + \frac{1}{4}\int_\tau^\infty (\partial_1 \Gamma_1')^2 d\tau' + \frac{1}{2}\partial_t \int_\tau^\infty P_1' d\tau'$$

$$+ m_1(C_2 - m_2) \tag{32c}$$

$$C_4 = m_4 - \frac{i}{2}\Lambda_1 + iP_1\partial_1\Gamma_1 + i\partial_t \int_\tau^\infty E_1' d\tau' + \frac{i}{4}\int_\tau^\infty (\partial_1 \Gamma_1')(\partial_1 P_1') d\tau'$$

$$+ m_1(C_3 - m_3) + (m_2 - m_1^2 - 2\partial_z m_1)(C_2 - m_2) \tag{32d}$$

and

$$\Gamma_1 = -\int_{-\infty}^{\infty} r q d\chi_2 \tag{33a}$$

(33b) $$\pi_1 = -\int_{\chi_2}^{\infty} d\chi_2(iq\partial_1 r - ir\partial_1 q)$$

(33c) $$\eta_1 = -\int_{\chi_2}^{\infty} d\chi_2[(\partial_1 q)(\partial_1 r) + (\partial_1\Delta_1^{(1)})(\partial_2\Delta_1^{(1)})]$$

(33d)
$$\lambda_1 = -\int_{\chi_2}^{\infty} d\chi_2[i(\partial_1 q)(\partial_1^2 r) - i(\partial_1^2 q)(\partial_1 r) + (\partial_2\Delta_1^{(1)})\partial_1\pi_1 + 2(\partial_1\Delta_1^{(1)})\partial_2\pi_1]$$

(33e) $$P_1 = \pi_1(\chi_2 \to -\infty)$$

(33f) $$E_1 = \eta_1(\chi_2 \to -\infty)$$

(33g) $$\Lambda_1 = \lambda_1(\chi_2 \to -\infty)$$

(33h) $$\Delta_1^{(1)} = -\int_{\chi_2}^{\infty} (rq)d\chi_2$$

One can easily show that since m_j satisfies (24), the integral in (31) does produce a constant of the motion, but it is simply the Fourier transform of the initial value of m_1. The second part of C_2 is nontrivial. One notes that the integrals are equivalent to integrals over χ_1. Then since this part is only a function of χ_1 and t, one can show that this part of C_2 can only give a constant in (31) if it vanishes. Thus

(34a) $$\partial_t \int_{\chi_1}^{\infty} \Gamma_1(s,t)ds = \frac{1}{4}P_1$$

similarly C_3 and C_4 gives

(34b) $$\partial_t \int_{\chi_1}^{\infty} P_1 ds = E_1 - \frac{1}{4}\partial_1^2\Gamma_1 - \Gamma_1\partial_1\Gamma_1 - \frac{1}{2}\int_{\chi_1}^{\infty} [\Gamma_1(s,t)]^2 ds$$

$$\partial_t \int_{\chi_1}^{\infty} E_1 ds = \frac{1}{2}\Lambda_1 - P_1\partial_1\Gamma_1 - \frac{1}{4}\int_{\chi_1}^{\infty} (\partial_1\Gamma_1)(\partial_1 P_1) ds \tag{34c}$$

It is obvious that (34a) gives a conservation law.(The corresponding one from R_2^2 is exactly the same). However (34b) or (34c) do not. Letting $\chi_1 \to -\infty$, since E_1, Λ_1 and Γ_1 vanish in this limit, we have

$$\partial_t \int_{-\infty}^{\infty} P_1 d\chi_1 = -\int_{-\infty}^{\infty} \Gamma_1^2 d\chi_1 < 0 \tag{35a}$$

$$\partial_t \int_{-\infty}^{\infty} E_1 d\chi_1 = -\frac{1}{4}\int_{-\infty}^{\infty} (\partial_1\Gamma_1)(\partial_1 P_1) d\chi_1 \tag{35b}$$

showing that the integral of P_1 will always decrease until the field vanishes in some norm.

So we do not obtain a set of conserved quantities. We obtain only one conserved quantity and an infinity of dynamical relations. The quantity R_1^1 is indeed a constant of the motion, but it is not a function of the dynamical variables. Instead it is only a function of the <u>initial data</u>. It is an open question as to how this structure for the transmission coefficients of DSI fits into known algebraic structure of integrable systems.

A final observation is that (35) does strongly resemble what one obtains in SIT[16,17] and cavitons[18]. There one has <u>local</u> conservation laws in the form of (5b). However spatial integration of (5b) then yields relations like (35) where the right-hand side results from boundary contributions. But in one-dimensional systems, such boundary contributions are exclusively restricted to integrable systems with singular dispersion relations. Here we see such occuring for <u>nonsingular</u> dispersion relations.

3. The Time-Dependent Schrödinger Equation in KPI

The KPI equation is

$$(u_t + 6u_x u + u_{xxx})_x = 3u_{yy} \tag{36}$$

whose Lax pair[6] is

$$iΨ_y = -Ψ_{xx} - uΨ \tag{37a}$$

$$Ψ_t = -4Ψ_{xxx} - 6uΨ_x - 3(u_x - i\int_{-\infty}^{x} u_y dx')Ψ \tag{37b}$$

The scattering problem in (37a) is simply again, the time-dependent Schrödinger equation, except that t has been replaced by y. The solution of the inverse scattering problem for (37a) requires the solution of a Fredholm integral equation of the second type[7]. One of the Fredholm alternatives is that homogeneous solutions can exist. And they do indeed exist[12]. Let

$$u = 2\partial_x^2 \ell n Δ \tag{38}$$

where

$$Δ = \frac{1}{η^2} + |x - ζy|^2 \tag{39}$$

with $ζ = ξ + iη$. Then u is the rational solution

$$u = \frac{4}{Δ^2}[\frac{1}{η^2} + η^2 y^2 - (x - ξy)^2] \tag{40}$$

The general solution of (37a) for $Ψ$ is

$$Ψ = e^{i(2λx - 4λ^2 y)} χ \tag{41a}$$

$$χ = 1 + \frac{iχ_+}{λ - ζ} + \frac{iχ_-}{λ - ζ^*} \tag{41b}$$

where

$$χ_+ = \frac{1}{Δ}[\frac{1}{η} + 2x - 2ζ^* y] \tag{42a}$$

$$\chi_- = \frac{1}{\Delta}\left[-\frac{1}{\eta} + 2x - 2\zeta y\right] \tag{42b}$$

and λ is an arbitrary parameter. However there is a problem with this solution. It is true that χ does have a bound state at $\lambda = \zeta$ or $\lambda = \zeta^*$. For these complex values of λ, χ has a pole and the residue is square integrable. These solutions and their scattering data are necessary for solving the inverse scattering problem. However, the function Ψ, given by (41a) is <u>not</u> square integrable for these complex values of λ, due to the phase factor in (41a). Thus for the quantum mechanical problem, such a Ψ does not belong to the Hilbert space and is therefore not an allowable solution. This leaves us with a very interesting situation and it is the following. The inverse scattering solution for (37a) requires scattering data[7] from these bound states ($\lambda = \zeta$ or ζ^*), if they exist. This means that information about these bound states are probably necessary and essential for the reconstruction of the potential. If so then these states would also be necessary and essential for closure. However, these states are not quantum mechanically allowed states, which would imply that the allowed quantum mechanical states would not be closed. In other words, for the potential, (40), the quantum mechanical Hilbert space would be incomplete. But a fundamental assumption of quantum mechanics is that the Hilbert space is indeed complete. This apparent contridiction would place the entire mathematical structure of quantum mechanics into question. However, let us note that there is another possible explanation, and that is that perhaps these bound states, (42), are really not fundamental. In reality the known inverse scattering solution for (37a) is actually based on an analysis of the solutions for χ, not Ψ, which satisfies a slightly different equation than (37a). As is clear from (41b), (42) are quite valid square integrable functions for χ. In other words, the known inverse scattering solution pertains to χ, not Ψ, and it may be that there is another inverse scattering solution for Ψ with different scattering data in which the scattering data for $\chi_\pm$ are not required. Another possibility could be that there is some still unknown identities which gives the scattering data for the states $\chi_\pm$ in terms of other scattering information. Clearly, more work needs to be done.

In any case, we see that there are some indeed quite interesting questions here to be answered, and the answers could have some quite interesting implications for quantum mechanics and certainly for the time-dependent Schrödinger equation.

4. Conclusion

The time-dependent Schrödinger equation is a key equation in two major IST's for multidimensional integrable equations. In the DSI, it serves as a time-evolution operator for the transmission coefficient. As a result of this evolution, the transmission coefficients no longer contain conserved quantities, which now sets DSI apart from other integrable systems.

In the KPI, it serves as the scattering problem. An analysis of this scattering problem has revealed that for certain time dependent potentials, scattering data for unnormalizable states are required. If this is true, there could be important consequences for quantum mechanics. If it is not, then there is a definite need for more analysis which would allow us to prove such.

These are interesting facets of the time-dependent Schrödinger equation and indicate that even more interesting algebraic structures may like underneath multidimensional integrable systems than those known to be present in one dimensional integrable systems.

Bibliography

[1] F.B. Estabrook and H.D. Wahlquist, J. Math. Phys. 16, 1 (1975).

[2] D.J. Kaup, Physica 1D, 391-411 (1980).

[3] M. Sato and Y. Sato, in Nonlinear Partial Differential Equations in Applied Science, edited by H. Fujita, P.D. Lax, and G. Strang (Kinokuniya/ North-Holland, Tokyo, 1983), 259.

[4] J. Matsukidaira, J. Satsuma and W. Strampp, J. Math. Phys. 31, 1426-1434 (1990).

[5] P.M. Santini and A.S. Fokas, Commun. Math. Phys. 115, 375-419 (1988).

[6] S.V. Manakov, Physica D3, 420 (1981).

[7] A.S. Fokas and M.J. Ablowitz, Stud. Appl. Math. 69, 211-228 (1983).

[8] V.G. Bukurov, Phys. Lett. A 160, 367-371 (1991).

[9] D.J. Kaup, Physica 1D, 45-67 (1980).

[10] A.S. Fokas, Phys. Rev. Lett. 51, 3-6 (1983).

[11] D.J. Kaup, (To appear in Inverse Problems).

[12] V. Zakharov and S. Manakov, Funct. Analysis and Appl. 19, 11-26 (1985).

[13] A.S. Fokas and M.J. Ablowitz, Phys. Rev. Lett. 51, 7-10 (1983).

[14] M.J. Ablowitz, D.J. Kaup, A.C. Newell and H. Segur, Stud. Appl. Math. 53, 249 (1974).

[15] D.J. Kaup and A.C. Newell, Adv. in Math. 31, 67-100 (1979).

[16] G.L. Lamb, Jr., Rev. Mod. Phys. 43, 99-124 (1971).

[17] D.J. Kaup, Phys. Rev. A 16, 704-718 (1977).

[18] D.J. Kaup, Phys. Fluids 31, 1465-1470 (1988).

DEPARTMENT OF MATHEMATICS, CLARKSON UNIVERSITY, POTSDAM, NY 13699-5815

E-mail address: kaup@sun.mcs.clarkson.edu

Contemporary Mathematics
Volume **160**, 1994

Models of q -Algebra Representations: Matrix Elements of $U_q(su_2)$

E.G. KALNINS, WILLARD MILLER, JR. AND SANCHITA MUKHERJEE

ABSTRACT. This paper continues a study of one and two variable function space models of irreducible representations of q-analogs of Lie enveloping algebras, motivated by recurrence relations satisfied by q-hypergeometric functions. Here we consider the quantum algebra $U_q(su_2)$. We show that various q-analogs of the exponential function can be used to mimic the exponential mapping from a Lie algebra to its Lie group and we compute the corresponding matrix elements of the "group operators" on these representation spaces. This "local" approach applies to more general families of special functions, e.g., with complex arguments and parameters, than does the quantum group approach. We show that the matrix elements themselves transform irreducibly under the action of the quantum algebra. We find an alternate and simpler derivation of a q-analog, due to Groza, Kachurik and Klimyk, of the Burchnall-Chaundy formula for the product of two hypergeometric functions ${}_2F_1$. It is interpreted here as the expansion of the matrix elements of a "group operator" (via the exponential mapping) in a tensor product basis in terms of the matrix elements in a reduced basis.

1. Introduction

This paper continues the study of function space models of irreducible representations of q-algebras [**1, 2**]. These algebras and models are motivated by recurrence relations satisfied by q-hypergeometric functions [**3**] and our treatment is an alternative to the theory of quantum groups. Here, we consider the irreducible representations of the quantum algebra $U_q(su_2)$. We replace the

1991 *Mathematics Subject Classification.* Primary 33D55, 33D45; Secondary 17B37, 81R50.

Key words and phrases. basic hypergeometric functions, q-algebras, quantum groups, Clebsch-Gordan series.

This paper is in final form and no version of it will be submitted for publication elsewhere.

The second author was supported in part by the National Science Foundation under grant DMS 91–100324

The third author was supported by the Study Abroad fellowship of the Government of India.

usual exponential function mapping from the Lie algebra to the Lie group by the q-exponential mappings E_q and e_q. In place of the usual matrix elements on the group (arising from an irreducible representation) which are expressible in terms of Jacobi polynomials, we find eight types of matrix elements expressible in terms of q-hypergeometric series. These q-matrix elements do not satisfy group homomorphism properties, so they do not lead to addition theorems in the usual sense. However, they do satisfy orthogonality relations. Furthermore, in analogy with true group representation theory we can show that each of the eight families of matrix elements determines a two-variable model for irreducible representations of $U_q(su_2)$. In §3 we show how this two-variable model leads to orthogonality relations for one class of matrix elements.

In §4 we find an alternate and simpler derivation of a q-analog, due to Groza, Kachurik and Klimyk [**4**], of the Burchnall-Chaundy formula for the product of two hypergeometric functions ${}_2F_1$. It is interpreted here as the expansion of the matrix elements of a "group operator" (via the exponential mapping) in a tensor product basis in terms of the matrix elements in a reduced basis.

Our approach to the derivation and understanding of q-series identities is based on the study of q-algebras as q-analogs of Lie algebras, [**5, 6**]. Essentially, we are attempting to find q-analogs of the theory relating Lie algebra and local Lie transformation groups [**7, 8**]. A similar approach has been adopted by Floreanini and Vinet [**9-12**]. This is an alternative to the elegant papers [**13-21**] which are based primarily on the theory of quantum groups. The main justification of the "local" approach is that it is more general; it applies to more general families of special functions than does the quantum group approach.

The notation used for q-series in this paper follows that of Gasper and Rahman [**22**].

2. Models of finite dimensional $U_q(su_2)$ representations

The quantum algebra $U_q(su_2)$ is the associative algebra generated by the elements H, E_+, E_- that obey the commutation relations

$$\begin{aligned}&[H,E_+]=E_+,\quad [H,E_-]=-E_-,\\ &[E_+,E_-]=\frac{q^H-q^{-H}}{q^{\frac12}-q^{-\frac12}}.\end{aligned}\tag{2.1}$$

Here q is a real parameter such that $0<q<1$. In the limit as $q\to 1$ relations (2.1) go to the usual commutation relations for the complexification of the Lie algebra su_2. Finite dimensional irreducible representations of $U_q(su_2)$ are determined by the integral or half-integral number u: $2u=0,1,2,\cdots$. The corresponding representation $D(2u)$ is defined on the $(2u+1)$-dimensional Hilbert space H_{2u} with orthonormal basis $\{e_m : m = -u, -u+1, \cdots, u\}$ such that

$$(2.2)\qquad \begin{aligned} E_+ e_m &= ([u-m]_q[u+m+1]_q)^{\frac{1}{2}}\, e_{m+1} \\ E_- e_m &= ([u+m]_q[u-m+1]_q)^{\frac{1}{2}}\, e_{m-1} \\ He_m &= me_m \end{aligned}$$

where

$$(2.3)\qquad [m]_q = \frac{q^{\frac{m}{2}} - q^{-\frac{m}{2}}}{q^{\frac{1}{2}} - q^{-\frac{1}{2}}} = q^{-\frac{m-1}{2}} \left(\frac{1-q^m}{1-q}\right).$$

On this Hilbert space $E_+ = (E_-)^*$ and $H^* = H$. A second convenient basis for H_{2u} is the set $\{f_n : n = 0, 1, \cdots, 2u\}$ such that

$$(2.4)\qquad \begin{aligned} E_+ f_n &= -q^{-1}[2u-n]_q f_{n+1} \\ E_- f_n &= -q[n]_q f_{n-1} \\ Hf_n &= (-u+n)f_n. \end{aligned}$$

Here

$$(2.5)\qquad f_n = \left[\frac{(-1)^n q^{(\frac{3}{2}-u)n}(q;q)_n}{(q^{-2u};q)_n}\right]^{\frac{1}{2}} e_{-u+n}.$$

Since the element

$$(2.6)\qquad C = E_+E_- + \frac{q^{H-\frac{1}{2}} + q^{-H+\frac{1}{2}}}{(q^{\frac{1}{2}} - q^{-\frac{1}{2}})^2}$$

commutes with each generator of $U_q(su_2)$ it corresponds to a multiple of the identity operator I on H_{2u}. Indeed,

$$(2.7)\qquad C = \frac{q^{u+\frac{1}{2}} + q^{-u-\frac{1}{2}}}{(q^{\frac{1}{2}} - q^{-\frac{1}{2}})^2} I.$$

Now we consider the discrete series $\uparrow_u$ of infinite dimensional representations of $U_q(su_2)$. This is defined on the Hilbert space H_0 with orthogonal basis $\{f_n : n = 0, 1, \cdots\}$ such that relations (2.4) hold. Now, however, u is a negative real number and

$$||f_n|| = \sqrt{\frac{q^{(\frac{3}{2}-u)n}(q;q)_n}{(q^{-2u};q)_n}}, \quad n = 0, 1, 2, \cdots.$$

On this Hilbert space $E_+ = -(E_-)^*$ and $H^* = H$. (In the limit as $q \to 1$ these representations correspond to the positive discrete series of unitary irreducible representations of the Lie algebra $su(1,1)$, **[18, 19]**.)

Note, however, that for each complex number u such that $2u \neq 0, 1, 2, \cdots$ expressions (2.4) define an algebraically irreducible representation $\uparrow_u$ of $U_q(su_2)$ on an infinite dimensional vector space K consisting of all finite linear combinations of the basis vectors $\{f_n\}$.

We introduce a convenient one variable model of $D(2u)$. Here the vector space H_{2u} consists of polynomials $f(z)$ of maximum order $2u$ in the complex variable z. The action of $U_q(su_2)$ is defined by the operators

$$\begin{aligned} E_+ &= \frac{q^{-\frac{1}{2}}z}{1-q}(q^u T_z^{-\frac{1}{2}} - q^{-u} T_z^{\frac{1}{2}}) \\ E_- &= \frac{q^{\frac{3}{2}}}{z(1-q)}(T_z^{\frac{1}{2}} - T_z^{-\frac{1}{2}}) \\ H &= -u + z\frac{d}{dz} \end{aligned} \tag{2.8}$$

where $T_z f(z) = f(qz)$. The basis functions $\{f_n = z^n : n = 0, 1 \cdots, 2u\}$ satisfy relations (2.4). Similarly, we can introduce a one variable model of $\uparrow_u$. A basis for the vector space consists of the functions $\{f_n(z) = z^n : n = 0, 1, 2 \cdots\}$ in the complex variable z. The action of $U_q(su_2)$ on functions $f(z)$ is again defined by expressions (2.8). See [**1**] for derivations of the inner products corresponding to these models.

In analogy with a standard relationship between special functions and the representations of Lie groups we shall compute the "matrix elements" of q-analogs of the group operators $e^{\beta E_+} e^{\alpha E_-}$ with respect to the $\{f_n\}$ basis, in both the representations $D(2u)$ and $\uparrow_u$. Of course there are many q-analogs of the exponential mapping, none of which have all the properties needed to ensure that there is a true "group" associated with the q-algebra. Among the q-analogs we shall limit ourselves to the two that are most important, [**22**]:

$$e_q(z) = \sum_{k=0}^{\infty} \frac{z^k}{(q;q)_k}, \quad E_q(z) = \sum_{k=0}^{\infty} \frac{q^{k(k-1)/2}}{(q;q)_k} z^k. \tag{2.9}$$

If z is a complex number, the first series converges to $1/(z;q)_\infty$ for $|z| < 1$ and the second series converges to $(-z;q)_\infty$ for all z.

Using the model (2.4) we can define eight q-analogs of the matrix elements.

$$\begin{aligned} (e+, e-):& \quad e_q(\beta E_+) e_q(\alpha E_-) f_n = \sum_{n'} T_{n'n}^{(e+,e-)}(\alpha, \beta) f_{n'}, \\ (e-, e+):& \quad e_q(\beta E_-) e_q(\alpha E_+) f_n = \sum_{n'} T_{n'n}^{(e-,e+)}(\alpha, \beta) f_{n'}, \\ (e+, E-):& \quad e_q(\beta E_+) E_q(\alpha E_-) f_n = \sum_{n'} T_{n'n}^{(e+,E-)}(\alpha, \beta) f_{n'}, \end{aligned}$$

$$(e-, E+): \quad e_q(\beta E_-) E_q(\alpha E_+) f_n = \sum_{n'} T_{n'n}^{(e-,E+)}(\alpha, \beta) f_{n'}, \tag{2.10}$$

$$\begin{aligned} (E+, e-):& \quad E_q(\beta E_+) e_q(\alpha E_-) f_n = \sum_{n'} T_{n'n}^{(E+,e-)}(\alpha, \beta) f_{n'}, \\ (E-, e+):& \quad E_q(\beta E_-) e_q(\alpha E_+) f_n = \sum_{n'} T_{n'n}^{(E-,e+)}(\alpha, \beta) f_{n'}, \end{aligned}$$

$$(E+,E-):\quad E_q(\beta E_+)E_q(\alpha E_-)f_n = \sum_{n'} T^{(E+,E-)}_{n'n}(\alpha,\beta)f_{n'},$$

$$(E-,E+):\quad E_q(\beta E_-)E_q(\alpha E_+)f_n = \sum_{n'} T^{(E-,E+)}_{n'n}(\alpha,\beta)f_{n'}.$$

For the representations $D(2u)$ the sum on n' is from 0 to $2u$ and $n = 0, \cdots, 2u$. For the representations $\uparrow_u$, the sum on n' is from 0 to ∞ and $n = 0, 1, \cdots$. (Relations (2.10) and the relations to follow are well defined as formal power series in the variables α, β. We must look at each case separately to determine the domains of pointwise convergence of the series.)

Since $E_+^* = E_-$ for $D(2u)$ and $E_+^* = -E_-$ for $\uparrow_u$ the following relationships hold:

$$
\begin{aligned}
T^{(e+,e-)}_{n'n}(\alpha,\beta)A_{n'n} &= T^{(e+,e-)}_{nn'}(-\beta,-\alpha),\\
T^{(e-,e+)}_{n'n}(\alpha,\beta)A_{n'n} &= T^{(e-,e+)}_{nn'}(-\beta,-\alpha),\\
T^{(e+,E-)}_{n'n}(\alpha,\beta)A_{n'n} &= T^{(E+,e-)}_{nn'}(-\beta,-\alpha),\\
T^{(e-,E+)}_{n'n}(\alpha,\beta)A_{n'n} &= T^{(E-,e+)}_{nn'}(-\beta,-\alpha),\\
T^{(E+,E-)}_{n'n}(\alpha,\beta)A_{n'n} &= T^{(E+,E-)}_{nn'}(-\beta,-\alpha),\\
T^{(E-,E+)}_{n'n}(\alpha,\beta)A_{n'n} &= T^{(E-,E+)}_{nn'}(-\beta,-\alpha).
\end{aligned}
\tag{2.11}
$$

Here

$$A_{n'n} = \frac{q^{(3/2-u)(n'-n)}(q;q)_{n'}(q^{-2u};q)_n}{(q;q)_n(q^{-2u};q)_{n'}}.$$

Since $e_q(z)E_q(-z) = 1$, we have the identities

$$
\begin{aligned}
&a)\ \sum_\ell T^{(e+,e-)}_{n'\ell}(\alpha,\beta)T^{(E-,E+)}_{\ell n}(-\beta,-\alpha) = \delta_{n'n}\\
&b)\ \sum_\ell T^{(e-,e+)}_{n'\ell}(\alpha,\beta)T^{(E+,E-)}_{\ell n}(-\beta,-\alpha) = \delta_{n'n}\\
&c)\ \sum_\ell T^{(E-,e+)}_{n'\ell}(\alpha,\beta)T^{(E+,e-)}_{\ell n}(-\beta,-\alpha) = \delta_{n'n}.
\end{aligned}
\tag{2.12}
$$

Again, the sum is finite or infinite, depending on the representation $D(2u)$ or $\uparrow_u$.

Using the model (2.4) to compute the matrix elements (which are model independent) we obtain the explicit results

$$
\begin{aligned}
&T^{(e+,e-)}_{n'n}(\alpha,\beta)\\
&= \frac{(q^{-2u};q)_{n'}\beta^{n'-n}q^{-(n'-n)(n'+n+1)/4}}{q^{u(n-n')}(q;q)_{n'-n}(q^{-2u};q)_n(1-q)^{n'-n}}\\
&\times {}_3\phi_1\left(\begin{matrix} q^{-n}, & q^{1-n+2u}, & 0 \\ q^{n'-n+1} & & \end{matrix};q;\frac{\alpha\beta q^{n-u+1/2}}{(1-q)^2}\right)\\
&= \frac{(q;q)_n(-\alpha)^{n-n'}q^{-(n-n')(n'+n-5)/4}}{(q;q)_{n-n'}(q;q)_{n'}(1-q)^{n-n'}}\\
&\times {}_3\phi_1\left(\begin{matrix} q^{-n'}, & q^{1-n'+2u}, & 0 \\ q^{n-n'+1} & & \end{matrix};q;\frac{\alpha\beta q^{n'-u+1/2}}{(1-q)^2}\right)
\end{aligned}
$$

$$
\begin{aligned}
&T^{(e-,e+)}_{n'n}(\alpha,\beta)\\
&= \frac{(q^{-2u};q)_{n'}\alpha^{n'-n}q^{-(n'-n)(n'+n+1)/4}}{q^{u(n-n')}(q;q)_{n'-n}(q^{-2u};q)_n(1-q)^{n'-n}}\\
&\times {}_3\phi_1\left(\begin{matrix} q^{n'+1}, & q^{n'-2u}, & 0 \\ q^{n'-n+1} & & \end{matrix};q;\frac{\alpha\beta q^{u-n'+1/2}}{(1-q)^2}\right)\\
&= \frac{(q;q)_n(-\beta)^{n-n'}q^{-(n-n')(n+n'-5)/4}}{(q;q)_{n-n'}(q;q)_{n'}(1-q)^{n-n'}}\\
&\times {}_3\phi_1\left(\begin{matrix} q^{n+1}, & q^{n-2u}, & 0 \\ q^{n-n'+1} & & \end{matrix};q;\frac{\alpha\beta q^{u-n+1/2}}{(1-q)^2}\right)
\end{aligned}
$$

$$
\begin{aligned}
&T^{(E+,e-)}_{n'n}(\alpha,\beta)\\
&= \frac{(q^{-2u};q)_{n'}\beta^{n'-n}q^{(n'-n)(n'-3n-3)/4}}{q^{u(n-n')}(q;q)_{n'-n}(q^{-2u};q)_n(1-q)^{n'-n}}\\
&\times {}_2\phi_1\left(\begin{matrix} q^{-n}, & q^{1-n+2u} \\ q^{n'-n+1} & \end{matrix};q;\frac{-\alpha\beta q^{n'-u+1/2}}{(1-q)^2}\right)\\
&= \frac{(q;q)_n(-\alpha)^{n-n'}q^{-(n-n')(n+n'-5)/4}}{(q;q)_{n-n'}(q;q)_{n'}(1-q)^{n-n'}}\\
&\times {}_2\phi_1\left(\begin{matrix} q^{-n'}, & q^{1-n'+2u} \\ q^{n-n'+1} & \end{matrix};q;\frac{-\alpha\beta q^{n'-u+1/2}}{(1-q)^2}\right)
\end{aligned} \tag{2.13}
$$

$$T^{(E-,e+)}_{n'n}(\alpha,\beta)$$
$$= \frac{(q^{-2u};q)_{n'}\alpha^{n'-n}q^{-(n'-n)(n'+n+1)/4}q^{u(n'-n)}(-\alpha\beta q^{n-u+1/2}/(1-q)^2;q)_\infty}{(q;q)_{n'-n}(q^{-2u};q)_n(1-q)^{n'-n}(-\alpha\beta q^{u-n'+1/2}/(1-q)^2;q)_\infty}$$
$$\times {}_2\phi_1\left(\begin{matrix} q^{-n}, & q^{1-n+2u} \\ q^{n'-n+1} & \end{matrix}; q; \frac{-\alpha\beta q^{n-u+1/2}}{(1-q)^2}\right)$$
$$= \frac{(q;q)_n(-\beta)^{n-n'}q^{(n-n')(n-3n'+3)/4}(-\alpha\beta q^{n-u+1/2}/(1-q)^2;q)_\infty}{(q;q)_{n-n'}(q;q)_{n'}(1-q)^{n-n'}(-\alpha\beta q^{u-n'+1/2}/(1-q)^2;q)_\infty}$$
$$\times {}_2\phi_1\left(\begin{matrix} q^{-n'}, & q^{1-n'+2u} \\ q^{n-n'+1} & \end{matrix}; q; \frac{-\alpha\beta q^{n-u+1/2}}{(1-q)^2}\right)$$

$$T^{(E-,E+)}_{n'n}(\alpha,\beta)$$
$$= \frac{(q;q)_n(-\beta)^{n-n'}q^{(n-n')(n-3n'+3)/4}}{(q;q)_{n-n'}(q;q)_{n'}(1-q)^{n-n'}}$$
$$\times {}_2\phi_2\left(\begin{matrix} q^{n-2u}, & q^{n+1} \\ q^{n-n'+1}, & 0 \end{matrix}; q; \frac{\alpha\beta q^{u-n'+1/2}}{(1-q)^2}\right)$$
$$= \frac{(q^{-2u};q)'_n\alpha^{n'-n}q^{(n'-n)(n'-3n-3)/4}}{q^{u(n-n')}(q;q)_{n'-n}(q^{-2u};q)_n(1-q)^{n'-n}}$$
$$\times {}_2\phi_2\left(\begin{matrix} q^{n'-2u}, & q^{n'+1} \\ q^{n'-n+1}, & 0 \end{matrix}; q; \frac{\alpha\beta q^{u-n+1/2}}{(1-q)^2}\right)$$

$$T^{(E+,E-)}_{n'n}(\alpha,\beta)$$
$$= \frac{(q^{-2u};q)_{n'}\beta^{n'-n}q^{-(n'-n)(n'-3n-3)/4}}{q^{u(n-n')}(q;q)_{n'-n}(q^{-2u};q)_n(1-q)^{n'-n}}$$
$$\times {}_2\phi_2\left(\begin{matrix} q^{-n}, & q^{2u-n+1} \\ q^{n'-n+1}, & 0 \end{matrix}; q; \frac{\alpha\beta q^{n'-u+1/2}}{(1-q)^2}\right)$$
$$= \frac{(q;q)_n(-\alpha)^{n-n'}q^{(n-n')(n-3n'+3)/4}}{(q;q)_{n-n'}(q;q)_{n'}(1-q)^{n-n'}}$$
$$\times {}_2\phi_2\left(\begin{matrix} q^{-n'}, & q^{2u-n'+1} \\ q^{n-n'+1}, & 0 \end{matrix}; q; \frac{\alpha\beta q^{n-u+1/2}}{(1-q)^2}\right)$$

For the finite dimensional representations $D(2u)$ all of these matrix elements are polynomials. For the infinite dimensional representations the elements $T^{(e-,e+)}$ diverge for $\alpha\beta \neq 0$, the $T^{(E-,e+)}$ converge for $|\alpha\beta q^{u-n'+1/2}/(1-q)^2| < 1$, and the remaining matrix elements are entire functions. Identities (2.12) are valid for all α, β corresponding to the finite dimensional representations $D(2u)$. Using the ratio test to determine the domains of convergence corresponding to $\uparrow_u$ we find that (2.12a) converges for all α, β, (2.12b) diverges, and (2.12c), multiplied by $(-\alpha\beta q^{u-n'+1/2}/(1-q)^2;q)_\infty$, converges for all α, β.

Each of these families of matrix elements determines models of the irreducible representations $D(2u)$ and $\uparrow_u$. This is a consequence of the commutation relations (2.1). To see this we make use of the following formal power series results for linear operators X, Y:

LEMMA 1.

$$E_q(\alpha X)Ye_q(-\alpha X)=\sum_{n=0}^{\infty}\frac{\alpha^n}{(q;q)_n}[X,Y]_n$$

where

$$[X,Y]_0=Y,\quad [X,Y]_{n+1}=X[X,Y]_nq^n-[X,Y]_nX,\ n=0,1,\cdots.$$

LEMMA 2.

$$e_q(\beta X)YE_q(-\beta X)=\sum_{n=0}^{\infty}\frac{\beta^n}{(q;q)_n}[X,Y]'_n$$

where

$$[X,Y]'_0=Y,\quad [X,Y]'_{n+1}=X[X,Y]'_n-q^n[X,Y]'_nX,\ n=0,1,\cdots.$$

As a consequence of these lemmas we have

(2.14)

$$\begin{aligned} a)\ E_q(\alpha E_-)E_+ &= E_+E_q(\alpha E_-)-\frac{q^{1/2}\alpha}{(1-q)^2}\left(q^{-H}E_q(\alpha E_-)-q^HE_q(\alpha qE_-)\right)\\ b)\ e_q(\beta E_+)E_- &= E_-e_q(\beta E_+)+\frac{q^{1/2}\beta}{(1-q)^2}\left(q^{-H}e_q(\beta E_+)-q^He_q(\frac{\beta}{q}E_+)\right). \end{aligned}$$

Note also the easily verified identities

$$\begin{aligned} E_q(-\beta E_+)q^{-H}e_q(\beta qE_+)&=q^{-H},\\ E_q(-\beta qE_-)q^{-H}e_q(\beta E_-)&=q^{-H}. \end{aligned}$$

For the matrix elements (e+,e-) the operator identities

(2.15)

$$\begin{aligned} a)\ e_q(\beta E_+)e_q(\alpha E_-)E_- &= \frac{1}{\alpha}(I-T_\alpha)e_q(\beta E_+)e_q(\alpha E_-)\\ b)\ e_q(\beta E_+)e_q(\alpha E_-)E_+ &= \frac{1}{\beta}(I-T_\beta)e_q(\beta E_+)e_q(\alpha E_-)\\ &+\frac{\alpha q^{1/2}}{(1-q^2)}(q^{-u+n'}T_\beta^{-1}-q^{n'-u}T_\beta T_\alpha^{-1})e_q(\beta E_+)e_q(\alpha E_-)\\ c)\ [H,e_q(\beta E_+)e_q(\alpha E_-)] &= (\beta\partial_\beta-\alpha\partial_\alpha)e_q(\beta E_+)e_q(\alpha E_-) \end{aligned}$$

imply

$$a)\ -q[n]_q T^{(e+,e-)}_{n',n-1}(\alpha,\beta) = \frac{1}{\alpha}(I - T_\alpha) T^{(e+,e-)}_{n'n}(\alpha,\beta)$$
$$b)\ -q^{-1}[2u-n]_q T^{(e+,e-)}_{n',n+1}(\alpha,\beta) =$$
$$\left(\frac{1}{\beta}(I - T_\beta) + \frac{\alpha q^{1/2}}{(1-q)^2}(q^{-u+n'}T_\beta^{-1} - q^{u-n'}T_\beta T_\alpha^{-1})\right) T^{(e+,e-)}_{n'n}(\alpha,\beta)$$
$$c)\ (-u+n) T^{(e+,e-)}_{n'n}(\alpha,\beta) = (-u+n'+\alpha\partial_\alpha - \beta\partial_\beta)\, T^{(e+,e-)}_{n'n}(\alpha,\beta)$$

where $T_\alpha f(\alpha,\beta) = f(q\alpha,\beta)$.

For the matrix elements (e-,e+) we find

$$a)\ -q[n]_q T^{(e-,e+)}_{n',n-1}(\alpha,\beta) =$$
$$\left(\frac{1}{\beta}(I - T_\beta) + \frac{\alpha q^{1/2}}{(1-q)^2}(q^{u-n'}T_\beta^{-1} - q^{-u+n'}T_\beta T_\alpha^{-1})\right) T^{(e-,e+)}_{n'n}(\alpha,\beta)$$
$$b)\ -q^{-1}[2u-n]_q T^{(e-,e+)}_{n',n+1}(\alpha,\beta) = \frac{1}{\alpha}(I - T_\alpha) T^{(e-,e+)}_{n'n}(\alpha,\beta)$$
$$c)\ (-u+n) T^{(e-,e+)}_{n'n}(\alpha,\beta) = (-u+n'-\alpha\partial_\alpha + \beta\partial_\beta)\, T^{(e-,e+)}_{n'n}(\alpha,\beta),$$

for the matrix elements (E+,e-) we find

$$a)\ -q[n]_q T^{(E+,e-)}_{n',n-1}(\alpha,\beta) = \frac{1}{\alpha}(I - T_\alpha) T^{(E+,e-)}_{n'n}(\alpha,\beta)$$
$$b)\ -q^{-1}[2u-n]_q T^{(E+,e-)}_{n',n+1}(\alpha,\beta) =$$
(2.16)
$$\left(\frac{q}{\beta}(T_\beta^{-1} - I) + \frac{\alpha q^{1/2}}{(1-q)^2}(q^{-u+n'}T_\beta^{-1} - q^{u-n'}T_\beta T_\alpha^{-1})\right) T^{(E+,e-)}_{n'n}(\alpha,\beta)$$
$$c)\ (-u+n) T^{(E+,e-)}_{n'n}(\alpha,\beta) = (-u+n'+\alpha\partial_\alpha - \beta\partial_\beta)\, T^{(E+,e-)}_{n'n}(\alpha,\beta),$$

for the matrix elements (E-,e+) we find

$$a)\ -q[n]_q T^{(E-,e+)}_{n',n-1}(\alpha,\beta) =$$
$$\left(\frac{q}{\beta}(T_\beta^{-1} - I) + \frac{\alpha q^{1/2}}{(1-q)^2}(q^{u-n'}T_\beta^{-1} - q^{-u+n'}T_\beta T_\alpha^{-1})\right) T^{(E-,e+)}_{n'n}(\alpha,\beta)$$
$$b)\ -q^{-1}[2u-n]_q T^{(E-,e+)}_{n',n+1}(\alpha,\beta) = \frac{1}{\alpha}(I - T_\alpha) T^{(E-,e+)}_{n'n}(\alpha,\beta)$$
$$c)\ (-u+n) T^{(E-,e+)}_{n'n}(\alpha,\beta) = (-u+n'-\alpha\partial_\alpha + \beta\partial_\beta)\, T^{(E-,e+)}_{n'n}(\alpha,\beta),$$

for the matrix elements (E-,E+) we find

$$
\begin{aligned}
&a)\ -q[n]_q T^{(E-,E+)}_{n',n-1}(\alpha,\beta) = \\
&\quad \left(\frac{q}{\beta}(T_\beta^{-1}-I)+\frac{\alpha q^{1/2}}{(1-q)^2}(q^{u-n'}T_\beta^{-1}T_\alpha - q^{-u+n'}T_\beta)\right)T^{(E-,E+)}_{n'n}(\alpha,\beta)\\
&b)\ -q^{-1}[2u-n]_q T^{(E-,E+)}_{n',n+1}(\alpha,\beta) = \frac{q}{\alpha}(T_\alpha^{-1}-I)T^{(E-,E+)}_{n'n}(\alpha,\beta)\\
&c)\ (-u+n)T^{(E-,E+)}_{n'n}(\alpha,\beta) = (-u+n'-\alpha\partial_\alpha+\beta\partial_\beta)\,T^{(E-,E+)}_{n'n}(\alpha,\beta),
\end{aligned}
$$

and for the matrix elements (E+,E-) we have

$$
\begin{aligned}
&a)\ -q[n]_q T^{(E+,E-)}_{n',n-1}(\alpha,\beta) = \frac{q}{\alpha}(T_\alpha^{-1}-I)T^{(E+,E-)}_{n'n}(\alpha,\beta)\\
&b)\ -q^{-1}[2u-n]_q T^{(E+,E-)}_{n',n+1}(\alpha,\beta) = \\
&\quad \left(\frac{q}{\beta}(T_\beta^{-1}-I)+\frac{\alpha q^{1/2}}{(1-q)^2}(q^{-u+n'}T_\beta^{-1}T_\alpha - q^{u-n'}T_\beta)\right)T^{(E+,E-)}_{n'n}(\alpha,\beta)\\
&c)\ (-u+n)T^{(E+,E-)}_{n'n}(\alpha,\beta) = (-u+n'+\alpha\partial_\alpha-\beta\partial_\beta)\,T^{(E+,E-)}_{n'n}(\alpha,\beta).
\end{aligned}
$$

These relations are equivalent to q-difference relations satisfied by various q-hypergeometric series. Furthermore, it is easy to verify from the series that the relations hold also for u and n' complex. Thus we have a wide variety of two variable models of algebraically irreducible representations of $U_q(su_2)$. We note that this approach is closely related to the factorization method of quantum mechanics **[23]**.

Let X and Y be linear operators such that $YX = qXY$. A straightforward formal induction argument using this property, **[15]**, **[22, page 28]**, yields

LEMMA 3.

$$
(Y+X)^k = \sum_{\ell=0}^{k} \frac{(q;q)_k}{(q;q)_\ell (q;q)_{k-\ell}} X^\ell Y^{k-\ell},
$$
$$
e_q(X+Y)=e_q(X)e_q(Y), \quad E_q(X+Y)=E_q(Y)E_q(X).
$$

Using these results and iterating relations (2.14) we find

$$
E_q\left(\frac{\alpha\beta q^{1/2+H}}{(1-q)^2}\right)E_q(\beta E_+)e_q(\alpha E_-) = e_q(\alpha E_-)E_q(\beta E_+)E_q\left(\frac{\alpha\beta q^{1/2-H}}{(1-q)^2}\right),
$$
$$
E_q\left(\frac{\alpha\beta q^{1/2-H}}{(1-q)^2}\right)E_q(\alpha E_-)e_q(\beta E_+) = e_q(\beta E_+)E_q(\alpha E_-)E_q\left(\frac{\alpha\beta q^{1/2+H}}{(1-q)^2}\right)
$$

or

$$
E_q\left(\frac{\alpha\beta q^{1/2-u+n'}}{(1-q)^2}\right)T^{(E+,e-)}_{n'n}(\alpha,\beta) = E_q\left(\frac{\alpha\beta q^{1/2+u-n}}{(1-q)^2}\right)T^{(e-,E+)}_{n'n}(\beta,\alpha),
$$
$$
E_q\left(\frac{\alpha\beta q^{1/2+u-n'}}{(1-q)^2}\right)T^{(E-,e+)}_{n'n}(\beta,\alpha) = E_q\left(\frac{\alpha\beta q^{1/2-u+n}}{(1-q)^2}\right)T^{(e+,E-)}_{n'n}(\alpha,\beta).
$$

Identities (2.12a-c) yield orthogonality and biorthogonality relations for q-hypergeometric functions. For example, (2.12c) can be written in the form

(2.17)

$$\sum_{\ell} \frac{(zq^{\ell+1};q)_\infty(-q^{2u}z)^\ell(q^{-2u};q)_\ell}{q^{\ell(n+n')-\frac{\ell^2}{2}}} p_{n'}^{(\ell-n',2u-n'-\ell)}(zq^\ell) p_n^{(\ell-n,2u-n-\ell)}(zq^\ell)$$
$$= (zq^{1-n+2u};q)_\infty(-z)^n q^{2un-n/2-3n^2/2}(q;q)_n(q^{-2u};q)_n\delta_{nn'}.$$

Here the

$$p_n^{(\alpha,\beta)}(z) = \frac{(aq;q)_n}{(q;q)_n} {}_2\phi_1\left(\begin{matrix} q^{-n}, q^{n+1}ab \\ aq \end{matrix}; qz\right)$$
$$n = 0, 1, 2, \cdots$$

are little q-Jacobi polynomials, [**24**], where $a = q^\alpha$, $b = q^\beta$. Note that the little q-Jacobi polynomials satisfy the identity

$$\frac{(q;q)_n q^{\frac{1}{2}(n'-n)(\frac{1}{2}-n)}}{(q^{\beta+1};q)_n(-z)^n} p_n^{(n'-n,\beta)}(z) = \frac{(q;q)_{n'} q^{\frac{1}{2}(n-n')(\frac{1}{2}-n')}}{(q^{\beta+1};q)_{n'}(-z)^{n'}} p_{n'}^{(n-n',\beta)}(z).$$

Since $F_n(y) = q^{-n\ell} p_n^{(\ell-n,2u-n-\ell)}(zq^\ell)$ is a polynomial of order n in the variable $y = q^{-\ell}$, we see that expression (2.17) is an orthogonality relation for this set of polynomials. See [**17**] for similar identities associated with q-Bessel functions.

A nontrivial extension of identity (2.12c) is

$$\sum_{\ell} T_{n'\ell}^{(E-,e+)}(\alpha,\gamma) T_{\ell n}^{(E+,e-)}(\beta,-\alpha) = S_{n'n}(\gamma,\beta)$$

where the matrix elements $S_{n'n}(\gamma,\beta)$ are defined by

$$e_q(\beta E_-)E_q(\gamma E_-)f_n = \sum_{n'} S_{n'n}(\gamma,\beta) f_{n'}.$$

Explicitly,

$$S_{n'n}(\gamma,\beta) = \begin{cases} \frac{\beta^{n-n'}(q^{-n};q)_{n-n'}}{(1-q)^{n-n'}(q;q)_{n-n'}} q^{(n-n')(7+n+n')/4}(-\frac{\gamma}{\beta};q)_{n-n'} & \text{if } n \geq n' \\ 0 & \text{if } n < n'. \end{cases}$$

In the special case $\gamma/\beta = -q^{n'-n+1}$ we find the result

$$\sum_{\ell} (zq^{n'+\ell+2};q)_\infty \frac{(q^{-2u};q)_\ell}{(q;q)_\ell} (-z)^\ell q^{\ell(\frac{3}{2}+2u-n)+\ell^2/2} p_{n'}^{(\ell-n',2u-n'-\ell)}(zq^{n'+\ell+1}) \times$$
$$p_n^{(\ell-n,2u-n-\ell)}(zq^{n+\ell}) = (zq^{2u+2};q)_\infty(-z)^n q^{n(\frac{3}{2}+2u)-\frac{3}{2}n^2}(q;q)_n(q^{-2u};q)_n\delta_{nn'}.$$

(In the case of the Lie algebra of the Euclidean group in the plane, the analogous identities are the Hansen-Lommel identities for q-Bessel functions, [**25, 13, 2**].) There are similar extensions of (2.12a-b).

3. Orthogonality relations for matrix elements

The matrix elements of the form (E, e) and (e, E) satisfy orthogonality relations analogous to those arising from integration with respect to the Haar measure on $SU(2)$ in the $q = 1$ case. Indeed these relations are directly derivable from the q-difference relations for the matrix elements $T^{(E+,e-)}_{n'n}(\alpha, \beta)$ in §2. We follow the derivation in [**3**] but now make clear the connection to the representation theory of $U_q(su_2)$.

Making the change of variable $z = -\alpha\beta q^{n'-u-1/2}/(1-q)^2$ and factoring out the dependence of β in the first equation (2.13) for $T^{(E+,e-)}_{n'n}(\alpha, \beta)$ and in the recurrence relations for these expressions, we obtain the (ordinary) difference equations

$$\tau^{(a,b)} p_n^{(\alpha,\beta)} = -q^{1-n}(1 - q^{n+1}ab) p_{n-1}^{(\alpha+1,\beta+1)} \tag{3.1}$$
$$\tau^{*(aq,bq)} p_{n-1}^{(\alpha+1,\beta+1)} = -(1 - q^n) p_n^{(\alpha,\beta)}$$

where the $p_n^{(\alpha,\beta)}(z)$ are little q-Jacobi polynomials, a and b are complex numbers, $a = q^\alpha$, $b = q^\beta$ and

$$\tau^{(a,b)} = \frac{1}{z}(I - T_z), \quad \tau^{*(aq,bq)} = (z-1)T_z^{-1} + (aq - abq^2 z).$$

Let $S_{a,b}$ be the space of all polynomials in z with the complex bilinear form

$$(f_1, f_2)_{a,b} = \frac{1}{2\pi i}\oint_C f_1(z) f_2(z) w_{a,b}(z)\,\frac{dz}{z}, \quad f_1, f_2 \in S_{a,b} \tag{3.2}$$

where $C = \{z : |z| = 1 + \epsilon, \quad \epsilon > 0\}$ and $w_{a,b}$ is a weight function to be determined. We consider τ and τ^* as mappings

$$\tau^{(a,b)} : S_{a,b} \to S_{aq,bq}, \quad \tau^{*(aq,bq)} : S_{aq,bq} \to S_{a,b}$$

and look for a weight function so that τ^* is the adjoint of τ:

$$(\tau f, g)_{aq,bq} = (f, \tau^* g)_{a,b}, \quad \forall f \in S_{a,b}, \quad g \in S_{aq,bq}. \tag{3.3}$$

This yields the necessary and sufficient conditions

$$w_{aq,bq}(z/q) = \frac{z}{q}(1-z) w_{a,b}(z), \quad w_{aq,bq}(z) = aqz(1 - bqz) w_{a,b}(z)$$

with the solution, unique to within a constant multiplier,

$$w_{a,b}(z) = \frac{(\frac{z}{a}; q)_\infty (\frac{qa}{z}; q)_\infty}{(zbq; q)_\infty (\frac{1}{z}; q)_\infty (aq; q)_\infty (\frac{1}{a}; q)_\infty (-aq; q)_\infty (-\frac{1}{a}; q)_\infty}. \tag{3.4}$$

By construction $\tau^*\tau : S_{a,b} \to S_{a,b}$ is self-adjoint and

$$\tau^*\tau p_n^{(\alpha,\beta)} = \lambda_n p_n^{(\alpha,\beta)}, \quad \lambda_n = -q(1 - q^{-n})(1 - q^{n+1}ab).$$

Clearly $\lambda_n \neq \lambda_m$ if $n \neq m$ so we must have

$$(p_n^{(\alpha,\beta)}, p_m^{(\alpha,\beta)})_{a,b} = 0 \quad \text{if } m \neq n.$$

We have proved orthogonality for the little q-Jacobi polynomials. It remains to compute the normalization of these functions. Setting $f = p_n^{(\alpha,\beta)}$, $g = p_{n-1}^{(\alpha+1,\beta+1)}$ in the adjoint relation (3.3) we obtain the recurrence

$$||p_n^{(\alpha,\beta)}||^2_{a,b} = \frac{q(1-q^{n+1}ab)}{(1-q^n)}||p_{n-1}^{(\alpha+1,\beta+1)}||^2_{aq,bq}.$$

From the relation $(p_1^{(\alpha,\beta)}, p_0^{(\alpha,\beta)})_{a,b} = 0$ and the explicit expressions for p_1, p_0 we find

$$||1||^2_{a,bq} = \frac{(1-bq)}{(1-abq^2)}||1||^2_{a,b}. \tag{3.5}$$

(Here we use the fact that $(1-zbq, 1)_{a,b} = ||1||^2_{a,bq}$.)

To completely determine the norm we use the second equation (2.13) for $T^{(E+,e-)}_{n'n}(\alpha,\beta)$. Here the induced action of the recurrence operators on the basis $\{p_n^{(\alpha,\beta)}\}$ is

$$\mu^{(a,b)}p_n^{(\alpha,\beta)} = (1-aq^n)p_n^{(\alpha-1,\beta+1)}, \quad \mu^{(aq^{-1},bq)}p_n^{(\alpha-1,\beta+1)} = \frac{q^{-n}}{a}(1-bq^{n+1})p_n^{(\alpha,\beta)}$$

where

$$\mu^{(a,b)} = 1 - aT_z, \quad \mu^{(a,b)} : S_{a,b} \to S_{aq^{-1},bq},$$

$$\mu^{*(aq^{-1},bq)} = \frac{z-1}{az}T_z^{-1} + \frac{1-bqz}{az}, \quad \mu^{*(aq^{-1},bq)} : S_{aq^{-1},bq} \to S_{a,b}.$$

It is straightforward to verify that

$$(\mu f, g)_{aq^{-1},bq} = (f, \mu^* g)_{a,b}, \quad \forall f \in S_{a,b},\ g \in S_{aq^{-1},bq}, \tag{3.6}$$

so μ^* is the adjoint of μ. Now set $f = p_0^{(\alpha,\beta)} = 1$, $g = p_0^{(\alpha-1,\beta+1)} = 1$ in (3.6) to obtain

$$||1||^2_{aq^{-1},bq} = \frac{(1-bq)}{a(1-a)}||1||^2_{a,b}. \tag{3.7}$$

The solution of the recurrences (3.5) and (3.7) is

$$||1||^2_{a,b} = \frac{(abq^2;q)_\infty K(q)}{(bq;q)_\infty(aq;q)_\infty(-\frac{1}{a};q)_\infty(-aq;q)_\infty}.$$

Thus we have

$$\frac{1}{2\pi i}\oint_C \frac{(\frac{z}{a};q)_\infty(\frac{qa}{z};q)_\infty}{(\frac{1}{a};q)_\infty(\frac{1}{z};q)_\infty(zbq;q)_\infty}\,dz = \frac{(abq^2;q)_\infty K(q)}{(bq;q)_\infty}.$$

To compute $K(q)$ we make particular choices of a and b such that the integral becomes trivial:

$$a > 1+\epsilon > 1 > qa, \quad abq = 1.$$

It follows easily that $K(q) = 1/(q;q)_\infty$. Thus we have derived the normalization for the little q-Jacobi polynomials entirely by elementary techniques and proved complex orthogonality with respect to the bilinear form (3.3), (3.4).

To obtain the orthogonality of the $\{p_n^{(\alpha,\beta)}\}$ with respect to a real positive definite inner product we expand the integral expression for the complex bilinear form $(p_n^{(\alpha,\beta)}, p_m^{(\alpha,\beta)})_{a,b}$ by residues at the poles $z = q^k$, $k = 0, 1, 2, \cdots$ of the weight function. The straightforward computation yields

(3.8)
$$\int_0^1 p_n^{(\alpha,\beta)}(x)p_m^{(\alpha,\beta)}(x)\rho_{\alpha,\beta}(x)\,d_qx = \frac{(aq)^n(abq^{n+1};q)_\infty(aq;q)_n(q;q)_\infty(1-q)}{(bq^{n+1};q)_\infty(aq;q)_\infty(q;q)_n(1-abq^{2m+1})}\delta_{m,n}$$

where

$$\int_0^1 f(x)\,d_qx = (1-q)\sum_{k=0}^{\infty} f(q^k)q^k$$

and $\rho_{\alpha,\beta}(x) = x^\alpha(xq;q)_\infty/(bxq;q)_\infty$. See **[26-28]** for related results on orthogonality of q-series and associated polynomials of a discrete variable.

4. A tensor product identity for $U_q(su_2)$ representations

Given the irreducible representations $D(2u_1)$ and $D(2u_2)$ on the spaces H_{2u_1} and H_{2u_2}, respectively, we define the tensor product representation $D(2u_1) \otimes D(2u_2)$ of $U_q(su_2)$ on the space $H_{2u_1} \otimes H_{2u_2}$ by the operators **[5]**,

$$\begin{aligned} F_+ &= \Delta(E_+) = E_+ \otimes q^{\frac{1}{2}H} + q^{-\frac{1}{2}H} \otimes E_+ \\ F_- &= \Delta(E_-) = E_- \otimes q^{\frac{1}{2}H} + q^{-\frac{1}{2}H} \otimes E_- \\ L &= \Delta(H) = H \otimes I + I \otimes H. \end{aligned} \tag{4.1}$$

The operators $F_\pm$, L satisfy the same commutation relations as the operators $E_\pm$, H:

$$[L, F_\pm] = \pm F_\pm, \quad [F_+, F_-] = \frac{q^L - q^{-L}}{q^{\frac{1}{2}} - q^{-\frac{1}{2}}}. \tag{4.2}$$

From Lemma 3 we have the formal identity

$$\begin{aligned} E_q(\beta F_+)e_q(\alpha F_-) &= E_q(\beta E_+ \otimes q^{\frac{1}{2}H})E_q(\beta q^{-\frac{1}{2}H} \otimes E_+) \\ &\times e_q(\alpha E_- \otimes q^{\frac{1}{2}H})e_q(\alpha q^{-\frac{1}{2}H} \otimes E_-) \\ &= E_q(\beta E_+ \otimes q^{\frac{1}{2}H})e_q(\alpha E_- \otimes q^{\frac{1}{2}H}) \\ &\times E_q(\beta q^{-\frac{1}{2}H} \otimes E_+)e_q(\alpha q^{-\frac{1}{2}H} \otimes E_-). \end{aligned}$$

Thus,

$$T_{n'm',nm}^{(E+,e-)}(\alpha,\beta) = T_{n'n}^{(E+,e-),u_1}(\alpha q^{\frac{m'}{2}}, \beta q^{\frac{m'}{2}})T_{m'm}^{(E+,e-),u_2}(\alpha q^{-\frac{n}{2}}, \beta q^{-\frac{n}{2}}) \tag{4.3}$$

where the functions $T_{n'm',nm}^{(E+,e-)}(\alpha,\beta)$ are the matrix elements of the operator $E_q(\beta F_+)e_q(\alpha F_-)$ in the tensor product basis $\{f_n^{u_1} \otimes f_m^{u_2}\}$. In other words, for this operator the matrix elements in a tensor product basis actually factor.

It is well known that the tensor product decomposes into a direct sum of irreducible representations [**5, 29, 1**],

$$D(2u_1)\otimes D(2u_2)\cong\sum_{v=|u_1-u_2|}^{u_1+u_2}\oplus D(2v).$$

Choosing canonical orthonormal basis vectors $\{e^v_m : m=v,v-1,\cdots,-v\}$ for the irreducible subspace corresponding to the representation $D(2v)$ we define the Clebsch-Gordan coefficients by the expansion formula

$$e^v_m=\sum_{n_1,n_2}\begin{bmatrix}u_1&u_2&v\\n_1&n_2&m\end{bmatrix}_q e^{u_1}_{n_1}\otimes e^{u_2}_{n_2}. \tag{4.4}$$

A generating function for the coefficients is given by [**29, 1**]

$$\begin{aligned}
&x_2^{u_3+u_2-u_1}(\frac{x_3}{x_2}q^{u_1-u_2-u_3};q)_{u_2+u_3-u_1}x_3^{u_3+u_1-u_2}(\frac{x_1}{x_3}q^{u_2-u_3-u_1};q)_{u_3+u_1-u_2}\\
&\quad\times x_1^{u_1+u_2-u_3}(\frac{x_2}{x_1}q^{1+u_3-u_1-u_2};q)_{u_1+u_2-u_3}\\
&=(-1)^{u_3+u_2-u_1}q^{\frac12(u_1+u_2-u_3)+u_3(u_1+u_2)-\frac12(u_1^2+u_2^2+u_3^2)}\times\\
&\quad\left[\frac{(-1)^{u_1+u_2-u_3}(q;q)_{u_1+u_2-u_3}(q^{-u_1-u_2-u_3-1};q)_{u_1+u_2-u_3}}{(q^{-2u_1};q)_{u_1+u_2-u_3}(q^{-2u_2};q)_{u_1+u_2-u_3}}\right]^{\frac12}\times\\
&\sum_{n_1,n_2}x_3^{u_3-m}x_1^{u_1+n_1}x_2^{u_2+n_2}q^{\frac12[m(u_2-u_1+u_3)+n_1(u_2-u_3+u_1)+n_2(u_3-u_1+u_2+2)]}\\
&\times(-1)^{u_3+m}\left[\frac{(q^{-2u_3};q)_{u_3+m}(q^{-2u_1};q)_{u_1+n_1}(q^{-2u_2};q)_{u_2+n_2}}{(q;q)_{u_3+m}(q;q)_{u_1+n_1}(q;q)_{u_2+n_2}(-1)^{u_1+u_2-u_3}}\right]^{\frac12}\begin{bmatrix}u_1&u_2&u_3\\n_1&n_2&m\end{bmatrix}_q.
\end{aligned}$$

It follows that

$$\begin{aligned}
&\begin{bmatrix}u_1&u_2&v\\n_1&n_2&m\end{bmatrix}_q= \\
&\left[\frac{(q^{-2u_1};q)_{u_1+u_2-v}(q^{-2u_2};q)_{u_1+u_2-v}(q^{-2u_1};q)_{u_1+n_1}(q^{-2u_2};q)_{u_2+n_2}}{(q;q)_{u_1+u_2-v}(q^{-u_1-u_2-v-1};q)_{u_1+u_2-v}(q^{-2v};q)_{v+m}(q;q)_{u_2+n_2}}\right]^{\frac12}\\
&\times\frac{[(q;q)_{v+m}(q;q)_{u_1+n_1}]^{\frac12}(q^{n_1+v-u_2+1};q)_\infty}{(q;q)_\infty(q^{-2u_1};q)_{u_1+u_2-v}}\\
&\times q^{\frac12(u_1-u_2)(u_1+u_2+n_1-n_2)}q^{\frac12[-u_1^2+u_2^2+2u_1u_2-2u_2v-n_1u_1+n_2u_1-n_1u_2-n_2u_2]}\\
&\times{}_3\phi_2\left(\begin{matrix}q^{v-u_1-u_2}, & q^{u_1-u_2+v+1}, & q^{-u_2-n_2}\\ q^{v+n_1-u_2+1}, & q^{-2u_2}\end{matrix};q;q^{-v+m}\right).
\end{aligned}\tag{4.5}$$

Note that the transformation [**1**]

$$x_j\to x_j^{-1},\quad u_j\to u_j,\quad q\to q^{-1},\quad j=1,2,3$$

followed by a multiplication by $x_1^{2u_1}x_2^{2u_2}x_3^{2u_3}$ maps the generating function to

$$(-1)^{u_1+u_2+u_3}q^{2(u_1^2+u_2^2+u_3^2)-\frac12(u_1+u_2+u_3)^2+\frac32u_3-\frac12u_1-\frac12u_2}$$

times itself. This leads to the transformation formula

$$
{}_3\phi_2\left(\begin{matrix} q^{-n}, & b, & c \\ d, & e & \end{matrix}; q; \frac{deq^n}{bc}\right) = \frac{(e/c;q)_n}{(e;q)_n}\,{}_3\phi_2\left(\begin{matrix} q^{-n}, & c, & d/b \\ d, & cq^{1-n}/e & \end{matrix}; q; q\right),
$$

[22], so that the following alternate expression for the coefficients holds [4, 29, 30].

$$
\begin{bmatrix} u_1 & u_2 & v \\ n_1 & n_2 & m \end{bmatrix}_q = \tag{4.6}
$$

$$
\left(\frac{[u_1-u_2+v]![v+m]![v-m]![u_1-n_1]![2v+1]}{[u_1+n_1]![u_1+u_2-v+1]![u_1+u_2-v]![v+u_2-u_1]![u_2+n_2]![u_2-n_2]!}\right)^{\frac{1}{2}}
$$

$$
\times q^{-\frac{1}{4}(v+u_1-u_2+1)(u_1+u_2-v)+\frac{1}{2}(v-u_2)n_2-\frac{1}{2}n_1u_2}\frac{(-1)^{v-u_1-u_2}[2u_2]!}{[v-u_2-n_1]!}
$$

$$
\times{}_3\phi_2\left(\begin{matrix} q^{v-u_1-u_2}, & q^{u_1-u_2+v+1}, & q^{-u_2+n_2} \\ q^{v-u_2-n_1+1}, & q^{-2u_2} & \end{matrix}; q; q\right),
$$

provided $v = u_1+u_2, u_1+u_2-1, \cdots, |u_1-u_2|$, $n_i = u_i, u_i-1, \cdots, -u_i$ and $m = n_1+n_2$. Otherwise, the coefficients vanish. Here,

$$
[k]! = \frac{q^{-k(k-1)/4}}{(1-q)^k}(q;q)_k.
$$

Defining normalized matrix elements for the representation $D(2u)$ by

$$
\begin{aligned}
S^{(E+,e-)}_{n_1n_2}(\alpha,\beta) &= \langle E_q(\beta E_+)e_q(\alpha E_-)e_{n_1}, e_{n_2}\rangle \\
&= \left[\frac{(-1)^{n-n'}(q;q)_{n'}(q^{-2u};q)_n}{q^{(3/2-u)(n-n')}(q;q)_n(q^{-2u};q)_{n'}}\right]T^{(E+,e-)}_{n'n}(\alpha,\beta),
\end{aligned} \tag{4.7}
$$

where $n_1 = -u+n$, $n_2 = -u+n'$, we note from (4.4), (4.6) and (4.7) that the following identity, relating the matrix elements of the operator $E_q(\beta F_+)e_q(\alpha F_-)$ in two distinct orthonormal bases, must hold:

$$
\begin{aligned}
S^{(E+,e-)}_{m_1m_2,n_1n_2}(\alpha,\beta) = \sum_v &\begin{bmatrix} u_1 & u_2 & v \\ n_1 & n_2 & n_1+n_2 \end{bmatrix}_q S^{(E+,e-),v}_{m_1+m_2,n_1+n_2}(\alpha,\beta) \\
&\times\begin{bmatrix} u_1 & u_2 & v \\ m_1 & m_2 & m_1+m_2 \end{bmatrix}_q,
\end{aligned}
$$

or,

$$
\begin{aligned}
&{}_2\phi_1\left(\begin{matrix} q^a, q^b \\ q^c \end{matrix}; q; x\right) {}_2\phi_1\left(\begin{matrix} q^\alpha, q^\beta \\ q^\gamma \end{matrix}; q; q^{a+b-c}x\right) = \\
&\sum_{r=0}^{\infty} q^{r(r+c-1)} \frac{(q^a;q)_r (q^b;q)_r (q^\gamma;q)_r x^r}{(q;q)_r (q^c;q)_r (q^{c+\gamma+r-1};q)_r} \times \\
&{}_3\phi_2\left(\begin{matrix} q^{-r}, q^\alpha, q^{1-r-c} \\ q^\gamma, q^{1-r-a} \end{matrix}; q; q\right) {}_3\phi_2\left(\begin{matrix} q^{-r}, q^\beta, q^{1-r-c} \\ q^\gamma, q^{1-r-b} \end{matrix}; q; q\right) \\
&\times {}_2\phi_1\left(\begin{matrix} q^{a+\alpha+r}, q^{b+\beta+r} \\ q^{c+\gamma+2r} \end{matrix}; q; x\right).
\end{aligned} \tag{4.8}
$$

The result (4.8) is established, intitially, only for negative integer values of the parameters, corresponding to the finite dimensional irreducible representations of $U_q(su_2)$. However, using the same arguments as in the interpretation of (2.12) we can establish it for all complex values of the parameters for which both sides of the equation are well defined. This result was first derived by Groza, Kachurik and Klimyk, [**4**] who used an argument in the spirit of quantum groups. The proof here is much simpler and the interpretation of the result is somewhat different.

As a final note we remark that Biedenharn and Tarlini, [**31**], have shown how to extend the notion of tensor operators for a Lie algebra to q-tensor operators for a quantum algebra in such a way that a generalized Wigner-Eckart theorem holds. This relates to our q-algebra models.

References

1. E.G. Kalnins, H.L. Manocha and W. Miller (1992), *Models of q-algebra representations: Tensor products of special unitary and oscillator algebras*, J. Math. Phys. **33**, 2365–2383.
2. E.G. Kalnins, S. Mukherjee and W. Miller (1993), *Models of q-algebra representations: The group of plane motions*, SIAM J. Math. Anal. (to appear).
3. W. Miller, Jr. (1989), in *q-Series and Partitions, D. Stanton, ed.*, IMA Volumes in Mathematics and its Applications, Vol. 18, Springer-Verlag, New York.
4. V.A. Groza, I.I. Kachurik and A.U. Klimyk (1990), *On matrix elements and Clebsch-Gordan coefficients of the quantum algebra $U_q(SU_2)$*, J. Math. Phys. **31**, 2769–2780.
5. M. Jimbo (1985), *A q-difference analogue of $U(g)$ and the Yang-Baxter equation*, Letters in Mathematical Physics **10**, 63–69.
6. A.K. Agarwal, E.G. Kalnins, and W. Miller (1987), *Canonical equations and symmetry techniques for q-series*, SIAM J. Math. Anal. **18**, 1519-1538.
7. W. Miller (1968), *Lie Theory and Special Functions*, Academic Press, New York.
8. N. Ja. Vilenkin (1968), *Special Functions and the Theory of Group Representations*, American Mathematical Society, Providence, Rhode Island.
9. R. Floreanini and L. Vinet (1990), *q-Orthogonal polynomials and the oscillator quantum group*, INFN preprint, Trieste **AE-90/23**.
10. R. Floreanini and L. Vinet (1991), *Quantum algebras and q-special functions*, Letters Math. Physics **22**, 45–54.
11. R. Floreanini and L. Vinet (1991), *Addition formulas for q-Bessel functions*, Université de Montréal preprint, Montréal **UdeM-LPN-TH60**.
12. L. Floreanini and L. Vinet (1990), *q-Analogues of the parabose and parafermi oscillators and representations of quantum algebras*, J. Phys. **A23**, L1019–L1023.

13. H.T. Koelink (1991), *On quantum groups and q-special functions*, thesis **University of Leiden**.
14. H.T. Koelink and T.H. Koornwinder (1989), *The Clebsch-Gordan coefficients for the quantum group $S_\mu U(2)$ and q-Hahn polynomials*, Proc. Koninklijke Nederl. Akad. Wetenschappen, Series A **92**, 443–456.
15. T.H. Koornwinder (1989), *Representations of the twisted $SU(2)$ quantum group and some q-hypergeometric orthogonal polynomials*, Nederl. Akad. Wetensch. Proc. Ser **A92**, 97–117.
16. T.H. Koornwinder (1991), *The addition formula for little q-Legendre polynomials and the $SU(2)$ quantum group*, SIAM J. Math. Anal. **22**, 295–301.
17. T.H. Koornwinder and R.F. Swarttouw (1993), *On q-analogues of the Fourier and Hankel transforms group*, (to appear in Trans. American Math. Soc.).
18. T. Masuda, K. Mimachi, Y. Nakagami, M. Noumi, Y. Saburi and K. Ueno (1990), *Unitary representations of the quantum groups $SU_q(1,1)$: Structure of the dual space of $\mathfrak{U}_q(sl(2))$*, Lett. Math. Phys. **19**, 187–194.
19. T. Masuda, K. Mimachi, Y. Nakagami, M. Noumi, Y. Saburi and K. Ueno (1990), *Unitary representations of the quantum groups $SU_q(1,1)$: II. Matrix elements of unitary representations and the basic hypergeometric functions*, Lett. Math. Phys. **19**, 194–204.
20. L.L. Vaksman and L.I. Korogodskiĭ (1989), *An algebra of bounded functions on the quantum group of the motions of the plane, and q-analogues of Bessel functions*, Soviet Math. Dokl. **39**, 173–177.
21. S. L. Woronowicz (1987), *Twisted $SU(2)$ group. An example of a noncommutative differential calculus*, Publ. RIMS, Kyoto Univ. **23**, 117–181.
22. G. Gasper and M. Rahman (1990), *Basic Hypergeometric Series*, Cambridge University Press, Cambridge.
23. W. Miller (1970), *Lie theory and q-difference equations*, SIAM J. Math. Anal. **1**, 171–188.
24. G.E. Andrews and R. Askey (1985), *Classical Orthogonal Polynomials*, Lecture Notes # 1171, Springer-Verlag, New York, Berlin, 36–62.
25. H.T. Koelink (1991), *Hansen-Lommel orthogonality relations for Jackson's q-Bessel functions*, Report W-91-11 **University of Leiden**.
26. A. F. Nikiforov, S.K. Suslov, and V.B. Uvarov (1985), *Classical Orthogonal Polynomials of a Discrete Variable*, Nauka, Moscow (in Russian).
27. A.F. Nikiforov and S.K. Suslov (1986), *Classical orthogonal polynomials of a discrete variable on nonuniform lattices*, Lett. Math. Phys. **11**, 27–34.
28. E.G. Kalnins and W. Miller (1989), *Symmetry techniques for q-series: Askey-Wilson polynomials*, Rocky Mtn. J. Math. **19**, 223–230.
29. I.I. Kachurik and A.U. Klimyk (1989), *On Clebsch-Gordan coefficients of quantum algebra $U_q(SU_2)$*, Preprint, Inst. for Theor. Phys., Kiev.
30. N. Ja. Vilenkin and A. U. Klimyk (1992), *Representations of Lie Groups and Special Functions, Volume 3 (Chapter 14) (russian translation)*, Kluwer, Dordrecht, The Netherlands.
31. L.C. Biedenharn and M. Tarlini (1990), *On q-tensor operators for quantum groups*, Letters in mathematical Physics **20**, 271–278.

Department of Mathematics and Statistics, University of Waikato, Hamilton, New Zealand

School of Mathematics and Institute for Mathematics and its Applications, University of Minnesota, Minneapolis, Minnesota 55455.

E-mail address: miller@ima.umn.edu

Institute for Mathematics and its Applications, University of Minnesota, Minneapolis, Minnesota 55455.

Contemporary Mathematics
Volume **160**, 1994

MANY-ELECTRON CORRELATION PROBLEM AND LIE ALGEBRAS

JOSEF PALDUS

ABSTRACT. A brief outline and a formulation of the many-electron correlation problem, arising in finite-dimensional model calculations of atomic and molecular electronic structure, is given. It is shown how various algebraic methods, based on representation theory of relevant symmetry and/or dynamical groups, in particular of the unitary group U(n), can be employed to gain insight into the underlying formalism, as well as to design efficient algorithms and codes enabling one to obtain at least approximate solutions to these challenging many-electron model problems. Extensions and generalizations of this so-called unitary group approach (UGA), pertaining to system partitioning and/or spin dependent Hamiltonians, and various methodological developments enabling an efficient construction of matrix representatives of relevant Hamiltonians within the chosen model space, as well as applications to multi-reference coupled cluster theories, are also briefly mentioned.

1. INTRODUCTION

In contrast to subatomic systems, the quantum mechanical Hamiltonian that describes atoms, molecules and solids (at the nonrelativistic level) is explicitly known and easily written down, since only Coulomb electrostatic forces are involved. Yet, there is little hope of finding exact analytical solutions of the pertinent Schrödinger equation even for very simple atoms having two or more electrons. For molecular systems the task is naturally yet more demanding. Even when we adopt the Born-Oppenheimer approximation that freezes the nuclear framework and treat nuclear coordinates as parameters (rather than variables) in the resulting *electronic* Schrödinger equation,*) the task remains as challenging as ever. For example, considering the benzene molecule – a small molecule by an organic chemist's standards – the electronic Schrödinger equation is a linear second order partial differential equation in 126 variables. It is quite apparent that one cannot rely on standard

1991 *Mathematics Subject Classification.* 81V55, 22E70.

Supported by the Natural Sciences and Engineering Research Council of Canada

*) Born-Oppenheimer approximation represents an excellent approximation for most molecular systems, except the so-called 'floppy' molecules, whose certain vibrational modes have very low frequency. However, even in these cases, Born-Oppenheimer or adiabatic approximation provides a useful starting point in the so-called Herzberg-Teller expansion.

This paper is in final form and no version of it will be submitted for publication elsewhere.

techniques and must design simpler models if one wishes to have any hope of success in finding at least approximate solutions to this type of problem.

The model that one employs in the so-called *ab initio* calculations of molecular electronic structure[1] (and very often for the electronic structure of atomic systems as well) is obtained by truncating the infinitely dimensional Hilbert space of quadratically integrable bound state solutions of an electronic Schrödinger equation to suitable subspaces of finite dimension. Such subspaces are usually constructed as tensor powers of finite-dimensional one-electron subspaces that are in turn obtained as a linear span of a suitable choice of atomic, or atomic-like, (spin)orbitals [A(S)O's] localized on the nuclei constituting a given molecular system. Instead of physically more appropriate Slater-type AO's, one employs nowadays a larger basis of Gaussian-type AO's, or rather a linear combination of Gaussian AO's (the so-called contracted Gaussians) with suitably optimized orbital exponents and expansion coefficients. Thus, each contracted Gaussian mimics a corresponding Slater-type AO. This is done for computational reasons, since the use of Gaussian orbitals drastically simplifies the evaluation of necessary three and four center two-electron integrals (i.e., sixfold integrals involving AO's localized on three or four different nuclei).

The smallest AO basis that is required, referred to as the *minimum basis set* (MBS), consists of those AO's that constitute complete atomic shells (K, L, M, etc.) that are occupied by at least one electron in the ground state of constituting atoms. Doubling or tripling this orbital number in order to achieve greater flexibility and thus more accurate description of the electronic wavefunction one obtains the so-called *double-zeta* (DZ) or *triple-zeta* (TZ) bases. For the same reason one extends these bases by the so-called *polarization* orbitals (orbitals with higher angular momentum quantum numbers), particularly when attempting to describe excited states or, generally, those properties that depend critically on the outermost part of the molecular electronic cloud. Similarly, for the properties depending on the electronic charge distribution near the nuclei, one extends the standard basis with orbitals having large exponents and thus localized near the nuclei.

In the so-called *semi-empirical* approaches, one employs a hypothetical AO basis, regarding the corresponding one- and two-electron integrals as parameters that are calculated by using various approximations and by fitting experimental data. Thus, while the ab initio models are defined by the choice of AO's employed, which are then used to evaluate the necessary one- and two-electron integrals, the semi-empirical approaches employ these integrals as parameters. Nonetheless, the general form of the Hamiltonian employed is formally the same in both cases.

In molecular orbital (MO) type approaches, the nonorthogonal AO basis just described is then replaced by an orthonormal MO basis. Since the latter orbitals are given as linear combinations of the former ones, one refers to this approach as the MO-LCAO method. Most often, these linear combinations are determined by the Rayleigh-Ritz procedure relying on the variational principle with the trial wave function represented by a single Slater determinant (configuration, antisymmetrized product), yielding the SCF MO's, or a linear combination of configurations, resulting in the MC-SCF MO's (see below). The single determinantal ansatz for the wave function characterizes the so-called *independent particle models* (IPM's) and, when optimizing the energy, leads to the Hartree-Fock equations. For the finite-dimensional models described, these reduce to an algebraic pseudo-eigenvalue problem that is solved iteratively, until the self-consistent field (SCF) representing the averaged potential from the remaining electrons is obtained.

The IPM usually provides an excellent approximation, yielding well over 98% or even 99% of the total energy. Unfortunately, most chemical processes of interest involve only outermost electrons and are characterized by very small energy differences, representing a tiny fraction of the total energy. Thus, while the total energy may amount to hundreds or thousands of electron volts, the required chemical accuracy cannot exceed a few hundreds of an eV (i.e., 1 or 2 kcal/mol), since in many problems the difference of a few kcal/mol is crucial (e.g., the relative stability of different stereoisomers). The binding energies themselves seldom exceed a couple of hundred kcal/mol; typical H-CH bond strength being about 100 kcal/mol (i.e., $\sim$ 4 eV). Thus, even when one makes an error of only 1% in the total energy, this is entirely inadequate for chemical purposes, where very small energy differences, as compared to the total energies, decide most phenomena of interest. It is, therefore, imperative to proceed beyond the IPM. The difference between the exact nonrelativistic energy and the HF (or restricted HF) energy is referred to as the *correlation energy*, since it arises due to the correlations of motions of individual electrons that properly account for singular behavior of the Coulomb potential at the origin and that are "averaged out" in the IPM Hartree-Fock type approaches. Thus, the main emphasis in the present day computational quantum chemistry is on post-Hartree-Fock methods that can properly account for correlation phenomena and provide sufficiently accurate wave functions and energies for at least medium sized molecular systems.

We should also briefly mention the older *valence bond* (VB) approaches,[1] originated by Heitler and London,[2] in which the AO-type basis is directly employed to build N-electron states that afford simple chemical and physical interpretation (covalent and ionic structures). Although this approach has great appeal to chemists, the nonorthogonality of the AO basis presents a considerable computational challenge, which was the main reason why this approach was eclipsed by the MO based methods. Lately, however, thanks to new theoretical advances and greatly improved computational capabilities, this method is enjoying its 'rennaisance'.[1,3]

In this account we would like to indicate how one can employ certain dynamical symmetries of both MO and VB based models to facilitate the solution of the many-electron correlation problem. In contrast to classical symmetries that are characterized by the so-called *invariance*, *symmetry* or *degeneracy group* of the Hamiltonian, we shall rely on more subtle symmetries that are related with relatively recently introduced *dynamical groups* or *spectrum generating algebras*. However, we will not insist that the group employed contains all the bound states of the system within a single irreducible representation (irrep), as is done in standard dynamical group applications. In fact we shall refer to a given Lie group as a dynamical group *in a weak sense* if the Hamiltonian may be expressed in terms of its generators. Thus, the most important groups (or rather their Lie algebras) that we shall employ are unitary groups $\mathrm{U}(2^n)$ and $\mathrm{U}(n)$. The algebraic formalism based on these dynamical groups and their representation theory proved to be very useful in the design of efficient algorithms that enable one to obtain very accurate solutions for small and medium size many-electron correlation problem.

2. Problem formulation

All quantum chemical models that describe the (nonrelativistic) electronic structure of atomic and molecular systems employ a spin-independent Hamiltonian H, involving at most two-body interactions and defined on an N-electron space $\mathcal{W}_N$ given by the totally antisymmetric component (Pauli principle) of the N-th rank tensor power of a finite-dimensional subspace $\mathcal{W}$ of a general one-electron Hilbert space $\mathcal{H}$, $\mathcal{W} \subset \mathcal{H}$, i.e.

$$\mathcal{W}_N = \mathcal{W}^{\wedge N} . \tag{1}$$

Since H is spin-independent, it is convenient to represent the one-electron space $\mathcal{W}$ as a direct product of orbital ($\mathcal{V}$) and spin (Σ) spaces,

$$\mathcal{W} = \mathcal{V} \otimes \Sigma , \tag{2}$$

where $\dim \Sigma = 2$, $\Sigma = \text{Span}\{|\eta\rangle : \eta = \pm\frac{1}{2}\}$, the spin states $|\eta\rangle \equiv |\frac{1}{2}\eta\rangle$ being the spin-up ($\eta = \frac{1}{2}$) and spin-down ($\eta = -\frac{1}{2}$) eigenstates spanning the fundamental representation of SU(2). The spin orbitals $|A\rangle$ forming an orthonormal basis for $\mathcal{W}$ are thus represented by monomials

$$|A\rangle = |a\rangle \otimes |\eta\rangle , \tag{3}$$

with the (orthonormal) orbitals $|a\rangle$ spanning $\mathcal{V}$. Clearly, if $\dim \mathcal{V} = n$, then $\dim \mathcal{W} = 2n$. Although $\mathcal{V}$ is usually defined as a linear span of a nonorthogonal set of AO's, one subsequently employs an equivalent set of orthonormal MO's (e.g., HF, SCF or MC-SCF orbitals), as noted earlier.

The model Hamiltonian H is best defined using the second quantization formalism (for mathematically rigorous formulation, see Ref. 4) that associates with the spin orbital basis $\{|A\rangle\}$ a finite-dimensional Grassmann-type algebra of annihilation ($X_A \equiv X_{a\eta}$) and creation ($X_A^\dagger \equiv X_{a\eta}^\dagger$) operators defined on the Fock space $\mathcal{F} = \bigoplus_{K=0}^{2n} \mathcal{W}_K$, with $\mathcal{W}_0 \approx \mathbb{C}$ being the one-dimensional space that is isomorphic with $\mathbb{C}$ and spanned by the vacuum state $|0\rangle$. These operators fulfill the following anticommutator relations

$$\{X_A, X_B\} = \{X_A^\dagger, X_B^\dagger\} = 0, \quad \{X_A^\dagger, X_B\} = \delta_{AB} , \tag{4}$$

and satisfy the so-called vacuum property

$$X_A|0\rangle = 0, \quad (\forall A) . \tag{5}$$

The electronic model Hamiltonian then takes the form

$$H = \sum_{a,b} \langle a|\hat{z}|b\rangle \sum_{\eta} X_{a\eta}^\dagger X_{b\eta} + \tfrac{1}{2} \sum_{a,b,c,d} \langle ab|\hat{v}|cd\rangle \sum_{\eta,\tau} X_{a\eta}^\dagger X_{b\tau}^\dagger X_{d\tau} X_{c\eta} , \tag{6}$$

where $\langle a|\hat{z}|b\rangle$ and $\langle ab|\hat{v}|cd\rangle$ are the one and two electron integrals, respectively, $\hat{z}$ representing the kinetic energy and the potential due to the clamped-nuclei framework and $\hat{v}$ the interelectronic Coulomb repulsion. Since all the forces involved are spin inndependent, H commutes with total spin operators

$$[H, S^2] = [H, S_z] . \tag{7}$$

Thus, once an orthonormal basis $\{|a\rangle\}$ for $\mathcal{V}$ is chosen, we can evaluate the one and two electron integrals $\langle a|\hat{z}|b\rangle$ and $\langle ab|\hat{v}|cd\rangle$, respectively, and regard them as given parameters, just as in the semi-empirical approaches (in which, however, their number is drastically reduced using various simplifying assumptions). Our objective is then to find the eigenvalues and eigenvectors of H (generally as a function of the nuclear framework). Since the dimension of $\mathcal{W}_N$ rapidly increases with n and N, only an approximate solution of this problem is possible once n and/or N exceed certain limit (see below).

3. Basic Formalism

The most straightforward way to determine the desired eigenvectors $|\Psi\rangle$ and eigenvalues E of the particle-number preserving Hamiltonian H, Eq. (6),

$$H|\Psi\rangle = E|\Psi\rangle , \tag{8}$$

is to choose a basis $\{|\Phi_k\rangle\}$ for $\mathcal{W}_N$, calculate the matrix representative $\mathbf{H}$ of H in this basis, and diagonalize $\mathbf{H}$. Unfortunately, this procedure becomes quickly unmanageable, since $\dim \mathcal{W}_N$ rapidly increases with the orbital number n.

A substantial reduction in the dimensionality, without invoking any approximations, can be achieved by exploiting various invariant properties (or symmetries) of H. Adapting the basis for $\mathcal{W}_N$ with respect to any group or group chain $G_1 \supset G_2 \supset \cdots \supset G_r$ that leaves H invariant, reduces $\mathcal{W}_N$ into a direct sum of mutually orthogonal subspaces $\mathcal{W}_N^{(\lambda_1,\lambda_2,\dots,\lambda_r)}$ that may be uniquely labelled by the irreps of the groups G_i involved in the chain,

$$\mathcal{W}_N = \bigoplus_{\lambda_1,\lambda_2,\dots,\lambda_r} \mathcal{W}_N^{(\lambda_1,\lambda_2,\dots,\lambda_r)} . \tag{9}$$

The spatial symmetry of rigid molecular systems is often of only a limited usefulness when determining the potential energy surfaces (i.e., E as a function of molecular geometry), since it pertains to only equilibrium geometries. However, a substantial reduction for any nuclear framework geometry is achieved by exploiting the spin independence of H, Eq. (7), based on the chain SU(2) $\supset$ SU(1), when

$$\mathcal{W}_N = \bigoplus_{S,S_z} \mathcal{W}_N^{(S,S_z)} . \tag{10}$$

Even this reduction, however, is insufficient to provide subspaces of manageable dimensionality once we proceed beyond very small systems. For example, even for such a small molecule as H_2O having 10 electrons ($N = 10$), the ab initio model involving 20 orbitals ($n = 20$) has the dimension of the singlet subspace exceeding 5×10^7. Although problems of this size can be nowadays handled, thanks to modern computer technology and efficient algorithms whose underlying principles will be outlined below, it is quite evident that one is able to find exact solutions (so-called full configuration interaction or FCI results) for only relatively small model problems and that further approximations must be invoked if larger systems are to be handled.

When truncating the N-electron space, it is clearly essential that a physically suitable basis be chosen so that the truncation has the least possible effect on the accuracy of resulting energies and wave functions. To achieve this goal within a configuration interaction (CI) formalism (referred to as the shell model in nuclear physics and respresenting a simple application of the Rayleigh-Ritz variational principle), one chooses the one-electron basis (MO's) that provides the energetically best description of the state considered with a single antisymmetrized product (Slater determinant) trial wave function [so-called Hartree-Fock (HF) MO's, also referred to as self-consistent field (SCF) MO's when finite dimensional model space is employed] or, in open shell cases, using an appropriate set of quasidegenerate configurations [so-called multi-configuration or MC SCF MO's]. Thus, formally designating the single antisymmetrized product of SCF MSO's $|I_k\rangle, k = 1, \dots, N$ by $|\Phi_0\rangle$,

$$|\Phi_0\rangle = X^{\dagger}_{[I_0]}|0\rangle\,, \tag{11}$$

where $[I_0]$ designates an ordered set of these spin orbitals $[I_0] \equiv \{I_1, I_2, \dots I_n : I_1 < I_2 < \cdots < I_N\}$ and where we defined[5]

$$X^{\dagger}_{[J]} = X^{\dagger}_{J_1} X^{\dagger}_{J_2} \cdots X^{\dagger}_{J_k} \tag{12}$$

for an ordered set $[J] = \{J_1, J_2, \dots, J_k : J_1 < \cdots < J_k\}$, so that

$$X_{[J]} = X_{J_k} \cdots X_{J_2} X_{J_1}\,. \tag{13}$$

An arbitrary state $|\Psi\rangle \in \mathcal{W}_N$ then takes the form

$$|\Psi\rangle = \hat{C}|\Phi_0\rangle\,, \tag{14}$$

where $\hat{C}$ designates an appropriate wave operator,

$$\hat{C} = \sum_{k=0}^{N} \hat{C}_k\,,$$

with

$$\hat{C}_0 = c_0 \hat{1}\,, \quad \hat{C}_k = \sum_{[I''],[I']} \langle [I''] | \hat{c}_k | [I'] \rangle X^{\dagger}_{[I'']} X_{[I']}, \quad (k > 0)\,, \tag{15}$$

where $[I'] \subset [I_0]$, $[I''] \subset [I] \backslash [I_0]$, with $[I]$ designating the set of all orthonormal MSO's spanning $\mathcal{W}$, card $[I] = 2n = \dim \mathcal{W}$. The cardinality $k =$ card $[I'] =$ card $[I'']$ defines the excitation order of configurations $|\Phi_K\rangle \equiv |\Phi^{[I'']}_{[I']}\rangle = X^{\dagger}_{[I'']} X_{[I']} |\Phi_0\rangle$ spanning $\mathcal{W}_N$ [or its appropriate spin and/or symmetry adapted component $\mathcal{W}_N^{(\lambda_1, \dots, \lambda_r)}$]. The most often employed truncation scheme when employing so-called limited CI (LCI) method is by the excitation order (e.g., CI-SD method employing singly ($k = 1$) and doubly ($k = 2$) excited configurations). Although such a truncation scheme often provides very valuable results, it has basic shortcomings due to its lack of size extensivity (i.e., non-linear energy dependence on the particle number when the interactions are 'switched off'; in fact, the CI-SD wave function can be shown to be orthogonal to the FCI ground state wave function when N, $n \to \infty$). Thus, when studying extended systems (polymers), or reactive systems in which particle number in subsystems is not preserved, one often exploits so-called *coupled cluster* (CC) *approaches* that employ the exponential form for the wave operator,[6,7]

$$\hat{C} = \exp(\hat{T}), \tag{16}$$

with $\hat{T}$ representing the *connected* component of $\hat{C}$ [in the sense of many-body perturbation theory (MBPT)] and implying that $c_0 = 1$ (the so-called *intermediate normalization*). The components of the cluster operator $\hat{T}$ are determined by solving an energy-independent nonlinear system of algebraic equations that essentially represent the recurrence formulas generating selected types of MBPT diagrams. By solving these equations one automatically generates these MBPT diagrams and sums their contributions to an infinite order (assuming, of course, the solution exists). Once the operator $\hat{T}$, or rather its approximation (usually consisting of its one- and two-body components), is known, the energy is calculated using the asymmetric energy formula $E = \langle\Phi_0|He^T|\Phi_0\rangle = \langle\Phi_0|H(T_1 + T_2 + \frac{1}{2}T_1^2)|\Phi_0\rangle$, yielding a size-extensive result. This method provides excellent results for closed shell ground states.[8] However, an extension of this method to quasidegenerate or open-shell systems has been the subject of intense research and is far from being straightforward.[7,9,10]

4. Basic structure of fermionic algebra

As indicated earlier, the spin independence of the Hamiltonian, Eq. (6), can be exploited to significantly reduce the dimensionality of the model problem. For this goal one may exploit various techniques based either on the SU(2) or S_N group algebras. For small dimensions it is indeed irrelevant which method is employed. In the late sixties[11] it was realized that the same result can be accomplished using the unitary group U(n) that may be regarded as a dynamical group (in a weak sense) for an n-orbital model of many-fermion systems. In the many-electron case, one exploits the group chain[12,13]

$$\mathrm{U}(2n) \supset \mathrm{U}(n) \otimes \mathrm{U}(2). \tag{17}$$

There is, in fact, a very close relationship between the S_N and U(n) approaches,[14] since the symmetric group plays a fundamental role in the representation theory of U(n). Likewise, the link with the well-known SU(2) formalism [via the chain (17)] proved to be very useful,[15,16] particularly when employing the so-called graphical methods of spin algebras.[17] More recent work also established the fundamental role played by Clifford algebras and the related U(2^n) dynamical group, relying on the group chain[18–20]

$$\mathrm{U}(2^n) \supset \mathrm{Spin}(m) \supset \mathrm{SO}(m) \supset \mathrm{U}(n), \tag{18}$$

where $m = 2n + 1$ or $m = 2n$. All these developments led not only to a very effective and general spin adapted formalism,[12,13,16,21–23] but also provided very efficient and universal algorithm for handling of large scale many-electron correlation problems.[22,24–31] Before we briefly outline some of the ideas on which these algorithms are based, we shall indicate the origins of the unitary group dynamical symmetry for many-fermion systems.

Within the second quantization Grassmann algebra, Eqs. (4) and (5), one defines the so-called *number operators* N_I,

$$N_I = X_I^\dagger X_I \tag{19}$$

and their complements

$$\bar{N}_I = 1 - N_I = X_I X_I^\dagger , \tag{19'}$$

representing a mutually commuting set of projection operators (Hermitian idempotents)

$$N_I^2 = N_I, \quad \bar{N}_I^2 = \bar{N}_I, \quad N_I^\dagger = N_I, \quad \bar{N}_I^\dagger = \bar{N}_I , \tag{20}$$

$$[N_I, N_J] = [\bar{N}_I, \bar{N}_J] = [N_I, \bar{N}_J] = 0, \quad (\forall I, J) . \tag{20'}$$

We have also that

$$N_I \bar{N}_I = 0 , \qquad (\forall I) . \tag{20''}$$

Using these number operators, we can define the following compound projectors

$$e_{\{I\}} \equiv e_{I_1 I_2 \cdots I_r} := N_{I_1} N_{I_2} \cdots N_{I_r} \bar{N}_{I_{r+1}} \cdots \bar{N}_{I_{2n}} , \tag{21}$$

where $(I_1 I_2 \cdots I_{2n})$ is some permutation of $(1\,2 \cdots (2n))$. Note that in contrast to definition (12) we employ curly brackets to designate the index sets, $\{I\} \equiv \{I_1, I_2, \ldots, I_r\}$, in order to emphasize the abelian character of the products involved, Eq (20′). For the special case of $\{I\} = \emptyset$ we write

$$e_{\{0\}} \equiv e_0 = \bar{N}_1 \cdots \bar{N}_{2n} . \tag{22}$$

It is easily seen that $e_{\{I\}}$ represent orthogonal projectors, since

$$e_{\{I\}}^2 = e_{\{I\}}, \quad e_{\{I\}}^\dagger = e_{\{I\}}, \quad e_{\{I\}} e_{\{J\}} = \delta_{\{I\}\{J\}} e_{\{I\}} , \tag{23}$$

that provide the resolution of the identity, since

$$\sum_{\{I\}} e_{\{I\}} = \prod_{J=1}^{2n} (N_J + \bar{N}_J) = e \equiv 1 . \tag{24}$$

The fermionic algebra of creation and annihilation operators, Eq. (4), is also easily seen to be isomorphic with the Clifford algebra C_{4n}, since defining

$$\begin{aligned} \alpha_I &= X_I + X_I^\dagger , \\ \alpha_{I+2n} &= i(X_I - X_I^\dagger), \qquad (I = 1, \ldots, 2n) \end{aligned} \tag{25}$$

implies that

$$\{\alpha_J, \alpha_K\} = 2\delta_{JK}, \quad (J, K = 1, \ldots, 4n) . \tag{26}$$

We can thus construct another set of matrix units by exploiting products of creation and annihilation operators, Eqs. (12) and (13). Indeed, since

(27) $$X_{[I]}e_0 = e_0X^\dagger_{[I]} = 0\,,$$

so that

(28) $$e_0X_{[I]}X^\dagger_{[J]}e_0 = \delta_{[I][J]}e_0\,,$$

we define

(29) $$e_{[I][J]} = X^\dagger_{[I]}e_0X_{[J]}\,.$$

It follows that

(30) $$e_{[I][J]}e_{[K][L]} = \delta_{[J][K]}e_{[I][L]}\,.$$

Since also $[\bar{N}_I, X_J] = \delta_{IJ}X_J$, we have in fact that

(31) $$e_{\{I\}} = e_{[I][I]}\,,$$

so that

(32) $$e = \sum_{[I]} e_{[I][I]}\,.$$

We can in fact identify e_0 with the vacuum projector (see Ref. 5)

(33) $$e_0 = |0\rangle\langle 0|\,,$$

so that

(34) $$e_{[I][J]} = X^\dagger_{[I]}|0\rangle\langle 0|X_{[J]} = |[I]\rangle\langle[J]|\,.$$

It is useful to distinguish two classes of these projectors, namely the particle number conserving ones (card $[I]$ = card $[J]$) and nonconserving ones. Further, the former ones can be divided into the $(2n+1)$ subsets according to the cardinality of $[I]$ or $[J]$, each subset forming a subalgebra of C_{4n}. In the special case when card $[I]$ = card $[J] = 1$, we can make the identification $[I] \equiv I$, $[J] \equiv J$, so that

(35) $$e_{IJ} = |I\rangle\langle J|\,.$$

These operators multiply as standard matric units

(36) $$e_{IJ}e_{KL} = \delta_{JK}e_{IL}\,,$$

and their associative algebra may be turned into a Lie algebra by defining a Lie product as a commutator,

(37) $$[e_{IJ}, e_{KL}] = \delta_{JK}e_{IL} - \delta_{IL}e_{KJ}\,.$$

In fact, the same holds for C_{4n} [see Eq. (30)]. In the first case, Eq. (37), the resulting Lie algebra is isomorphic with the Lie algebra of GL$(2n)$ or, in view of the Hermitian property of e_{IJ}'s,

(38) $$e^\dagger_{IJ} = e_{JI} \qquad \text{or} \qquad e^\dagger_{[I][J]} = e_{[J][I]}\,,$$

with the Lie algebra of U$(2n)$, while in the latter case with U(2^{2n}).

The disadvantage of operators (29) or (34), (35) in practical applications is their high particle rank (by the particle rank of a particle-number conserving operator we understand the number of creation-annihilation pairs that are required in its second quantization representation). For this reason one introduces analogous low particle rank operators. Thus, one usually realizes the U($2n$) generators by one-particle operators

$$e^I_J = X^\dagger_I X_J \,. \tag{39}$$

These again possess the same structure constants as the generators (35),

$$\left[e^I_J, e^K_L\right] = \delta^K_J e^I_L - \delta^I_L e^K_J \,, \tag{40}$$

as well as the Hermitean property

$$(e^I_J)^\dagger = e^J_I \,. \tag{41}$$

However, they no longer form an associative algebra, since

$$e^I_J e^K_L = e^{IK}_{JL} + \delta^K_J e^I_L \,, \tag{42}$$

where

$$e^{IK}_{JL} = X^\dagger_I X^\dagger_K X_L X_J \,, \tag{43}$$

are of a particle rank two and cannot be expressed as linear combinations of one-particle operators e^I_J. Thus, while both operator sets $\{e_{IJ}\}$ and $\{e^I_J\}$ span isomorphic Lie algebras U($2n$), only the former one also forms a corresponding associative algebra.

The definition (39) may be generalized to an arbitrary particle rank,

$$e^{[I]}_{[J]} = X^\dagger_{[I]} X_{[J]} \,. \tag{44}$$

These operators, referred to as the *(spinorbital) replacement operators*,[5] as well as their spin-free counterparts (see below) play a useful role in many-electron theories. In fact, they represent a very convenient basis of the universal enveloping algebra of U($2n$) that is distinct from the standard Poincaré-Birkhoff-deWitt basis. If we allow the index sets $[II_1], [II_1I_2]$, etc. represent unordered spin orbital sets (so that the corresponding operators will differ at most by a phase from the ordered case), we can represent the replacement operators (44) in terms of the operators (29) or (34) as follows

$$e^{[I]}_{[J]} = e_{[I][J]} + \sum_{I_1}{}' e_{[II_1][JI_1]} + \sum_{I_1<I_2}{}' e_{[II_1I_2][JI_1I_2]} + \cdots \,, \tag{45}$$

where the prime indicates that the summation is restricted to spin orbitals that are not included in $[I]$ or $[J]$. This relationship, or its inverse, indicate the equivalence of operator sets $\{e_{[I][J]}\}$ and $\{e^{[I]}_{[J]}\}$ as two distinct bases of U(2^{2n}). However, since these bases are not related by a unitary transformation, the commutators of general replacement operators are not as simple as those for $e_{[I][J]}$, when

$$\left[e_{[I][J]}, e_{[K][L]}\right] = \delta_{[J][K]} e_{[I][L]} - \delta_{[L][I]} e_{[K][J]} \,. \tag{46}$$

In fact, only the one particle replacement operators e^I_J, Eq. (39), are closed with respect to the commutator product, Eq. (40), forming a basis for the spin orbital Lie algebra U($2n$). Equation (45) expresses then an important relationship between the U($2n$) and U(2^{2n}) generators that implies the imbedding of U($2n$) in U(2^{2n}) via SO($4n+1$), [for corresponding orbital chain, see Eq. (18)], that is employed in the so-called Clifford algebra unitary group approach (CAUGA).[18–20]

5. Spin adaptation

In order to exploit the spin-independence of the model Hamiltonians employed in molecular electronic structure calculations, we rely on the group chain (17). Considering the spin orbital group U($2n$) generators $e^I_J \equiv e^{i\eta}_{j\tau}$, Eq. (39), satisfying relationships (40) and (41), it is easy to realize that the corresponding partial traces

(47) $$E^i_j = \sum_{\sigma=\pm\frac{1}{2}} e^{i\sigma}_{j\sigma} = e^{i(1/2)}_{j(1/2)} + e^{i(-1/2)}_{j(-1/2)},$$

(47′) $$E^\sigma_\tau = \sum_{i=1}^{n} e^{i\sigma}_{i\tau},$$

form the generators of the corresponding orbital U(n) and spin U(2) subgroups,

(48) $$\left[E^i_j, E^k_\ell\right] = \delta^k_j E^i_\ell - \delta^i_\ell E^k_j, \qquad (E^i_j)^\dagger = E^j_i$$

(48′) $$[E^\sigma_\tau, E^\mu_\nu] = \delta^\mu_\tau E^\sigma_\nu - \delta^\sigma_\nu E^\mu_\tau, \qquad (E^\sigma_\tau)^\dagger = E^\tau_\sigma$$

that mutually commute

(49) $$\left[E^i_j, E^\sigma_\tau\right] = 0.$$

Since H, Eq. (6), is spin independent, it is expressible solely in terms of orbital generators,

(50) $$H = \sum_{a,b} \langle a|\hat{z}|b\rangle E^a_b + \tfrac{1}{2} \sum_{a,b,c,d} \langle ab|\hat{v}|cd\rangle E^{ab}_{cd},$$

where

(51) $$E^{ab}_{cd} = E^a_c E^b_d - \delta^b_c E^a_d$$

is the two-particle *orbital replacement operator*, representing a spin-free analogue of (43) [cf., also, Eq. (42)]. Thus, we can formulate our model problem within the U(n) rather than the full U($2n$), labeling the U(n) subproblems by the total spin quantum numbers S and S_z. In fact, in view of the Pauli principle, only totally antisymmetric states are admissible in our problem [see also Eq. (1)]. For an N-electron system, these are characterized by the irrep $(1^N\dot{0}) \equiv (1^N 0^{n-N})$. Since the irreps of U(2) can involve at most two-rowed Young tableaux, i.e. $(\sigma_1\sigma_2)$, it is not difficult to show that the corresponding U(n) irreps $(\lambda_1\lambda_2\cdots\lambda_n)$ can have at most two columns, since only products of mutually conjugate irreps of U(n) and U(2) can yield a single column irrep of U($2n$), Eq. (17). Since the total particle number N is conserved,

$$N = \sum_{i=1}^{n} E_i^i = \sum_{\sigma=-1/2}^{1/2} E_\sigma^\sigma = \sum_{i,\sigma} e_{i\sigma}^{i\sigma}\,, \tag{52}$$

the U(2), and the corresponding U(n), irreps are uniquely labeled by the total spin S, given by the largest possible value of S_z,

$$S_z = \tfrac{1}{2}\left(E_{+1/2}^{+1/2} - E_{-1/2}^{-1/2}\right)\,. \tag{53}$$

Designating the general U(n) irrep having at most two columns as $(2^a 1^b 0^c)$, we find easily that[12]

$$\begin{aligned} n &= a + b + c \\ N &= 2a + b \\ 2S &= b\,. \end{aligned} \tag{54}$$

Thus, for an n-orbital model of an N-electron system, the states with spin multiplicity $(2S+1)$ may be obtained by considering the carrier space of the U(n) irrep $(2^a 1^b 0^c) \equiv [a, b, c]$ with

$$a = \tfrac{1}{2}N - S, \quad b = 2S, \quad c = n - a - b\,. \tag{55}$$

This is the basis of the so-called UGA approach that will be briefly described in the next section.

Even though at most two-body operators are involved in our Hamiltonian, Eqs. (6) or (50), certain approaches employ higher than two-body operators (e.g., CC methods, where 3 or 4 body cluster components are often considered). To formulate the spin-adapted versions of such approaches, it is often convenient to employ higher than two-particle replacement operators, Eq. (51). These high-rank spin-free orbital replacement operators (also sometimes referred to as 'excitation operators'[32] or 'spin-free reduced density operators'[33]) may be defined recursively, similarly as in Eq. (51), namely

$$E_{b_1 b_2 \cdots b_k}^{a_1 a_2 \cdots a_k} = E_{b_1 b_2 \cdots b_{k-1}}^{a_1 a_2 \cdots a_{k-1}} E_{b_k}^{a_k} - \sum_{i=1}^{k-1} \delta_{b_i}^{a_k} E_{b_1 b_2 \cdots b_k \cdots b_{k-1}}^{a_1 a_2 \cdots a_i \cdots a_{k-1}}\,. \tag{56}$$

These operators are very useful in formulating various many-body theories. In fact, they represent the spin-adapted equivalents of 'normal product' ordered operators that are essential in the MBPT and CC methodologies[34] and are closely related to density matrices. We briefly list some of their useful properties:

(i) The replacement operators $E_{ij\cdots}^{ab\cdots}$ are invariant to an arbitrary simultaneous permutation of their lower and upper indices,

$$E_{b_1 b_2 \cdots b_k}^{a_1 a_2 \cdots a_k} = E_{b_{p_1} b_{p_2} \cdots b_{p_k}}^{a_{p_1} a_{p_2} \cdots a_{p_k}}\,, \tag{57}$$

where $P : i \to p_i (i = 1, \ldots, k)$ is any permutation from S_k.

(ii) The E-operators may be easily symmetry adapted, since the permutation group S_k acts naturally on a general operator (57) as on any tensor. In view of property (i), it is immaterial whether S_k acts on subscripts or superscripts. For example, for $k = 2$, we define

$$
\begin{aligned}
{}^{[2]}E^{ab}_{cd} &\equiv {}^{1}E^{ab}_{cd} = E^{ab}_{cd} + E^{ab}_{dc} \\
{}^{[1^2]}E^{ab}_{cd} &\equiv {}^{3}E^{ab}_{cd} = E^{ab}_{cd} - E^{ab}_{dc} ,
\end{aligned}
\tag{58}
$$

where we label the symmetry adapted operators either by the S_2 irrep symbols or by the corresponding spin multiplicities $(2S+1)$. For $k > 2$, the partial symmetrization with respect to S_ℓ, $\ell < k$, is often useful.[5]

(iii) The mean value of $E^{a_1a_2\cdots a_k}_{b_1b_2\cdots b_k}$ in $|\Phi\rangle$ vanishes, i.e.

$$
\langle\Phi|E^{a_1a_2\cdots a_k}_{b_1b_2\cdots b_k}|\Phi\rangle = 0 ,
\tag{59}
$$

unless all the orbitals $a_1a_2\cdots a_kb_1b_2\cdots b_k$ are occupied in $|\Phi\rangle$. The same holds for spin orbital replacement operators, where in fact

$$
\langle\Phi|E^{A_1A_2\cdots A_k}_{B_1B_2\cdots B_k}|\Phi\rangle = \Delta^{A_1A_2\cdots A_k}_{B_1B_2\cdots B_k} ,
\tag{59'}
$$

with

$$
\Delta^{A_1A_2\cdots A_k}_{B_1B_2\cdots B_k} = \sum_{P\in S_k} (-1)^p \delta^{A_1}_{Bp_1}\delta^{A_2}_{Bp_2}\cdots\delta^{A_k}_{Bp_k} ,
\tag{60}
$$

the sum extending over all permutations $P \in S_k$, $P : i \to p_i (i = 1, \dots k)$ and p designating the parity of P.

(iv) Product of replacement operators is given by the Wick's theorem (see, e.g., Ref. 34) type rule (with all contracted terms having the positive phase), e.g.

(61)

$$
\begin{aligned}
E^{abc}_{ijk}E^{de}_{\ell m} &= E^{abcde}_{ijk\ell m} + \delta^d_i E^{abce}_{\ell jkm} + \delta^d_j E^{abce}_{i\ell km} + \delta^d_k E^{abce}_{ij\ell m} + \delta^e_i E^{abcd}_{mjk\ell} + \delta^e_j E^{abcd}_{imk\ell} \\
&\quad + \delta^e_k E^{abcd}_{ijm\ell} + \delta^{de}_{ij} E^{abc}_{\ell mk} + \delta^{de}_{ji} E^{abc}_{m\ell k} + \delta^{de}_{ik} E^{abc}_{\ell jm} + \delta^{de}_{ki} E^{abc}_{mj\ell} + \delta^{de}_{jk} E^{abc}_{i\ell m} + \delta^{de}_{kj} E^{abc}_{im\ell} ,
\end{aligned}
$$

where $\delta^{ab\cdots}_{ij\cdots}$ designates the unit tensor $\delta^{ab\cdots}_{ij\cdots} = \delta^a_i\delta^b_j\cdots$. Thus, in analogy to Wick's theorem, the 5-body term corresponds to the normal product, the 4-body terms to single contractions (or pairings), and the 3-body terms to double contractions. One contracts the superscripts of the right hand side operator with the subscript of the left hand side one. This rule also gives us the commutators of replacement operators that are often required in many-body theory.

(v) The symmetry adapted operators that correspond to irreps characterized by Young diagrams having more than two columns must vanish in view of Pauli principle (this rule may be of course generalized to an arbitrary spin). For example, for at most two-column irreps, we must have that

$$
\sum_{P\in S_3} E^{a_1a_2a_3}_{bp_1bp_2bp_3} = 0 .
\tag{62}
$$

This rather trivial result seems to be equivalent[35] to Green's polynomial identities[36] for U(n) generators (for a general irrep this is only a conjecture).

(vi) Let us note, finally, that the expectation value of rank k replacement operators yields the k-th order reduced density matrices.[5,33,35]

6. Unitary group approach

We now present a brief account of the UGA,[6,13,21,23] particularly as implemented in large scale CI calculations.[24–31] As shown in the preceding section, for any n-orbital model of an N-electron system, the U(n) irrep $(2^a 1^b 0^c) \equiv [a\,b\,c]$, Eq. (55), characterizes a subspace of $\mathcal{W}_N$ with spin multiplicity $(2S+1) = (b+1)$. Thus, once we choose an appropriate basis for the carrier space of this irrep, we can construct the matrix representatives of U(n) generators and thus of the Hamiltonian H, Eq. (50).

Mathematically, the most convenient basis for U(n) irrep carrier spaces is that adapted to the Gel'fand-Tsetlin chain,[37] written schematically as

$$\mathrm{U}(n) \supset \mathrm{U}(n-1) \supset \cdots \supset \mathrm{U}(2) \supset \mathrm{U}(1)\,. \tag{63}$$

The basis vectors for a given irrep are then uniquely labeled by the triangular Gel'fand tableaux $[\mathbf{m}] = [m_{ij}]$, $(i \leq j, j = 1, 2, \ldots, n)$, with the rows $\mathbf{m}_i = (m_{1i}\, m_{2i} \cdots m_{ii})$ representing the highest weights of the irreps of the subgroups in the chain and satisfying the so-called 'betweenness conditions' of Weyl,[38] $m_{ij} \geq m_{i,j-1} \geq m_{i+1,j}$. Thus the top row $\mathbf{m}_n$ is identical in all tableau that are associated with a given irrep of U(n). Although the general formulas giving an explicit form of generator matrix representatives in this basis are known,[37,39] their complexity limits their direct applicability and usefulness in actual applications.[13] It is thus important to inspect the general structure of U(n) representations for the specific many-electron case and to design a suitable notation for this case. (In fact, this notation proved to be useful[40] even for irreps involving more than two columns in the corresponding Young or Weyl tableaux that are required in applications to systems of fermions with higher spin than one half, or in the nuclear correlation problem, where both spin and isospin must be taken into account and where, in fact, UGA originated.[11])

One immediately observes that for at most two column irreps the Gel'fand tableaux notation is greatly redundant. We thus replace the highest weight irrep labels $\mathbf{m}_i = (m_{1i}\, m_{2i} \cdots m_{ii})$, that in our case are of the form $(2^{a_i} 1^{b_i} 0^{c_i})$, by the triplet of integers $[a_i\, b_i\, c_i]$, $a_i + b_i + c_i = i$. Thus, instead of Gel'fand tableaux with $\frac{1}{2}n(n+1)$ integers, we employ the so-called ABC tableaux[6] (nowadays often referred to as Paldus tableaux[21–27,41–45]) with $3n$ integers,

$$[\mathbf{m}] = [a_i\, b_i\, c_i], \quad (i = 1, \ldots, n) \tag{64}$$

where

$$a_i + b_i + c_i = i\,. \tag{65}$$

It is often even more convenient to employ the difference ΔABC tableau $[\Delta a_i\, \Delta b_i\, \Delta c_i]$, with entries

$$\Delta x_i = x_i - x_{i-1} \quad (x = a, b, c; i = 1, \ldots, n; x_0 \equiv 0)\,, \tag{66}$$

where

$$\Delta a_i + \Delta b_i + \Delta c_i = 1 \tag{67}$$

and

$$0 \le \Delta a_i \le 1, \quad 0 \le \Delta c_i \le 1 . \tag{68}$$

We can thus conveniently label our spin-adapted states (i.e., GT basis vectors) by arrays of quaternary *step numbers* d_i,

$$d_i = 1 + 2\Delta a_i - \Delta c_i = 2\Delta a_i + \Delta \bar{c}_i = (\Delta a_i \Delta \bar{c}_i)_2 , \tag{69}$$

where the bar designates a binary complementation: $\bar{0} = 1, \bar{1} = 0$. In other words, the step numbers represent binary values of the row of the $\Delta A\bar{C}$ tableaux $[\Delta a_i\, \Delta \bar{c}_i]$ (for their direct relationship with Weyl tableaux, see Ref. 46). The orbital occupancies n_i, given by the diagonal elements of weight generators,

$$\langle [\mathbf{m}'] | E_{ii} | [\mathbf{m}] \rangle = \delta_{[\mathbf{m}'],[\mathbf{m}]}\, n_i([\mathbf{m}]) , \tag{70}$$

are equal to

$$n_i \equiv n_i([\mathbf{m}]) = \Delta a_i + \Delta \bar{c}_i = 2\Delta a_i + \Delta b_i . \tag{71}$$

Clearly,

$$0 \le n_i \le 2 . \tag{72}$$

It is thus straightforward to construct the N-electron spin-adapted basis for an n-orbital model that is associated with the U(n) irrep $[a\,b\,c]$, Eq. (55). We simply start with the highest weight $[a_n\, b_n\, c_n] = [a\,b\,c]$ and subtract at each level the four possible step-rows $[0\,0\,1]$, $[0\,1\,0]$, $[1\ {-1}\,1]$ and $[1\,0\,0]$, corresponding to the step numbers 0, 1, 2 and 3, respectively, while keeping only states with non-negative entries. The total number of these states is given by Weyl's dimension formula, which in the present case reads[12]

$$\dim[a\,b\,c] = \frac{b+1}{n+1} \binom{n+1}{a} \binom{n+1}{c} . \tag{73}$$

As already indicated, the dimension of the irreps involved rapidly increases with increasing n and N. On the other hand, there is a great deal of an identical 'local structure' when we consider a Gel'fand-Tsetlin basis as a whole. Realizing that there is only a limited number of possible irreps or rows in Gel'fand or ABC tableaux that are associated with a given irrep, we collect all identical rows $[a_i\, b_i\, c_i]$ at each level, together with four chaining indices j_{ik} $(i = 0, 1, 2, 3)$ corresponding to the four possible step numbers, and construct the so-called *distinct row table* (DRT).[21] Using this information, we can easily determine the lexical index of each ABC or $\Delta A\bar{C}$ tableau and vice versa. Thus, the entire basis set for a given irrep is encoded in the DRT, involving $3+4$ integers for each distinct row (or $3+4+4$ if also chaining indices for reverse tableaux generation are included[21]). Moreover, Shavitt[21] introduced a convenient graphical representation of the DRT through an expanded branching diagram[47] that represents the entire Gel'fand-Tsetlin basis as a two-rooted graph with vertices representing the distinct rows, and edges the step numbers. Each state or configuration vector corresponds then to a walk on this graph, extending from the top root, representing a given U(n) irrep $[a\,b\,c]$, to the bottom root, representing the fictitious zero-level irrep $[0\,0\,0]$. This global representation of the Gel'fand-Tsetlin basis is very useful in actual implementations of UGA. To indicate the efficiency of the DRT representation we note that, e.g., for the triplet states of H_2O described by a DZ-type basis ($N = 10$, $n = 20$, $S = 1$), we have $\dim[4\,2\,14] = 99, 419, 400 \approx 10^8$, while the DRT contains only 355 distinct rows (i.e., at most $\sim 4 \times 10^3$ integers). Similarly, for a TZ-type model for singlets ($N = 10$, $n = 30$, $S = 0$), in which case $\dim[5\,0\,25] = 4, 035, 556, 161$, while the number of distinct rows is 511. This enables a very efficient computer representation of even huge N-electron basis sets.

The crucial problem is then an efficient evaluation of generator matrix elements and, particularly, matrix elements of generator products that are required for the calculation of the two-electron component of the Hamiltonian [it is, in fact, more convenient to calculate directly the matrix elements of pertinent 2-body replacement operators (51)]. This problem received considerable attention of numerous researchers and the combination of various viewpoints and approaches resulted in versatile algorithms for this problem. In fact, the computation of these matrix elements is so fast that only certain precursors (so called segment values) are stored while the required matrix elements are calculated 'on the fly'. This is particularly the case in the so-called 'integral driven' approaches.

For the elementary raising (or lowering) generators E^i_{i+1} (or E^{i+1}_i), one obtains in fact simple closed expressions.[12] The remaining generator matrix elements (say the raising ones) may be, in principle, obtained from the recursion relations

$$E^i_{j+1} = [E^i_j, E^j_{j+1}]\,. \tag{74}$$

When relatively small dimensions are involved and FCI is considered, this is in fact very efficient. Unfortunately, in most realistic calculations one must employ truncated basis sets, so that this method cannot be used, since some of the required intermediate states will be missing. The same problem arises when computing matrix elements of generator products as a product of matrix representatives of individual generators (not to mention the problems of storage requirements when using this simple approach). Consequently, it is the direct evaluation of matrix elements and of their products that is called for.

The present approaches rely on the fact[15,16,21] (best seen from the SU(2) representation) that the generator matrix elements can be factored into the segment values W_r (given essentially by the SU(2) $6j$-symbols), namely

$$\langle[\mathbf{m}']|E^i_j|[\mathbf{m}]\rangle = \prod_{r\in\Omega} W_r\,, \tag{75}$$

where

$$\Omega \equiv \Omega(E^i_j) = \{k : \min(i,j) \le k \le \mathbf{m}ax(i,j)\}\,. \tag{76}$$

A similar technique may also be exploited for matrix elements of generator products, where at most two such product terms are involved (see Refs. 15,16 for details). This structure of general one- and two-body matrix elements, and the required explicit expressions for segments values, may be derived using very different approaches, ranging from a heuristic matrix element factorization,[23b] SU(2) based approaches[15,16] exploiting graphical methods of spin algebras,[17] Gould and Chandler's approach[45] based on Green's polynomial identities[36] for generators and the implied projection operators,[48] as well as tensor operator based approaches[49] that emulate a spin-adapted form of the second quantization formalism. Similar approaches may also be exploited in the S_N based formalism.[14,15] In fact, the mutual 'cross-fertilization' of various formulations proved to be most beneficial, since each offers a distinct viewpoint revealing different aspects of the complex algebraic structure involved. It is, however, beyond the scope of this overview to describe these various approaches in any detail. We only mention that many of these approaches may be extended to handle the system partitioning,[19,47,49,50] when one employs a basis set adapted to the chain

(77)
$$\mathrm{U}(n_1+n_2+\cdots+n_r)\supset \mathrm{U}(n_1)\otimes \mathrm{U}(n_2+\cdots+n_r)\supset\cdots\supset \mathrm{U}(n_1)\otimes\cdots\otimes \mathrm{U}(n_2)\,,$$

or spin-dependent Hamiltonians,[51,52] when we need to compute matrix elements of general spin-orbital generators $e^{i\sigma}_{j\tau}$ in the $\mathrm{U}(n)\otimes\mathrm{U}(2)$ basis, etc.

7. Clifford algebra UGA

We have seen earlier the fundamental role played by Clifford algebras in finite-dimensional quantum-chemical models of many-electron systems (Section 4). From numerous possibilities offered by the CAUGA chain, Eq. (18), we shall only briefly consider the final imbeding $\mathrm{U}(2^n)\supset\mathrm{U}(n)$ (for the role of intermediate groups in the chain (18) and the uniqueness of the state labeling problem, see Refs. 20 and 53).

Designating now the $\mathrm{U}(n)$ generators by Λ^i_j in order to distinguish them from $\mathrm{U}(2^n)$ generators E^p_q, both satisfying the same commutation relations, we employ the spanning set of Clifford algebra monomials that are uniquely labeled by the orbital occupation numbers m_i, $m_i=0$ or 1, $(i=1,\ldots,n)$, to obtain a realization of $\mathrm{U}(2^n)$ generators E^p_q. Exploiting Eq. (35) we thus write

$$E^p_q = |p)(q|\,, \tag{78}$$

where the 'multi-particle' states $|p)$ are in a one-to-one correspondence with the orbital Clifford algebra C_{2n} monomials $(m_1\,m_2\cdots m_n)$ just mentioned. A simple and convenient labeling of these monomials is achieved by interpreting the occupation number arrays as binary integers, defining

$$p\equiv p\{m_i\} = 2^n-(m_1\,m_2\cdots m_n)_2\,. \tag{79}$$

Thus, E^p_q represent the orbital analogues of spin orbital 'multi-particle' generators $e_{[I][J]}$ used earlier.

With this notation it is now easy to show[18] that we can represent any $\mathrm{U}(n)$ generator Λ^i_j as a simple linear combination of $\mathrm{U}(2^n)$ generators E^p_q, all coefficients being $+1$ or -1 (for elementary generators, all coefficients are $+1$),

$$\Lambda^i_j = (-1)^{j-i+1}\sum_{p,q,r}(-1)^{S_2(q)}\,E^{p+r}_{p+r'}\,,\qquad (i\neq j) \tag{80}$$

with

$$r \equiv \tau_{ij}(r,q), \quad r' \equiv \tau_{ji}(q,r)$$

(81) $$\tau_{k\ell}(u,v) = 2^{n-\ell}(2^{\ell-k+1}u + 2v + 1)\,,$$

$$1 \le p \le 2^{n-j}, \quad 0 \le q \le 2^{j-i-1} - 1, \quad 0 \le r \le 2^{i-1} - 1\,,$$

and $S_2(q)$ designates the digital sum of q_2 (the binary representation of q). For the weight generators we get simply

(82) $$\Lambda_i^i = \sum_{p=1}^{2^n} m_i^p E_p^p\,,$$

m_i^p being the occupation number of the i-th orbital in the p-th monomial. It is important to note that while the above representation contains, generally, 2^{n-2} generators, only

(83) $$\left[\begin{pmatrix} n-2 \\ a-1 \end{pmatrix} + \begin{pmatrix} n-2 \\ c-1 \end{pmatrix}\right] \frac{1}{1+\delta_{b,0}}$$

of these are needed when considering the U(n) irrep $[a\,b\,c]$.

The great advantage of this representation lies in the fact that any U(n) spin adapted state can be represented as a linear combination of U(2^n) two-box states constituting totally symmetric irrep $(2\dot{0})$. Even though the U(2^n) $\downarrow$ U(n) subduction is not generally multiplicity free, the subgroup chain (18) can provide additional labels.[20,53] (Moreover, the SO($2n$) group involves two-body generators of the type $X_I^\dagger X_J^\dagger$ or $X_I X_J$ and SO($2n$+1) group also single particle creators and annihilators, so that they may be employed even for particle-number nonconserving systems.) In fact, this property holds in general, since one can show[20,53,54] that any p-column irrep of U(n) is contained at least once in the totally symmetric p-box irrep $(p\,\dot{0})$ of U(2^n). Since the matrix elements of E_q^p generators within the totally symmetric U(2^n) irreps are easily obtained, the problem of calculating the U(n) representations has thus been greatly simplified. For example, in the 2-box case that is relevant to many electron problem , we simply have

(84) $$E_q^p(i|j) = \delta_{qi}(p|j) + \delta_{qj}(i|p)\,,$$

where we employed the unnormalized states

(85) $$(i|j) = \sqrt{1+\delta_{ij}}[i|j]\,,$$

with $[i|j]$ designating the normalized two-box states of $(2\,\dot{0})$ irrep of U(2^n), $[i|j] \equiv |\boxed{i|j}\rangle$.

In contrast to standard UGA, which is based on Gel'fand-Tsetlin states that correspond to Yamanouchi-Kotani coupling scheme, the CAUGA is not associated with any particular coupling scheme. In fact, using directly the orthonormal two-box tableau $[i|j]$, one obtains a partially spin-adapted scheme[31] that is equivalent to using S_z-adapted Waller-Hartree bi-determinantal states.[55] These have been successfully exploited in FCI-type codes that can nowadays handle problems involving up to $\sim 10^8$ such states.[28,31] Although these highly vectorized codes are essentially based on classical Slater determinant-like N-electron basis functions, and thus could have been, in principle, implemented a long time ago (at least in a non-vectorized version), it was not until the UGA was well in place, and the formulation of the problem in terms of unitary group generators became widely employed, that these codes materialized. Moreover, the most efficient version of these codes, in which the number of intermediate states is substantially reduced, was inspired by the CAUGA formalism.[31]

Another interesting feature of the CAUGA scheme is the fact that it naturally leads to the nonorthogonal Weyl-Rumer basis that is characteristic of the *valence bond* (VB) method,[1] initiated by Heitler and London's pioneering work.[2] In fact, the VB basis naturally arises through a canonical construction that may be best illustrated in terms of the so-called harmonic level excitation diagram (HLED) scheme:[56] starting from the highest (lowest) weight state, representing the zeroth level of the diagram, we successively apply to each state all the elementary lowering (raising) generators, starting always with Λ^n_{n-1}, until the entire basis is generated (unless we truncate the procedure as in the LCI approaches). The ordering of the resulting Weyl-Rumer states (also called Clifford-Weyl states when orthonormal one-electron basis is employed[57]) is such that their Schmidt orthogonalization yields Gel'fand-Tsetlin basis (that may also be obtained directly using the $\mathrm{U}(n)$ Clebsch-Gordan coefficients[58] or S_N outer product reduction coefficients[19]). There is also a direct link of the HLED scheme[59] with the so-called Verma bases that were explored by Li *et al.*[60] Thus, the CAUGA scheme can be very useful in implementing VB-type methodology,[61,62] although the nonorthogonality problems still represent a computational bottleneck in spite of the progress made in recent years.[63]

To conclude this sketchy outline of CAUGA we summarize its main features and advantages:

(i) CAUGA effectively reduces an N-electron problem to a number of two-boson problems;

(ii) CAUGA enables an exploitation of an arbitrary coupling scheme, being particularly suited for VB methods (recall that UGA is restricted to a rather unphysical Gel'fand-Tsetlin basis);

(iii) CAUGA can be applied to particle-number non-conserving operators;

(iv) CAUGA drastically simplifies the algorithm for the evaluation of matrix representatives of arbitrary generators or their products;

(v) CAUGA easily extends to systems of fermions with an arbitrary spin (i.e., to multi-column $\mathrm{U}(n)$ irreps);

(vi) CAUGA is amenable to a general point group symmetry adaptation and system partitioning;

(vii) CAUGA can exploit many UGA features, such as the DRT concept, so that it can be regarded as an extension, rather than a replacement, of UGA.

8. Valence universal coupled cluster ansatz

In order to indicate the general applicability of UGA and/or CAUGA formalism, we briefly indicate how it relates to CC methodology. As mentioned earlier, single-reference CC methods exploit the exponential ansatz for the wave operator, Eq. (16), embodying the connected and linked cluster theorems (see, e.g., Ref. 7) and yielding the desired size-extensivity of the resulting energies. This method, in several of its variants, proved to be highly successful in describing nondegenerate closed-shell ground states and is nowadays extensively used[8] [one of its variants is also referred to as quadratic CI (QCI)[64,65]]. Unfortunately, an extension of this methodology to quasidegenerate or degenerate open-shell systems is far from straightforward and the past two decades witnessed numerous controversies and pitfalls in this regard.

It is well beyond the scope of this article to provide even a cursory outline of the underlying theory (effective Hamiltonian formalism, Bloch equations, etc.) or concepts (valence universality *vs.* state universality, intruder state problem, etc.). We will thus only attempt to indicate the essence of possible generalizations of the single-reference CC ansatz, Eq. (16), to the multi-reference (MR) case and the role played by UGA and/or CAUGA formalisms. In contrast to the single-reference case, when the (spin) orbitals are naturally split into the occupied (I_0) and unoccupied or virtual ($I \backslash I_0$) subsets, once the reference state $|\Phi_0\rangle$ is chosen, Eq. (11), the MR approaches require a partitioning of the one-electron space into three mutually orthogonal subspaces. This is achived by partitioning the MO (or spinorbital) basis into the core (C), active or valence (A) and excited or virtual (E) sets. The N-electron model space $\mathcal{M}_0$ is then defined as a linear span of configurations with all n_C = card (C) core spin orbitals occupied, none of the excited spin orbitals occupied and $(N - n_C)$ active spin orbitals occupied in all possible ways.

The great step forward in the development of MR-CC theories proved to be Lindgren's postulate of the normal product cluster ansatz.[66,67] Nonetheless, a considerable effort was required before a deeper understanding of the so-called valence universality, implied by this ansatz, was achieved. Of equal importance was the state universal ansatz due to Jeziorski and Monkhorst.[68] The valence universality essentially implies (see Refs. 7,10,69-71) that the whole class of model spaces $\mathcal{M}_0^{(n)}$ with $n = 0, 1, \ldots, n_A$ valence electrons must be simultaneously considered, n_A = card (A), $\mathcal{M}_0^{(n)} \subset \mathcal{W}_{(n_C+n)}$, and that there exists[71] a *unique* valence universal wave operator Ω from the so-called model algebra $\mathcal{A}$ (see below), whose restriction to $\mathcal{M}_0^{(n)}$ yields appropriate wave operators U_n for $\mathcal{M}_0^{(n)}$, $n = 0, 1, \ldots, n_A$. In practical terms this implies that in the so-called valence universal or Fock space CC approaches one has to consider, along with the studied system, also its various ions, starting with the system involving only core electrons, while in the state universal or Hilbert space approaches, only a given system with a fixed electron number need be considered. However, while the Fock space approaches rely on a single valence universal wave operator Ω, the Hilbert space approaches require a separate wave operator for each reference configuration.

To achieve a rigorous formulation of Fock space CC approaches and of the underlying valence universality theorem,[71] it is advantageous to employ the formalism based on replacement operators (see Section 4), the latter providing a convenient basis for the universal enveloping algebra of U(n), as noted earlier. To simplify the notation, we shall designate the spin orbitals belonging to the C, A and E subsets by the lower case letters from the beginning $(a, b, c, d, \ldots)$, the middle $(i, j, k, \ell, \ldots)$ and the end $(r, s, t, u, \ldots)$ of the Latin alphabet, respectively. Further, we shall use lower case Greek letters as follows: $\alpha, \beta, \gamma, \delta \in C \cup A$; $\rho, \sigma, \tau, \omega \in A \cup E$ and, finally, $\lambda, \kappa, \mu, \nu \in C \cup A \cup E$ for the generic labels. With this notation we define[71] a *model algebra* $\mathcal{A}$ as the operator algebra generated by the identity ($e \equiv \hat{1}$) and one-body replacement generators of the type e^{ρ}_{α}, using an ordinary product. Recalling the basic properties of replacement operators, the following set provides a basis for $\mathcal{A}$,

$$\{e,\, e^{\rho}_{\alpha},\, e^{\rho\sigma}_{\alpha\beta},\, e^{\rho\sigma\tau}_{\alpha\beta\gamma}, \dots\}\,. \tag{86}$$

In contrast to the closed shell case, when $n_A = 0$, $\mathcal{A}$ is not abelian, although $\mathcal{A}$ still leaves $\mathcal{W}_N$ invariant, since it only contains the particle number preserving operators.

We observe that only active labels can appear both as sub- and super-scripts on the replacement operators constituting $\mathcal{A}$. Consequently, we have that[71]

$$\mathcal{A} = \mathcal{A}_{\text{int}} \oplus \mathcal{A}_{\text{ext}}, \quad \mathcal{A} = \text{Span}\{\mathcal{A}_{\text{int}}, \mathcal{A}_{\text{ext}}\}, \quad A_{\text{int}} \cap A_{\text{ext}} = \emptyset\,, \tag{87}$$

where $\mathcal{A}_{\text{int}} = \text{Span}\{e,\, e^{i}_{j},\, e^{ij}_{k\ell}, \dots\}$ and $\mathcal{A}_{\text{ext}} = \text{Span}\{e^{\rho}_{\alpha},\, e^{\rho\sigma}_{\alpha\beta}, \dots\}$ with each replacement operator carrying at least one nonvalence label. In fact, $\mathcal{A}_{\text{ext}}$ is a two-sided ideal of $\mathcal{A}$ and maps $\mathcal{M}_0$ into $\mathcal{M}_0^{\perp}$, while $\mathcal{M}_0$ is invariant under $\mathcal{A}_{\text{int}}$. Moreover, every element of $\mathcal{A}_{\text{ext}}$ is nilpotent. All these algebras can also be considered as Lie algebras with commutator taken as a Lie product.

In order to show the connection with the standard second quantization based formalism, we define another algebra by replacing an ordinary product with a commutative 'normal product' that we designate by an asterisk, namely[71]

$$e^{\rho}_{\alpha} * e^{\sigma}_{\beta} = e^{\rho\sigma}_{\alpha\beta}, \qquad e^{\rho\sigma}_{\alpha\beta} * e^{\tau}_{\gamma} = e^{\rho\sigma\tau}_{\alpha\beta\gamma}, \text{ etc.} \tag{88}$$

We observe that a $*$-product of p- and q-body replacement operators (that are in the normal product form with respect to the the true vacuum $|0\rangle$[34]) yields a $(p+q)$-body operator that is again in the normal product form. It can then be shown, that the desired generalized valence universal wave operator Ω takes the form

$$\Omega = \mathbf{e}^{S}\,, \tag{89}$$

where $\mathbf{e}$ designates the 'normal ordered' exponential function defined as follows

$$\mathbf{e}^{X} = \hat{1} + X + \frac{1}{2!} X * X + \frac{1}{3!} X * X * X + \cdots\,, \tag{90}$$

with $X \in \mathcal{A}$, and $S \in \mathcal{A}_{\text{ext}}$ designates the valence universal cluster operator. Since $S \in \mathcal{A}_{\text{ext}}$, the series defining $\mathbf{e}^{S}$ terminates, and $\Omega = \hat{1} + \hat{\Omega}, \hat{\Omega} \in \mathcal{A}_{\text{ext}}$.

In closing we note that the concept of the model algebra enables a rigorous formulation of the valence universality theorem for general operators[71] and provides a useful and efficient formalism for the derivation of working equations for various MR-CC methods.[5,10,71] Moreover, when using the orbital versions of replacement operators,[5,32] the spin-adapted formulation of the theory may be achieved. We finally note that the UGA formalism is also indispensable for CC approaches employing a single but multiconfigurational reference state.[72–74] These approaches represent very practical, though in general only approximately size-extensive, methods.

In this very incomplete account of an algebraic approach to the many-electron correlation problem that is based on the unitary group U(n) and U(2^n) representation theory, we tried to convey the simplicity, elegance, power and universality of this approach as well as its usefulness and power in designing the required methodology. The versatility of UGA is perhaps best exemplified by the fact that it admits very diverse computational implementations, so far primarily exploited in large scale CI packages. These various computer codes exploit the integral driven,[24,30] loop-driven,[25a,75] split graph,[76] shape[25b] or interaction-block driven[75] UGA as well as matrix element driven approaches[29] based on the above mentioned HLED scheme.[56] An interesting implementation of UGA in atomic shell [or weak ligand field] theory, exploiting the Racah group chain $U(7) \supset SO(7) \supset G_2 \supset SO(3)$ [$\supset$ G], was recently put forward by Wen *et al.*[77]

The advancement of UGA also led[24b] to the development of partially spin-adapted FCI algorithms (preserving S_z but not S^2) employing Slater determinantal (or, rather, Waller-Hartree double-determinantal[55]) basis that are particularly suitable for modern vector processors.[28,78] The insight afforded by the CAUGA chain, Eq. (18), led to additional significant improvements.[31] With a massive investment in supercomputer resources (particularly very large memory), these types of code are capable of surpassing the one-billion limit in FCI calculations as demonstrated by Olsen *et al.*,[79] while computations involving 10^7 to 10^8 determinants may be carried out on mini-supercomputers. Although the FCI calculations are limited to relatively small systems (8-10 electrons with 30-20 orbitals), they provide useful benchmarks for testing various approximate theories (cf., e.g., Refs. 80 and 81). For general applications, the design of suitable truncation schemes is essential.[82]

The above-mentioned multitude of existing UGA based CI programs (at both the FCI and MR CI-SD levels) is nowadays exploited in various practical applications. It is well beyond the scope of this report to review these numerous and diverse exploitations. We shall thus restrict ourselves to a biased list of a few illustrative examples that have been carried out during the last few years and that employ either the original or improved loop-driven code by Brooks and Shaefer[25a] or the monumental COLUMBUS Program System.[30] The former ones, carried out primarily by the Strasbourg group, (i) explored the electron density in iron (II) porphyrin bis(water), establishing ^5E_g as the ground state of iron (II) tetraphenylporphyrin bis (tetrahydrofuran),[83] (ii) determined all electronic states of Na_3^+ when dissociating into $2Na(3s) + Na^+$ and elucidated the role of the static Jahn-Teller effect in this process,[84] (iii) analyzed chemical bonding in the collinear SrSC molecule in its lowest $^4\Sigma^-$, $^4\Pi$, $^4\Delta$ and $^4\Phi$ states[85] and in $CrCO_2$ and $TiCO_2$,[86] (iv) studied the appropriateness of widely used pseudopotentials for group 1 (K, Rb, Cs), 2(Ca, Sr, Ba) and 8(Fe, Ru, Os) atoms[87] and, finally, (v) examined high-spin chemical bonding in the MCS *vs* MCO molecules with M = Sc, Ti, V, Cr and Cu.[88] The COLUMBUS Program System (we concentrate on papers exploiting its CI component) was employed, among others, to study (i) the protonation and deprotonation energies of NH_3, H_2O, HF, PH_3, H_2S, HCl and HCN,[89] (ii) the reaction pathways for the $CH_2 + H \rightleftharpoons CH_3{}^* \rightleftharpoons CH + H_2$ reaction,[90] (iii) the electronic structure of undoped and doped copper oxides that play an important role in high-T_c superconductors,[91] (iv) the symmetry breaking in the first excited state of *trans*-N_2H_2,[92] (v) the molecular structure of neutral, protonated and deprotonated trimers of HF and HCl[93] and (vi) the electronic structure of the diuranium[94] and AlCu[95] molecules. Another interesting application involved the interpretation of resonance-enhanced multiphoton ionization (REMPI) mass spectra of the OH radical in the gas phase[96] and the determination of the *ab initio* potential energy surface for the dynamical calculations of photodissociation of FNO in the first absorbtion band $S_0 \to S_1$.[97] Recent applications enabled the challenging problem of determining accurate electron affinities[98] and electronic and molecular structure of various products of the $Sc^+ + H_2S$ reaction[99] to be addressed.

However incomplete and biased this list of various applications may be, it should provide ample illustration of the usefulness and versatility of various UGA and CAUGA based codes for solving diverse and topical problems in chemistry and molecular physics.

In addition to highly efficient algorithms that enabled the design of the above-mentioned codes for large-scale CI calculations,[24–30,75–77] the UGA formalism proved also useful in electron propagator theory,[100] the CC approaches,[5,23,44,71–74,101–104] many-body perturbation theory,[29,44] MC SCF approaches,[30b,105] and nuclear magnetic resonance spectroscopy.[106] A very important extension of UGA based CI methodology pertains to the computation of analytic energy gradients with respect to nuclear coordinates, enabling an efficient exploration of reaction paths and a determination of local minima on multidimensional potential energy hypersurfaces,[107] recently also implemented in the COLUMBUS Program System.[108] Another interesting development was a formulation of an efficient method for calculating molecular radiative intensities in the vacuum UV and soft X-ray wavelength regions.[109] The UGA formalism was also exploited in calculations of frequency dependent polarizabilities and van der Waals coefficients using time-dependent coupled Hartree-Fock theory[110] and of second order properties, particularly the two-photon cross sections and absorbtion coefficients.[111] A UGA-type formalism was also employed to study the partitioning of the spin-independent Hamiltonian for open-shell systems providing a simple derivation of the generalized Brillouin theorem, Hartree-Fock equations and random phase approximation.[112]

Lately, much work was also carried out for applications to spin-dependent Hamiltonians as required when handling the relativistic effects[52,113,114] (particularly the spin-orbit interaction[51]). We also mention the possibility of exploiting the parastatistics[53,115] (with para-fermions of order 2) and recent applications to density matrices.[35,116] The most sorely needed progress is required in handling of nonorthogonal basis sets[117] and a simpler and more straightforward handling of point group symmetries,[118] whenever present, would also be helpful. There are considerable possibilities in automating the algebraic derivations using suitable symbolic languages software, particularly in CC type applications to complex open-shell systems.[72,119,120]

For a more detailed account of UGA, CAUGA and related algebraic approaches, the interested reader is encouraged to survey some existing reviews on this topic.[14,22,23,30b,41–44,46,56,121]

ACKNOWLEDGEMENTS

The author wishes to express his sincere thanks and gratitude to Professors N. Kamran and P. Olver for giving him the opportunity to take part in their most stimulating Special Session of the 873rd AMS Meeting. Continued support by NSERC is also gratefully acknowledged.

REFERENCES

1. For basic concepts and terminology, see e.g. R. McWeeny, *Methods of Molecular Quantum Mechanics*, 2nd edition, Academic Press, London, 1989; A. Szabo and N.S. Ostlund, *Modern Quantum Chemistry: Introduction to Advanced Electronic Structure Theory*, McGraw Hill, New York, 1989; G. Náray-Szabó, P.R. Surján and J.G. Ángyán, *Applied Quantum Chemistry* (Reidel, Dordrecht, 1987); R. Zahradník and R. Polák, *Elements of Quantum Chemistry*, Plenum, New York, 1980; H. Primas and U. Müller-Herold, *Elementare Quantenchemie*, Teubner, Stuttgart, 1984.
2. W. Heitler and F. London, Z. Phys. **44** (1927), 455.
3. D.L. Cooper, J. Gerratt and M. Raimondi, Adv. Chem. Phys. **69** (1987), 319.
4. F.A. Berezin, *The Method of Second Quantization*, 2nd edition, Nauka, Moscow, 1986 (in Russian); English translation of the 1st edition was published by Academic Press, 1966.
5. J. Paldus and B. Jeziorski, Theor. Chim. Acta **73** (1988), 81.
6. J. Čížek, J. Chem. Phys. **45** (1966), 4256; Adv. Chem. Phys. **14** (1969), 35.
7. J. Paldus, in '*Methods in Computational Molecular Physics*', S. Wilson and G.H.F. Diercksen, eds., NATO ASI Series, Plenum, New York, 1992, pp. 99-194.
8. R.J. Bartlett, Annu. Rev. Phys. Chem. **32** (1981), 359; J. Phys. Chem. **93** (1989), 1697, and references therein.
9. D. Mukherjee and S. Pal, Adv. Quantum Chem. **20** (1989), 292.
10. J. Paldus, in '*Relativistic and Correlation Effects in Molecules and Solids*', G.L. Malli, ed., NATO ASI Series, Plenum, New York, 1993, in press.
11. M. Moshinsky, *Group Theory and the Many-Body Problem*, Gordon and Breach, New York, 1968; M. Moshinsky and T.H. Seligman, Ann. Phys. (NY) **66** (1971), 311.
12. J. Paldus, J. Chem. Phys. **61** (1974), 5321.
13. J. Paldus in '*Theoretical Chemistry: Advances and Perspectives*', Vol. 2, D. Henderson and H. Eyring, eds., Academic Press, New York, 1976, pp. 131-290.
14. W. Duch and J. Karwowsky, Comp. Phys. Rep. **2** (1985), 93; W. Duch, J. Mol. Structure (Theochem) **234** (1991), 27; J. Karwowsky, in '*Methods in Computational Molecular Physics*', S. Wilson and G.H.F. Diercksen, eds., NATO ASI Series, Plenum, New York, 1992, pp. 65-98.
15. G.W.F. Drake and M. Schlesinger, Phys. Rev. A**15** (1977), 1990.
16. J. Paldus and M.J. Boyle, Phys. Scr. **21** (1980), 295.
17. E. El Baz and B. Castel, *Graphical Methods of Spin Algebras in Atomic, Nuclear and Particle Physics*, Dekker, New York, 1972 and references therein.
18. J. Paldus and C.R. Sarma, J. Chem. Phys. **83** (1985), 5135.
19. J. Paldus, M.J. Gao and J.Q. Chem, Phys. Rev. A**35** (1987), 3197.
20. C.R. Sarma and J. Paldus, J. Math. Phys. **26** (1985), 1140.

21. I. Shavitt, Int. J. Quantum Chem., Symp. **11** (1977), 131; Symp. **12** (1978), 5.
22. J. Hinze, (ed.), *The Unitary Group for the Evaluation of Electronic Energy Matrix Elements*, (Lecture Notes in Chemistry, Vol. 22), Springer-Verlag, Berlin, 1981.
23. (a) J. Paldus, in '*Mathematical Frontiers in Computational Chemical Physics*', IMA Series, Vol. 15, D.G. Truhlar, ed., Springer-Verlag, Berlin, 1988, pp. 262-299; (b) I. Shavitt, *ibid.*, pp. 300-349.
24. (a) P.E.M. Siegbahn, J. Chem. Phys. **70** (1979), 5391; **72** (1980), 1647; (b) Chem. Phys. Lett. **109** (1984), 417.
25. (a) B.R. Brooks and H.F. Schaefer, J. Chem. Phys. **70** (1979), 5092; (b) P. Saxe, D.J. Fox, H.F. Schaefer and N.C. Handy, J. Chem. Phys. **77** (1982), 5584.
26. H. Lischka, R. Shepard, F. Brown and I. Shavitt, Int. J. Quantum Chem., Symp. **15** (1981), 91.
27. V.R. Saunders and J.H. van Lenthe, Mol. Phys. **48** (1983), 923.
28. P.J. Knowles and N.C. Handy, Chem. Phys. Lett. **111** (1984), 315.
29. M.J. Downward and M.A. Robb, Theor. Chim. Acta **46** (1977), 129; D. Hegarty and M.A. Robb, Mol. Phys. **38** (1979), 1795; in Ref. 22, pp. 100-107.
30. (a) R. Shepard, I. Shavitt, R.M. Pitzer, D.C. Comeau, M. Pepper, H. Lischka, P.G. Szalay, R. Ahlrichs, F.B. Brown and J.-G. Zhao, Int. J. Quantum Chem., Symp. **22** (1988), 149; (b) R. Shepard, in '*Relativistic and Electron Correlation Effects in Molecules and Solids*', G.L. Malli, ed., NATO ASI Series, Plenum, New York, 1993, in press.
31. S. Zarrabian, C.R. Sarma and J. Paldus, Chem. Phys. Lett. **155** (189), 183; R.J. Harrison and S. Zarrabian, *ibid.* **158** (1989), 393.
32. W. Kutzelnigg, J. Chem. Phys. **82** (1985), 4166 and references therein.
33. J. Hinze and J.T. Broad, in Ref. 22, pp. 332-344.
34. J. Paldus and J. Čížek, Adv. Quantum Chem. **9** (1975), 105.
35. M.D. Gould, J. Paldus and G.S. Chandler, J. Chem. Phys. **93** (1990), 4142.
36. H.S. Green, J. Math. Phys. **12** (1971), 2106; A.J. Bracken and H.S. Green, *ibid.* **12** (1971), 2099.
37. I.M. Gel'fand and M.L. Tsetlin, Dokl. Akad. Nauk SSSR **71** (1950), 825,1070.
38. H. Weyl, *Gruppentheorie und Quantenmechanik*, Hirzel, Leipzig, 1928 (English translation: "*The Theory of Groups and Quantum Mechanics*", Dover, New York, 1964).
39. G.E. Baird and L.C. Biedenharn, J. Math. Phys. **4** (1963), 1449.
40. R.D. Kent and M. Schlesinger, Phys. Rev. A **42** (1990), 1155; J. Math. Phys. **32** (1991), 1102 and references therein.
41. M.A. Robb and U. Niazi, Comp. Phys. Rep. **1** (1984), 127; Reports in Molecular Theory **1** (1990), 23.
42. R. Pauncz, *Spin Eigenfunctions: Construction and Use*, Plenum, New York, 1979, Chap. 9.
43. S. Wilson, *Electron Correlation in Molecules*, Clarendon, Oxford, 1984, Chap. 5.

44. F.A. Matsen and R. Pauncz, *The Unitary Group in Quantum Chemistry*, Elsevier, Amsterdam, 1986.
45. M.D. Gould and G.S. Chandler, Int. J. Quantum Chem. **25** (1984), 553,603,1089; **26** (1984), 441; **27** (1985), 878(E).
46. J. Paldus, Phys. Rev. A**14** (1976), 1620; in '*Symmetries in Science II*', B. Gruber and R. Lenczewski, eds., Plenum, New York, 1986, pp. 429-446.
47. J. Paldus and M.J. Boyle, Phys. Rev. A**22** (1980), 2299; M.J. Boyle and J. Paldus, *ibid.* **22** (1980), 2316.
48. M.D. Gould, J. Math. Phys. **21** (1980), 444; **22** (1981), 15; J. Aust. Math. Soc. B **26** (1984), 257.
49. X. Li and J. Paldus, J. Math. Chem. **4** (1990), 295, and Parts II and III, *ibid.*, in press.
50. M.D. Gould and J. Paldus, Int. J. Quantum Chem. **30** (1986), 327; M.D. Gould, *ibid.* **30** (1986), 364; P.J. Burton and M.D. Gould, J. Chem. Phys. **96** (1992) 5261; H.-L. Lin, Theor. Chim. Acta **79** (1991), 43,53.
51. G.W.F. Drake, in Ref. 22, pp. 243-259.
52. M.D. Gould and J. Paldus, J. Chem. Phys. **92** (1990), 7394 and references therein.
53. M.D. Gould and J. Paldus, J. Math. Phys. **28** (1987), 2304.
54. M.D. Gould and J. Paldus, Phys. Rev. A **34** (1986), 804.
55. I. Waller and R.D. Hartree, Proc. Phys. Soc. (London) A **124** (1929), 119.
56. J. Paldus, in '*Electrons in Finite and Infinite Structures*', P. Phariseau and L. Scheire, eds., Plenum, New York, 1977, pp. 411-429.
57. J. Paldus, S. Rettrup and C.R. Sarma, J. Mol. Structure (Theochem) **199** (1989), 85.
58. X. Li and J. Paldus, Int. J. Quantum Chem. **36** (1989), 127.
59. J. Paldus and J. Planelles, unpublished results.
60. S.-P. Li, R.V. Moody, N. Nicolescu and J. Patera, J. Math. Phys. **27** (1986), 668.
61. X. Li and J. Paldus, J. Mol. Structure (Theochem) **229** (1991), 249; Int. J. Quantum Chem. **41** (1992), 117.
62. J. Paldus and X. Li, Israel J. Chem. **31** (1991), 351; in '*Group Theory in Physics*', AIP Conference Proceedings No. 266, A. Frank, T.H. Seligman and K.B. Wolf, eds., American Institute of Physics, New York, 1992, pp. 159-178.
63. J. Verbeek, Ph.D. Thesis, University of Utrecht, The Netherlands, 1990.
64. J.A. Pople, M. Head-Gordon and K. Raghavachari, J. Chem. Phys. **87** (1987), 5968; **90** (1989), 4635.
65. J. Paldus, J. Čížek and B. Jeziorski, J. Chem. Phys. **90** (1989), 4356; **93** (1990), 1485.
66. I. Lindgren, Int. J. Quantum Chem., Symp. **12** (1978), 33.
67. I. Lindgren and J. Morrison, *Atomic Many-Body Theory*, Springer-Verlag, Berlin, 1982.
68. B. Jeziorski and H.J. Monkhorst, Phys. Rev. A **24** (1981), 1668.
69. I. Lindgren and D. Mukherjee, Phys. Rep. **151** (1987), 93.
70. D. Mukherjee and S. Pal, Adv. Quantum Chem. **20** (1989), 292.
71. B. Jeziorski and J. Paldus, J. Chem. Phys. **90** (1989), 2714.
72. C.L. Janssen and H.F. Schaefer, Theor. Chim. Acta **79** (1991), 1 and

references therein.
73. J. Paldus and X. Li, in '*Symmetries in Science VI: From the Rotation Group to Quantum Algebras*', B. Gruber, ed., Plenum, New York, 1993, in press.
74. B. Jeziorski, P. Jankowski and J. Paldus, unpublished results.
75. Y. Wang, Z. Wen, Z. Zhang and Q. Du, J. Comput. Chem. **13** (1992), 187.
76. P.-Å. Malmquist, A. Rendell and B.O. Roos, J. Phys. Chem. **94** (1990), 5477.
77. Z. Wen, Y. Wang and Z. Zhang, Int. J. Quantum Chem. **37** (1990), 631; Z. Wen, Y. Wang, Y. Dou and Z. Liu, Theor. Chim. Acta **76** (1989), 269.
78. N.C. Handy, Int. Rev. Phys. Chem. **7** (1988), 351.
79. J. Olsen, P. Jørgensen and J. Simons, Chem. Phys. Lett. **169** (1990), 463.
80. M. Barysz, H.J. Monkhorst and L.Z. Stolarczyk, Theor. Chim. Acta **80** (1991), 483.
81. S.A. Kucharski, A. Balková, P.G. Szalay and R.J. Bartlett, J. Chem. Phys. **97** (1992), 4289.
82. R.J. Harrison, J. Chem. Phys. **94** (1991), 5021.
83. M.-M. Rohmer, Inorg. Chem. **28** (1989), 4574.
84. G.-H. Jeung, M. Broyer and P. Labastie, Chem. Phys. Lett. **165** (1990), 494.
85. G.-H. Jeung, Chem. Phys. Lett. **176** (1991), 233.
86. G.-H. Jeung, Mol. Phys. **65** (1988), 669; **67** (1989), 747.
87. T. Leininger, G.-H. Jeung, M.-M. Rohmer and M. Pélissier, Chem. Phys. Lett. **190** (1992), 342.
88. G.-H. Jeung, J. Amer. Chem. Soc. **114** (1992), 3211.
89. J.E. Del Bene and I. Shavitt, J. Phys. Chem. **94** (1990), 5514.
90. M. Aoyagi, R. Shepard, A.F. Wagner, T.H. Dunning, Jr. and F.B. Brown, J. Phys. Chem. **94** (1990), 3236.
91. S. Yamamoto, K. Yamaguchi and K. Nasu, Phys. Rev. B**42** (1990), 266.
92. J.E. Del Bene, K. Kim and I. Shavitt, Can. J. Chem. **69** (1991), 246.
93. J.E. Del Bene and I. Shavitt, J. Mol. Structure (Theochem) **234** (1991), 499.
94. M. Pepper and B.E. Bursten, J. Amer. Chem. Soc. **112** (1990), 7803.
95. M. Bär and R. Ahlrichs, Chem. Phys. Letters **178** (1991), 241.
96. R. Forster, H. Hippler, K. Hoyermann and G. Rohde, Chem. Phys. Letters **183** (1991), 465.
97. H.U. Suter, J.R. Huber, M. von Dirke, A. Untch and R. Schinke, J. Chem. Phys. **96** (1992), 6727.
98. R.A. Kendall, T.H. Dunning, Jr. and R.J. Harrison, J. Chem. Phys. **96** (1992), 6796.
99. J.L. Tilson and J.F. Harrison, J. Phys. Chem. **96** (1992), 1667.
100. G. Born and I. Shavitt, J. Chem. Phys. **76** (1982), 558; G. Born, Int. J. Quantum Chem., Symp. **16** (1982), 633; **28** (1985), 335.
101. A. Banerjee and J. Simons, Int. J. Quantum Chem. **19** (1981), 207.
102. H. Baker and M.A. Robb, Mol. Phys. **50** (1983), 1077.
103. M.R. Hoffmann and J. Simons, J. Chem. Phys. **88** (1988), 993; **90** (1989), 3671.

104. D. Mukhopadhyay, Jr. and D. Mukherjee, Chem. Phys. Letters **163** (1989), 171.
105. P. Siegbahn, A. Heiberg, B. Roos and B. Levy, Phys. Scr. **21** (1980), 323.
106. R.D. Kent and M. Schlesinger, Phys. Rev. B **27** (1983), 46; P.S. Ponnapalli, M. Schlesinger and R.D. Kent, *ibid.* **31** (1985), 1258; R.D. Kent, M. Schlesinger and P.S. Ponnapalli, *ibid.* **31** (1985), 1264.
107. B.R. Brooks, W.D. Laidig, P. Saxe, J.G. Goddard, Y. Yamaguchi and H.F. Schaefer III, J. Chem. Phys. **72** (1980), 4652.
108. R. Shepard, H. Lischka, P.G. Szalay, T. Kovar and M. Ernzerhof, J. Chem. Phys. **96** (1992), 2085.
109. H. Ågren, A. Flores-Riveros and H.J. Aa. Jensen, Phys. Scr. **40** (1989), 745.
110. H. Hettema and P.E.S. Wormer, J. Chem. Phys. **93** (1990), 3389.
111. G.L. Bendazzoli, P. Palmieri and S. Rettrup, J. Chem. Phys. **91** (1989), 5518.
112. J. Koutecký and C. Scheuch, Int. J. Quantum Chem. **37** (1990), 373.
113. R.D. Kent, M. Schlesinger and I. Shavitt, Int. J. Quantum Chem. **41** (1992), 89; A.V. Titov, *ibid.* **42** (1992), 1711.
114. J.S. Battle and M.D. Gould, Chem. Phys. Lett. **201** (1993), 284.
115. M.D. Gould and J. Paldus, Phys. Rev. A **34** (1986), 804.
116. J. Paldus and M.D. Gould, Theor. Chim. Acta **83** (1993), in press.
117. S. Rettrup, T. Thorsteinsson and C.R. Sarma, Int. J. Quantum Chem. **40** (1991), 719.
118. M.J. Gao, J.Q. Chen and J. Paldus, Int. J. Quantum Chem. **32** (1987), 133.
119. P. Jankowski, unpublished results.
120. X. Li and J. Paldus, unpublished results.
121. I. Shavitt, in '*New Horizons in Quantum Chemistry*', P.-O. Löwdin and B. Pullman, eds., Reidel, Dordrecht, The Netherlands, pp. 279-293.

Department of Applied Mathematics, University of Waterloo, Waterloo, Ontario N2L 3G1, Canada

E-mail address: PALDUS@WATDCS.UWaterloo.ca

Contemporary Mathematics
Volume **160**, 1994

QUASI-EXACTLY-SOLVABLE SPECTRAL PROBLEMS AND CONFORMAL FIELD THEORY

MIKHAIL A. SHIFMAN

ABSTRACT. We discuss recent progress in quasi-exactly solvable spectral problems. A few attempts of extending this method beyond the known algebraic structure are briefly reviewed. Remarkable parallels between the quasi-exactly solvable problems and conformal field theory in two dimensions is a new rapidly developing direction in this field which we dwell upon in the second half of this talk.

1. INTRODUCTION

This talk is devoted to a special class of spectral problems – admitting partial or complete algebraization – and intriguing parallels revealed recently between these problems and two-dimensional conformal field theories (CFT). When I mention spectral problems I always keep in mind Quantum Mechanics and Schrödinger-type equations, although, certainly, by no means the approach is limited to Quantum Mechanics; as you will see, it is quite universal and is applicable to any spectral problem.

A unique feature of the Hamiltonians falling in the special class mentioned above is that they can be represented as bilinear combinations of generators of finite-dimensional Lie groups, the so called Lie-algebraic Hamiltonians. All exactly solvable (ES) quantal problems belong to this class, but actually it is much wider. Under certain circumstances we can get "quasi-exactly solvable" problems (QES), i.e. those in which only a part of the spectrum is calculable algebraically. It is important that the algebraic sector can include any given (but fixed) number of states, and they totally decouple from the rest of the spectrum – no analytic continuation in parameters can lead us outside the algebraic sector. (Notice: not all Lie-algebraic Hamiltonians represent physically reasonable exactly or quasi-exactly solvable problems).

The history of the development of ideas which culminated in the discovery of QES is a remarkable example of how closely different branches of physics and mathematics are intertwined at present: basically one and the same approach has emerged independently and almost simultaneously in physical and mathematical environments. The approach has been formulated in a general and systematic form in

1991 *Mathematics Subject Classification.* 81C05, 81C40, 17B15.
Supported in part by DOE under the contract No. DOE-AC02- 83ER40105
This paper is in final form and no version will be submitted for publication elsewhere

refs. [1, 2, 3] [1], although specific examples of QES were found empirically long ago. The pioneering work [1] revealed a hidden algebraic structure in the known examples and outlined a routine allowing one to construct new QES Hamiltonians on an industrial basis. Essentially the same algebraic procedure is formulated in the paper [3] which appeared a little bit later. The approach proposed in [2], *per se*, is not based on the algebraic principles, see below. A representative list of works published before 1989 is given in ref. [4] which might be considered as an introductory survey of the subject. Only the physical line of evolution is traced there since at the time this paper has been written I was unaware of the mathematical investigations in this field; ref. [3] by obvious reasons escaped my attention.

Recent progress goes mainly in two directions: (i) classification of multidimensional QES problems [5, 6, 7, 8] (for one-dimensional systems the existing classification is complete); and (ii) revealing relations between QES and CFT which might be mutually beneficial allowing one to use quantal methods in CFT and vice versa. The observation of a natural connection between QES in quantum mechanics and two-dimensional CFT has been first made in ref. [8]. A new impetus in this line of research is given by two recent works [9, 10]. The first of them explicitly demonstrates that the so called decoupling equations for the conformal blocks in a special class of CFT identically coincide with the equations for the QES wave functions derived in [2]. Moreover, the computation of the conformal weights is explicitly reduced to that of the QES eigenvalues. And finally, probably the most remarkable observation concerns a quantal analog of the operator product expansion (fusion rules) of CFT. The fact that there should exist a relation between the QES wave functions playing the same role as the fusion rules in CFT has been conjectured in [8]. The issue, however, has not been traced in detail, and the conjectured form of the relation turned out to be unrealistic. Ref. [9] proves that the QES analog of the CFT fusion rules is a set of fusion rules in the parameter space which gives a limiting expression for the wave function in a given QES model in terms of the wave function of a daughter ES system.

The first half of this talk describes results on QES spectral problems obtained mostly after 1989. The second half is devoted to the issue of QES *versus* CFT.

This is my first experience in purely mathematical audience, and I bring my apologies for the fact that my language will not be completely appropriate to the case. Hopefully it will still be understandable. A professional mathematical excursion in the quasi-exactly-solvable problems has been undertaken yesterday by Prof. Olver [11]. By comparing his review talk to the present one you can easily establish a dictionary facilitating the translation of the physical terminology in mathematical and vice versa. Further explanations can be found in the literature cited below.

2. Partial Algebraization in Quantum Mechanics

2.1. Basic idea

In this subsection we will discuss a simplest example illustrating the basic ideas of the algebraic (Turbiner- Kamran-Olver[2]) approach. To begin with let us recall

[1]The original papers of Ushveridze appeared in Russian in 1988 as Lebedev Institute preprints and then were published in *Kratkie soobscheniya po fizike*, a journal which is hardly accessible in most of the western universities. Therefore I cite here a very detailed review paper [2] of the author which summarizes all basic results.

[2]TKO in what follows

in more detail what the term *quasi-exactly solvable* actually means. Consider a quantum-mechanical system with the Hamiltonian $\mathcal{H}$ and assume for definiteness that the spectrum of the Hamiltonian is discrete. (This is not necessary, though). To find the eigenvalues and eigenfunctions in a normal way one solves a differential equation of the second order, supplemented by certain boundary conditions. The procedure is equivalent to diagonalization of an infinitely-dimensional matrix, which is obviously a non-algebraic operation. Everybody knows, however, that there exist exceptional cases, like e.g. the famous harmonic oscillator, where one can construct an algebraic algorithm yielding all spectrum. The algebraization of the spectrum in this problem is based on the fact that the Hamiltonian is representable as a bilinear combination of the raising and lowering operators,

$$\mathcal{H} = \frac{1}{2}(a^+ a + \omega) \tag{1}$$

obeying the Heisenberg algebra,

$$[aa^+] = 2\omega, \quad where \quad a^+ = -d_x + \omega x, \quad a = d_x + \omega x. \tag{2}$$

The algebraic structure encoded in eqs. (1) and (2) is sufficient to ensure that all eigenvalues and eigenfunctions are calculable, step by step, purely algebraically.

The new class of special spectral problems, with partial algebraization, is much wider. The basic idea remains the same – construct $\mathcal{H}$ in terms of generators of some Lie algebra. If these generators are realized in such a way that there exist finite dimensional representations, then we automatically get in the given spectral problem an algebraic sector whose dimension is equal to the dimension of the representation considered. The QES Hamiltonians contain numerical parameters, and finite-dimensional representations appear only for specific values of these parameters, which are, thus, quantized. Shortly I will explain the difference between the QES systems and the standard Lie-algebraic Hamiltonians, like, for instance, the conventional quantum top, which are exactly solvable. Here I would like just to illustrate the idea by a simplest example.

Consider the one-dimensional anharmonic oscillator

$$\mathcal{H} = \frac{1}{2}[p^2 + (x^6 - (8j+3)x^2)] \tag{3}$$

where x is the coordinate, p is the momentum, $p = -id/dx$, j is a numerical parameter. If j is half-integer the spectral problem with the Hamiltonian (3) have an algebraic sector of dimension $2j+1$.

To see that this is indeed the case let us make a step back and start from what I call a "gauge" transformed Hamiltonian

$$\mathcal{H}_{"G"} = -J_0 J_- - (2j+1)J_- - 2J_+ \tag{4}$$

where J_0 and $J_\pm$ are the generators of the $SL(2,R)$ group,

$$J_0 = \xi\frac{d}{d\xi} - j \ , \ J_+ = \xi^2\frac{d}{d\xi} - 2j\xi, \quad J_- = \frac{d}{d\xi}, \tag{5}$$

j is a numerical parameter, and ξ is an auxiliary variable. The generators J act in the space of polynomials of ξ. It is easy to check that for half-integer values of j

there exists a finite dimensional irreducible representation,

$$R_j = \langle \xi^0, \xi^1, \dots, \xi^{2j} \rangle \tag{6}$$

containing $(2j+1)$ elements. Therefore, any Hamiltonian built from J's is guaranteed to have an algebraic sector of dimension $2j+1$, a trivial observation stemming from the fact that the generators acting on the elements of the representation leave us inside the representation. What is less trivial, is a possibility to reduce $\mathcal{H}_{"G"}$ to a physically reasonable quantal Hamiltonian (3). By reasonable I mean a system with the hermitean Hamiltonian and normalizable wave functions.

To establish the connection between (4) and (3) notice that we are free to do change of variables *plus* quasigauge transformations. In the case at hand

$$\xi = x^2,$$

and in terms of the new "physical" variable x

$$\mathcal{H}_{"G"}(x) = -\frac{1}{2}\frac{d^2}{dx^2} + x^3\frac{d}{dx} - 4jx^2 + const. \tag{7}$$

The new Hamiltonian $\mathcal{H}_{"G"}(x)$ is still non-hermitean due to the presence of the $x^3 d/dx$ term. This non-hermiticity can be gauged away, however. Indeed, let us do the quasigauge transformation,

$$\mathcal{H}_{"G"}(x) \to \mathcal{H}(x) = e^{-a(x)}\mathcal{H}_{"G"}(x)e^{a(x)} \tag{8}$$

with $a(x) = x^4/4$. If $a(x)$ were a purely imaginary phase this would be a normal unitary gauge transformation which by no means could transform a non-hermitean operator into a hermitean one. Our transformation is non-unitary since $a(x)$ is real – that is why it is called quasigauge – and it kills the non-hermitean term. As usual, under this transformation

$$\frac{d}{dx} \to \frac{d}{dx} + \mathcal{A}(x) \quad where \quad \mathcal{A}(x) = \frac{da}{dx}. \tag{9}$$

Substituting the "gauge" transformed derivative in eq. (7) we immediately arrive to the Hamiltonian (3). The last step is to verify that the eigenfunctions obtained after the "gauge" transformation are normalizable. The fact that they can be normalized is quite obvious since all eigenfunctions belonging to the algebraic sector have the form

$$\psi(x) \sim \mathcal{P}_{2j}(x^2)e^{-a(x)}$$

where $\mathcal{P}_k(x)$ stands for a polynomial of the k-th order, and $a(x) = x^4/4$.

The Hamiltonian (4) yields one of the simplest examples of physically reasonable QES system. It is obvious that eq. (4) can be substituted by a generic $SL(2,\mathbf{R})$-algebraic Hamiltonian of the form

$$\mathcal{H}_{"G"} = \sum_{+,-,0} (C^{ab}J_aJ_b + C^aJ_a) \tag{10}$$

where the generators J_a are defined in eq. (5) while C^{ab}, C^a are numerical coefficients. Choosing different sets of the coefficients C we arrive at different quantal systems. (Some of them are reducible to each other, though; others are physically meaningless, see below). A full classification of one-dimensional QES systems based on $SL(2,\mathbf{R})$ is worked out in refs. [1, 2, 3]. Ref. [2] gives also general conditions under which the "gauge" transformation leads to physically reasonable quantal Hamiltonians – stable vacuum, the absence of "wrong" singularities in the potential, etc. The normalizability of the eigenfunctions of the generic $SL(2,\mathbf{R})$ QES systems is discussed in a recent work [12]. In the previous investigations of the issue (e.g. [2]) the requirement of normalizability has been analyzed only with respect to a particular (although wide) class of QES systems, namely those where the physical domain is a finite interval of the x axis, (x_1, x_2), such that (i) $V(x) < \infty$ at $x \in (x_1, x_2)$; (ii) $V(x_1) = V(x_2) = +\infty$; (iii) $\psi(x_1) = \psi(x_2) = 0$. In this case the situation is rather transparent and simple. Ref. [12] analyses a more complicated situation when the physical interval is $(-\infty, +\infty)$ and the wave functions $\in \mathcal{L}_2(R^1)$. Conditions on the coefficients C are found ensuring the finite norm for all eigenfunctions from the algebraic sector of an arbitrary $SL(2,\mathbf{R})$ system living on $x \in (-\infty, +\infty)$.

In the remainder of this talk the normalizability conditions are not touched upon at all. In many instances it may well happen that the QES systems with a well-defined algebraic sector are physically inappropriate since the corresponding wave functions are not normalizable.

2.2 Ushveridze formalism

Having built an $SL(2,\mathbf{R})$-algebraic Hamiltonian admitting partial algebraization, one can make a step further – diagonalize the finite-dimensional algebraic sector. In the general case this can be done by reducing the problem to an analog problem from classical planar electrostatics. Such a reduction lies in the basis of the program [2] which, originally, has been formulated in non-algebraic terms [3].

Let us remind basic elements. The most general form of the $SL(2,\mathbf{R})$ QES Hamiltonian is as follows [1]:

$$\mathcal{H}_{"G"}(\xi) \equiv e^{a}\mathcal{H}e^{-a} = -\frac{1}{2}\mathcal{P}_4(\xi)\frac{d^2}{d\xi^2} + \mathcal{P}_3(\xi)\frac{d}{d\xi} + \mathcal{P}_2(\xi). \tag{11}$$

The coefficients of polynomials $\mathcal{P}_i(\xi)$ are not completely independent; certain constraints are imposed by the $SL(2,\mathbf{R})$-algebraic structure. Moreover, some of them may even be redundant since making appropriate $SL(2,\mathbf{R})$ transformations (they reduce to a specific change of variables) we can always get rid of three parameters. In other words, there should exist a standard form of $\mathcal{H}_{"G"}$. What form is to be considered as standard is a matter of convenience, and we will not go into further detail here. A few additional remarks are presented in subsect. 2.3. Now we just note that generically $\mathcal{P}_4(\xi)$ can be parametrized in terms of five parameters,

$$\mathcal{P}_4(\xi) = -2C_{++}(\xi - a_1)(\xi - a_2)(\xi - a_3)(\xi - a_4), \tag{12}$$

[3] As a matter of fact, the relation between the algebraic TKO approach and that of Ushveridze is not fully understood so far. In particular, according to [2], the so called finite series of QES problems can be constructed, with the number of states in the spectrum belonging to the algebraic sector not exceeding 7. It remains to be seen whether such series can be obtained starting from the Lie-algebraic approach. If we stick literally to the procedure outlined in the previous subsection there is certainly no place for the finite series of ref. [2].

where C_{++} is the coefficient in front of J^+J^+ and a_i $(i = 1,2,3,4)$ are numbers. Then the polynomials $\mathcal{P}_3(\xi)$ and $\mathcal{P}_2(\xi)$ are representable in the following form:

$$\mathcal{P}_3(\xi) = \mathcal{P}_4(\xi) \sum_{\alpha=1}^{4} \frac{b_\alpha}{\xi - a_\alpha},$$

$$\mathcal{P}_2(\xi) - E = \mathcal{P}_4(\xi) \sum_{\alpha=1}^{4} \frac{c_\alpha}{\xi - a_\alpha}. \tag{13}$$

The parameters a_α are arbitrary. They can be either real or, if complex, should appear as pairs of complex-conjugate numbers. The same is applicable to b_α. Unlike a_α's the parameters b_α are constrained,

$$\sum_{\alpha=1}^{4} b_\alpha = 2j - 1. \tag{14}$$

As far as the parameters c_α are concerned, they must satisfy three conditions,

$$\sum_{\alpha=1}^{4} c_\alpha = 0, \ \sum_{\alpha=1}^{4} c_\alpha a_\alpha = -j(2j-1), \ \sum_{\alpha=1}^{4} c_\alpha a_\alpha^2 = -2j \sum_{\alpha=1}^{4} b_\alpha a_\alpha, \tag{15}$$

where the constraint (14) is taken into account. The remaining freedom in c_α's reflects the fact that we can have different energy eigenvalues.

Now, the eigenvalue equation $\mathcal{H}_{"G"}\psi_{"G"} = E\psi_{"G"}$ is equivalent to

$$-\frac{1}{2} \sum_{m,n=1,m\neq n}^{2j} \frac{1}{\xi - \xi_m}\frac{1}{\xi - \xi_n} + \sum_{\alpha=1}^{4} \frac{b_\alpha}{\xi - a_\alpha} \sum_{m=1}^{2j} \frac{1}{\xi - \xi_m} + \sum_{\alpha=1}^{4} \frac{c_\alpha}{\xi - a_\alpha} = 0 \tag{16}$$

provided that the eigenfunction is chosen in the form

$$\psi_{"G"}(\xi) = (\xi - \xi_1)(\xi - \xi_2)...(\xi - \xi_{2j}) \tag{17}$$

where ξ_1, ξ_2, ..., ξ_{2j} are free parameters to be determined from eq. (16). The possibility to choose $\psi_{"G"}$ in this form stems from the algebraic structure of the Hamiltonian: the set $\{1, \xi, ..., \xi^{2j}\}$ forms a finite-dimensional representation of $SL(2,\mathbf{R})$ in the same realization as the generators J, eq. (5). Reducing the left-hand side to the sum of simple poles and requiring all residues to vanish we get the following equation on ξ_k:

$$-\sum_{m=1,m\neq k}^{2j} \frac{1}{\xi_k - \xi_m} - \sum_{\alpha=1}^{4} \frac{-b_\alpha}{\xi_k - a_\alpha} = 0, \tag{18}$$

$$k = 1, 2, ..., 2j.$$

As was noted in [2] $(2j+1)$ conditions on the parameters ξ_k displayed in eq. (18) can be interpreted as an equilibrium condition for an analog system of $(2j+1)$ charged particles on the plane (see fig. 1). Namely, assume that we have $(2j+1)$ particles situated along the real axis on the complex ξ plane. All particles have electric charges equal to 1 and repel each other according to the standard Coulomb law (the Coulomb potential is logarithmic on the plane). All these particles can move freely. In addition to the moving particles there are four fixed force centers located at a_α. If all a_α are real the force centers lie on the same line (fig. 1a), and the corresponding values of b_α are also real. Otherwise, the force centers are at complex-conjugate points (fig. 1b), with the complex-conjugate values of b_α. In the first case, real a_α and b_α, the force centers are to be treated as nailed Coulomb charges $(-b_1)$, $(-b_2)$, etc.; they can either repel or attract the moving particles, depending on the sign of b_α. If the force centers are at the complex-conjugate points, they are to be treated as nailed dyons, carrying both, electric and magnetic charges, $(-Reb_\alpha)$ and $(-Imb_\alpha)$, respectively. The magnetic charges of the dyons create a vortex-like electrostatic field which on the real axis is directed parallel to the axis. Therefore, if the moving particles are situated appropriately, an (unstable) equilibrium is possible. The equilibrium configurations provide solutions to eqs. (18) and, thus, fully determine the eigenfunctions and the eigenvalues.

This rather lengthy description is summarized by one formula for the potential energy of the analog Coulomb system [2]:

$$W(\xi_k) = -\sum_{m,n=1,m\neq n}^{2j} \log \mid \xi_m - \xi_n \mid - \sum_{m=1}^{2j}\sum_{\alpha=1}^{4}(-Reb_\alpha)\log \mid \xi_m - a_\alpha \mid$$

$$-\sum_{m=1}^{2j}\sum_{\alpha=1}^{4}(-Imb_\alpha)\Phi(\xi_m - a_\alpha) \tag{19}$$

where $\Phi(x)$ is the phase of the complex number x,

$$\Phi(x) = \arctan\frac{Imx}{Rex}.$$

The last term describes the electrostatic energy of the magnetic charge living on a plane, $rot\vec{E} = -(Imb)$.

It is worth noting that a similar reduction – from a quantum mechanical wave function to an analog electrostatic system – is described in the literature in different contexts long ago, see e.g. [13].

For future comparison with the decoupling equation in CFT it is convenient to write down here the $SL(2,\mathbf{R})$ spectral problem in the "pole" form stemming from eq. (13),

$$\mathcal{P}_4(\xi)\left[-\frac{1}{2}d^2 + \sum_{\alpha=1}^{4}\frac{b_\alpha}{\xi - a_\alpha}d + \sum_{\alpha=1}^{4}\frac{c_\alpha}{\xi - a_\alpha}\right]\psi_{"G"} = 0 \tag{20}$$

where $d \equiv d/d\xi$. Below we will also need a gauge transformed equivalent of this equation, with the term linear in d gauged away,

$$\left[-d^2 + \sum_{\alpha=1}^{4}\frac{b_\alpha(b_\alpha+1)}{(\xi - a_\alpha)^2} + \sum_{\alpha=1}^{4}\frac{2\tilde{c}_\alpha}{\xi - a_\alpha}\right]\tilde{\psi}_{"G"} = 0 \tag{21}$$

where

$$\tilde{c}_\beta = c_\beta + \sum_{\gamma=1,\gamma\neq\beta}^{4}\frac{b_\beta b_\gamma}{a_\beta - a_\gamma}. \tag{22}$$

The reduction to the analog electrostatic problem is only one side of the Ushveridze approach to building QES systems. The approach reveals a surprising relation of QES systems to a totally different algebraic structure. It has been shown recently [14, 27] that the QES quantal problems are generated, alternatively, by a reduction of integrable models of magnets on discretized lattices, the so called Gaudin models [13]. The latter are characterized by infinitely-dimensional algebras and are exactly solvable by the Bethe ansatz.

2.3. General analysis of QES systems. Multidimensional case. [4]

A class of the Lie-algebraic Hamiltonians has been studied in quantum mechanics long time ago under the name of the generalized quantum tops. The standard construct which we will discuss in brief below leads to ES systems, with the infinite number of disentangled finite-dimensional blocks. The discussion of the generalized quantum tops will help us to understand what new has been introduced with the discovery of QES problems. We will emphasize those points which have to be changed in passing to QES systems.

The standard procedure starts from choosing a *compact* finite-dimensional Lie group G of dimension D and rank r. (Notice: QES Hamiltonians are obtained from *non-compact* groups which are usually not considered in this context). Given the group we write down the Hamiltonian as a bilinear combination of the generators of the group,

$$\mathcal{H} = -\frac{1}{2}\sum_{a,b=1}^{D} C^{ab} J_a J_b \tag{23}$$

where C^{ab} is a constant symmetric matrix, and possible linear terms in eq. (23) are omitted for simplicity (they are not of principle importance in the case at hand, unlike the QES systems, and can be easily added, if necessary).

Now, we assume that the generators J_a are realized as vector fields on a homogeneous space $\mathcal{M} = G/H$, H being a stationary subgroup. In other words, the orbit of the group G – it is denoted by $\mathcal{M}$ – is treated as the *configuration space* of a quantal system we are going to build. (Notice: for QES system special degenerate orbits are to be chosen which are treated, then, as the *phase space*.)

In the generic case there exists a G-invariant metric g_0 on $\mathcal{M}$. Such a metric definitely exists if H is compact.

If the coordinates on $\mathcal{M}$ are denoted by ξ^μ ($\mu = 1, 2, ..., d$ where d stands for the dimension of $\mathcal{M}$) the generators of G take the form

$$J_a = \mathcal{J}_a^\mu(\xi)\frac{\partial}{\partial \xi^\mu} \tag{24}$$

where possible inhomogeneous parts are suppressed. They can be readily generated by "gauge" transformations

$$J_a \to e^{a(\xi)} J_a e^{-a(\xi)} \tag{25}$$

[4] Discussion in this subsection essentially follows ref. [8].

without any change in the algebra of J_a's. It is obvious that under the "gauge" transformations (25) we get the generators in a new, equivalent form, with inhomogeneous parts included,

$$J_a = \mathcal{J}_a^\mu(\xi)\frac{\partial}{\partial\xi^\mu} + E_a(\xi) \tag{26}$$

where

$$E_a(\xi) = -\mathcal{J}_a^\mu\partial_\mu a(\xi). \tag{27}$$

Once the coordinates on $\mathcal{M}$ are introduced and the vector fields $\mathcal{J}_a^\mu(\xi)$ are found the invariant metric g_0 can be written explicitly,

$$(g_0)^{\mu\nu} = \sum_{a=1}^{D}\mathcal{J}_a^\mu(\xi)\mathcal{J}_a^\nu(\xi). \tag{28}$$

The algorithm for constructing the vector fields $\mathcal{J}_a^\mu(\xi)$ on $\mathcal{M}$ for compact groups is worked out in detail in the mathematical literature since it is extensively used within the Kirillov-Kostant procedure of building the representations of the Lie groups [16] (an outline of the algorithm is given, e. g. in [8]).

Moreover, it is well-known that the operators iJ_a are hermitean with respect to the G-invariant scalar product on $\mathcal{M}$. This property is crucial for the proof of the following assertion (see e.g. [8]): any Hamiltonian of the form (23) where J_a are the generators given in eq.(24) is reducible to

$$-2\mathcal{H} = e^\phi(\Delta_g - 2V)e^{-\phi} \tag{29}$$

where $g^{\mu\nu}(\xi)$ is the deformed metric,

$$g^{\mu\nu}(\xi) = \sum_{a,b}C^{ab}\mathcal{J}_a^\mu(\xi)\mathcal{J}_b^\nu(\xi), \tag{30}$$

$$g = \frac{1}{det[g^{\mu\nu}]}, \tag{31}$$

and Δ_g stands for the covariant Laplace operator,

$$\Delta_g = (g)^{-1/2}\partial_\mu(g)^{1/2}g^{\mu\nu}\partial_\nu. \tag{32}$$

Moreover, the "imaginary phase" ϕ is expressed in terms of the ratio g/g_0,

$$\phi = \frac{1}{2}\log\frac{\sqrt{g}}{\sqrt{g_0}}, \tag{33}$$

while the potential V reduces to

$$V = \frac{1}{2}g^{\mu\nu}\partial_\mu\phi\partial_\nu\phi - \frac{1}{2}\Delta_g\phi. \tag{34}$$

(g_0 introduced in eq. (32) is the same as in eq. (30) with the metric $g^{\mu\nu}$ substituted by the invariant metric $(g_0)^{\mu\nu}$.)

Summarizing, we see that the Hamiltonian $\mathcal{H} = -\frac{1}{2}\sum C^{ab}J_aJ_b$ is nothing else than the scalar Laplace operator on a curved space with the metric (29) plus a potential, modulo a quasigauge transformation.

It is clear that the Hamiltonian $\mathcal{H}$ is hermitean and physically "reasonable" for any set of the coefficients C^{ab} – not surprisingly since the operators iJ_a are hermitean in the case at hand. (Notice: this is *not* the case for QES systems.)

In the spectral problem with the Hamiltonian (23) the full spectrum can be constructed algebraically, a characteristic feature of the systems whose configuration space coincides with an orbit of the compact Lie group. To prove that the algebraization is complete it is sufficient to observe that the quadratic Casimir operator $\mathcal{H}_0 = \sum J_a J_a = \Delta_{g_0}$ is not degenerate and, moreover, it commutes with the Hamiltonian (23).

The Hilbert space of functions on $\mathcal{M}$ then splits into a sum of eigen-subspaces of the hermitean operator $\mathcal{H}_0 = \Delta_{g_0}$. Since $\mathcal{H}_0$ commutes with J_a every such subspace R_λ (corresponding to the $\mathcal{H}_0$ eigenvalue λ) forms a finite-dimensional representation of the group G. By construction, its quadratic Casimir coefficient is λ. Once the coordinates ξ^μ are fixed, the concrete form of the elements of the representation is known (generalized spherical harmonics!). In the basis of the generalized spherical harmonics the Hamiltonian (23) obviously reduces to the block form, each block completely disconnected from all others. The dimension of each block is $\{\ \mathrm{dim} R_\lambda\} \times \{\mathrm{dim} R_\lambda\}$ (fig. 2). Diagonalization of each finite-dimensional block can be carried out separately. It is important, of course, that the Hamiltonian (23) does not depend on the quantum numbers of the specific representations R_λ. (Notice: for QES Hamiltonians such a dependence does appear.)

After this brief excursion in the theory of the standard quantum tops let us proceed to discussion of QES problems. The QES systems emerge in those degenerate cases when the invariant metric $g_0^{\mu\nu}$ does not exist, or, in other words, the quadratic Casimir operator $\mathcal{H}_0 = \sum J_a J_a$ is totally degenerate: $\sum J_a J_a$ reduces to a numerical constant. Then, there are infinitely many eigenfunctions with the same eigenvalue of $\mathcal{H}_0$. As a result, the argument presented above, demonstrating the complete algebraization of $\mathcal{H}$, is not applicable. It remains to add that $\mathcal{H}_0$ is degenerate for some non-compact groups and specific orbits.

Thus, proceeding from the "old-fashioned" Lie-algebraic Hamiltonians to the QES problems we start from the assumption that no invariant metric g_0 exists on $\mathcal{M} = G/H$. Moreover, we would like to consider $\mathcal{M}$ as something like a phase space of a quantal system. Generically, the action of the group on $\mathcal{M}$ is not transitive, and the phase space does not exactly coincide with any of the orbits of the group G – some points are singular. Of special interest, however, is the situation when a dense orbit in the phase space does exist. Then it is symplectic and, hence, corresponds to a co-adjoint orbit of G. Moreover, using the standard Kirillov-Kostant procedure [16] one can obtain explicit expressions for the generators of the group in the form (26) (see [17] for details). Let us outline in gross features the steps crucial for constructing QES systems with the phase space $\cong$ orbit.

Given a semi-simple group[5] G one first analyses the orbits of the group in the adjoint representation, which can be of different types. For any orbit the dimension of the corresponding space is even, $dim\mathcal{M} = d = 2\delta$ where δ is an integer. In the

[5]The choice of the semi-simple Lie algebras is only due to historical reasons; this restriction was accepted in the first works, see e.g. ref. [8]. The requirement on G to be semi-simple is neither necessary nor essential, as was noticed recently in [18] where new examples of two-dimensional QES systems are found.

general case

$$(35) \qquad 2\delta = dimG - rankG.$$

As a rule, there exist also degenerate orbits with dim$\mathcal{M}$ < dim$G$ - rank$G$, all the way to dim$\mathcal{M}$=0. For instance, for $SL(3, \mathbf{R})$ the non-degenerate orbits have dimension 6; at the same time it is not difficult to explicitly construct degenerate orbits, say, of dimension 4.

Furthermore, we embed the orbit $\mathcal{M}$ in the D-dimensional space where D= dim$G$. The set of coordinates on $\mathcal{M}$ splits into δ pairs (p_i, q_i) $(i = 1, 2, ..., \delta)$. Each point $(p_i, q_i) \in \mathcal{M}$ considered as a point from the D- dimensional space is characterized by a vector $J_a(p_i, q_i)$, $a = 1, ..., D$. All q_i's are to be treated as quantum mechanical coordinates while all p_i's as canonical momenta, to be realized as $p_i = -i\partial/\partial q_i$. Now, requiring

$$(36) \qquad [p_i, q_j] = -i\delta_{ij},\ [p_i, p_j] = 0,\ [q_i, q_j] = 0$$

we get the group law

$$(37) \qquad [J_a, J_b] = if_{abc}J_c$$

where f_{abc} are the structure constants. Moreover, the generators J_a take the form of the linear differential operators (26). For the orbits under discussion J_a turn out to be non-hermitean. There is a subtle question which deserves mentioning – the question of ordering of p's and q's in J_a. The expressions pq and qp are equivalent at the classical level, of course. When we pass to the quantum realization the ordering becomes important. One can show that the requirement of the group law (36) fixes the quantum form of J_a almost unambiguously. The residual trivial ambiguity is due to the fact that instead of $p_i = -i\partial/\partial q_i$ one can always choose $p_i = -i\partial/\partial q_i + \partial_i\Phi(q)$, where $\Phi(q)$ is a quasiphase, without altering the group commutation relations. This is nothing else than the "gauge" freedom repeatedly discussed above.

The orbit $J_a(p, q)$ is characterized by a set of numerical parameters $\{j\}$, see e.g. eq. (5) for $SL(2, \mathbf{R})$. These parameters determine the value of the quadratic Casimir operator. Finite-dimensional representations of G – and we are interested only in those – appear only for discrete values of the numerical parameters. In other words, we get a quantization condition. For instance, in the case of $SL(2, \mathbf{R})$

$$(38) \qquad J_+J_- - J_0J_0 = j(j+1),$$

and the quantization implies that j is half-integer. (See also [7, 11] where the quantization in algebras leading to two-dimensional quantum mechanics are considered. A general discussion of the phenomenon of quantization of the cohomology parameters in the QES Hamiltonians is given in [19].) From eq. (37) we also explicitly see that the quadratic Casimir operator is degenerate for the $SL(2, \mathbf{R})$ orbit we consider.

The general expression for the QES Hamiltonian is a quadratic combination

$$(39) \qquad \mathcal{H} = \sum C^{ab}J_aJ_b + C^aJ_a.$$

Clearly not all terms here are independent. First of all, one bilinear term can be eliminated thanks to the condition (37). Second, using the freedom of motions along the orbit we can eliminate, in addition, D other terms. For instance, in the case of $SL(2,\mathbf{R})$ the change of variables

$$z = \frac{\alpha\xi + \beta}{\gamma\xi + \delta}\ , \ \alpha\delta - \beta\gamma = 1$$

where $\alpha, \ldots, \gamma$ are parameters, realizes the motion along the orbit. The new generators are linear combinations of the old, modulo a quasigauge transformation. Choosing three independent parameters from the set $\{\alpha, \beta, \gamma, \delta\}$ in a special way, we can eliminate some (at most, three) of the coefficients C.

Assume now, that a degenerate orbit is chosen, numerical parameters in J_a quantized, and a finite-dimensional representation is built. The central question is: "Can one reduce $\mathcal{H}$, eq. (38), to an ordinary Schrödinger-type operator?" The word "ordinary" essentially means that $\mathcal{H}$ is hermitean (see below).

The issue of hermiticity of the Lie-algebraic Hamiltonians in the case of non-compact groups and degenerate orbits is the most difficult problem which defies general solution so far. In the one-dimensional quantum mechanics based on $SL(2,\mathbf{R})$ the intrinsic non-hermiticity of the QES Hamiltonians resides in a single function – the coefficient in front of $d/d\xi$ – which can be always gauged away by an appropriate quasigauge transformation. The quantal system emerging in this way is not necessarily stable from the quantum-mechanical point of view, but at least it is hermitean. In the multi-dimensional case of higher groups and higher-dimensional orbits the general classification of the hermitean Lie-algebraic Hamiltonians is a challenging and basically open problem. It is clear that not all Hamiltonians (38) are equivalent to hermitean systems, but in certain instances the hermiticity can be achieved by a special choice of the coefficients C [5] (plus the freedom of the "gauge" transformation, of course). The work on classification of two-dimensional QES problems has already started [6]. All relevant four-dimensional orbits are found admitting finite-dimensional representations. The issue of hermiticity is not solved in its general form even for two-dimensional QES systems.

What little is known about this aspect of the higher dimensional case can be summarized in a few remarks. First of all, the formulation of the problem. We start from a generic Lie-algebraic Hamiltonian (38) (corresponding to non-compact groups and degenerate orbits) and try to transform it to a Schrödinger type hermitean operator, modulo "gauge" transformations. Now, the Schrödinger-type spectral equation has the following generic form:

$$-\frac{1}{2}\left[\frac{1}{\sqrt{g}}\partial_\mu(\sqrt{g}g^{\mu\nu}\partial_\nu\psi)\right] + V\psi = E\psi, \tag{40}$$

where $g^{\mu\nu}$ is metric. The equation describes a particle in the potential V on a spatial manifold with the contravariant (positive definite) metric $g^{\mu\nu}$. Allowing additionally for an external magnetic field we would get the Schrödinger operator of a more general form. Let us assume for the time being that the magnetic field is absent.

(Notice: in one-dimensional problems the metric could always be made flat). The question is: "Under what conditions is the equivalence between (38) and (39) possible?"

To answer the question, let us first make a "gauge" transformation with the quasiphase a. Then, eq. (39) becomes

(41)
$$[-\frac{1}{2}g^{\mu\nu}\partial_\mu\partial_\nu - \frac{1}{2}(\partial_\mu g^{\mu\nu})\partial_\nu - \frac{1}{2}(\partial_\mu \log\sqrt{g})g^{\mu\nu}\partial_\nu + g^{\mu\nu}\mathcal{A}_\mu\partial_\nu + \delta V]\psi_{"G"} = E\psi_{"G"}$$

where

$$\mathcal{A}_\mu = \partial_\mu a = \partial a/\partial\xi^\mu,$$

and δV is defined in terms of the original potential V and $\mathcal{A}_\mu$,

(42)
$$V = \frac{1}{2}g^{\mu\nu}\mathcal{A}_\mu\mathcal{A}_\nu - \frac{1}{2}(\partial_\mu\log\sqrt{g})g^{\mu\nu}\mathcal{A}_\nu - \frac{1}{2}g^{\mu\nu}(\partial_\mu\mathcal{A}_\nu) - \frac{1}{2}(\partial_\mu g^{\mu\nu})\mathcal{A}_\nu + \delta V.$$

Along with the "vector potential" $\mathcal{A}_\mu$ it is convenient to introduce a new auxiliary "potential"

(43)
$$\mathcal{B}_\mu = \partial_\mu(\log\sqrt{g}) - 2\mathcal{A}_\mu = \partial_\mu(\log\sqrt{g} - 2a).$$

Then the spectral equation (40) takes the form

(44)
$$-2\mathcal{H}_{"G"}\psi_{"G"} \equiv [g^{\mu\nu}\partial_\mu\partial_\nu + (\partial_\mu g^{\mu\nu})\partial_\nu + g^{\mu\nu}\mathcal{B}_\mu\partial_\nu - 2\delta V]\psi_{"G"} = -2E\psi_{"G"}.$$

Now let us compare the expression in the square brackets with the bilinear combination (38). Assume that the coefficients C^{ab} are chosen in a certain way. The set of the coefficients C^{ab} uniquely fixes the metric $g^{\mu\nu}$, see eq. (29). Moreover, the bilinear piece $C^{ab}J_aJ_b$ produces also some terms linear in ∂_ν as well. Next, we add the terms C^aJ_a and require the linear in ∂_ν part of $\sum[C^{ab}J_aJ_b + C^aJ_a]$ to be equal to

(45)
$$[(\partial_\mu + \mathcal{B}_\mu)g^{\mu\nu}]\partial_\nu.$$

Since $g^{\mu\nu}$ is already fixed, this requirement gives us $d/2 = \delta$ equations on $\mathcal{B}_\mu$,

(46)
$$(\partial_\mu + \mathcal{B}_\mu)g^{\mu\nu} = M^\nu,\ \ k = 1, 2, ..., \delta \equiv \frac{1}{2}dim\mathcal{M},$$

where M^ν are the coefficients in front of ∂_ν in $\sum[C^{ab}J_aJ_b + C^aJ_a]$ (these coefficients are built from polynomials of ξ^μ and depend on the coefficients C.)

For non-degenerate metrics $g^{\mu\nu}$ eq. (45) can always be solved,

(47)
$$\mathcal{B}_\mu = g_{\mu\nu}(M^\nu - \partial_\gamma g^{\gamma\nu}),\ \ \mu = 1, 2, ..., \delta.$$

Eq. (46) certainly ensures the desired form of $\mathcal{H}_{"G"}$, eq. (39), but one must remember that the functions $\mathcal{B}_\mu$ must be pure gauge, eq. (42). This requirement imposes the following constraints on $\mathcal{B}_\mu$:

(48)
$$\partial_\nu\mathcal{B}_\mu - \partial_\mu\mathcal{B}_\nu = 0,$$

to be considered as self-consistency conditions. If and only if the functions $\mathcal{B}_\mu$ given in eq. (46) are pure gauge can we speak of the QES problem of the Schrödinger type, eq. (39).

The constraint on $\mathcal{B}_\mu$ could, in principle, be somewhat relaxed should we decide to consider external magnetic fields, see the remark after eq. (39). In this case the physical Hamiltonian can contain additional terms linear in the first derivatives that can not be gauged away, and still be hermitean. The issue has not been elaborated in the context of the QES systems, and I will not touch upon it.

Both $g^{\gamma\nu}$ and M^{ν} in eq. (46) are explicit functions of the coefficients C; so is $\mathcal{B}_{\mu}$. Hence, the consistency conditions (47) are a system of $\frac{\delta}{2}(\delta-1)$ equations on $\frac{1}{2}D(D-1)$ independent coefficients C. (I remind that D is the dimension of the group G while $d=2\delta$ is that of the orbit considered). This system definitely admits some relatively simple solutions, as stems from examples found in [5]. The structure of the general solution is unknown.

2.4. Extensions of the Lie-algebraic QES problems.

The algebraic approach to constructing QES outlined above is most developed and transparent but by no means unique. Several works are devoted either to alternative approaches or to generalizations. First of all, let me mention the work [2] where an alternative (Ushveridze) formalism was introduced. Originally it has been developed with no reference to the underlying algebraic structure at all. Later it has been shown [4], however, that most of the results of [2] can be successfully reformulated in the algebraic language. At present it is not quite clear whether the Ushveridze formalism actually allows one to go beyond the algebraic TKO method. In ref. [2] it is asserted that there exist certain QES systems – the so called finite series – which seem to be irreducible to the Lie-algebraic QES. The question of whether they can be actually obtained after appropriate generalizations remains open. Some additional remarks concerning this issue can be found in a recent review [15] which, among other things, illustrates the possibilities of the analytic method [2] in relating the QES systems to other well-known problems of mathematical physics. The Gaudin spin-magnets on discretized lattices (completely integrable within the Bethe ansatz) are shown to give rise to equations of the type considered in [2].

The most straightforward generalizations of the standard algebraic procedure were worked out long ago; see e.g. [5]. (In the given context "long ago" means in 1988-89.) I assume that the "old" works containing suggestions how one could generalize the original procedure [1, 3] are already known to this or that extent, and will not dwell on them. The idea which immediately comes to one's mind is to consider higher Lie algebras and/or to substitute the "normal" Lie algebras by their extensions: graded Lie algebras (called also superalgebras) or quantum algebras. The first option – superalgebras – leading to *matrix* QES equations have been considered in [5]. The second topic – QES operators based on quantum algebras – surfaced later [20]. Speaking in the most general terms one may say that for quantum algebras instead of the differential operators one gets finite-difference operators, with the corresponding spectral problem. Serious questions remain unresolved so far, however, and the corresponding paper [20] remains unpublished. The both topics are even in more infant stage than other aspects of the theory of QES. Serious new investments are obviously needed, especially from the mathematical side, before we will be able to say that the situation is well understood and all relevant details are worked out.

A few words about other – less straightforward – proposals concerning how the parent QES systems can be transformed to generate daughter Hamiltonians with new properties. Ref. [21], treats the Coulomb potential which, as such, does not fall in the standard classification. A method of algebraization suggested there can, perhaps, be called "functional reproduction" based on $SL(2, \mathbf{R})$.

Ref. [22] shows, that starting from a QES system with N eigenstates in the algebraic sector and consistently applying Witten's supersymmetry [23], step by step, one can easily build $N-1$ daughter QES systems. At each step we generate a new QES system and loose one level from the algebraic sector. In the mathematical language this method is called the Darboux transformation. Related ideas are also discussed in [24, 25, 26].

Finally, it is in order to mention two quite recent publications [27, 28].

In [27] it is shown that the classical Gelfand-Levitan method [29] allows one to build algebraically, starting from the given QES potential, a *wider* class of QES problems with the same number of states in the spectrum in the algebraic sector as in the original problem. Start from a given QES Hamiltonian and assume that the eigenfunctions and eigenvalues belonging to the algebraic sector are established (N levels). Next, using the Gelfand-Levitan method construct a class of new Schrödinger equations with the same N eigenvalues – what happens with the rest of the spectrum is of no concern to us. The observation of Ref. [27] is as follows. One can always choose the kernel of the Gelfand-Levitan equation in such a way that this equation is algebraically soluble. This implies, in turn, that the new potential and the new eigenfunctions are algebraically expressible in terms of the original ones. The new potential contains more free parameters (up to N additional parameters).

In general, the corresponding problem – building daughter potentials with the given part of the spectrum – is being discussed in the mathematical literature in the general context (with no reference to QES) for a long time. It is called "isospectral problem", see e.g. [30, 31, 32]. Ref. [27] merely adapts general results in the QES environment.

Another interesting question is raised and solved in Ref. [28]. In a free translation from mathematical I would formulate the question in the following way. Assume we have linear differential operators (Hamiltonians) of higher orders (for instance, of the fourth or sixth order). Can they be equivalent to the standard QES system with the Hamiltonian quadratic in derivatives, and if yes, under what conditions?

Without even attempting to present the results of [28] in their full rigor and generality, let me illustrate the construction suggested using $SL(2, \mathbf{R})$ as an example – again, as many times above.

Relevant realization of the $SL(2, \mathbf{R})$ generators is given in eq. (5). We will introduce now new generators,

$$\tilde{J}_a = \exp(-\frac{1}{2}tJ_+^2)J_a\exp(\frac{1}{2}tJ_+^2), \quad a = 0, +, -$$

where t is a numerical parameter. It is obvious that the commutation relations of the new generators are the same good old $SL(2, \mathbf{R})$ relations of the old generators,

$$[\tilde{J}_0\tilde{J}_+] = \tilde{J}_+, \ [\tilde{J}_0\tilde{J}_-] = -\tilde{J}_-, \ [\tilde{J}_+\tilde{J}_-] = -2\tilde{J}_0.$$

Therefore, the new generators are as good for building $SL(2, \mathbf{R})$ QES Hamiltonians as the old ones. On the other hand, being expressed in terms of the old generators

(5) $\tilde{J}_a$ have the form

$$\tilde{J}_+ = J_+, \quad \tilde{J}_0 = J_0 + tJ_+^2, \quad \tilde{J}_- = J_- + t(J_+J_0 + J_0J_+) + t^2J_+^3. \tag{49}$$

In deriving eq. (48) we used the standard commutation relations for J_a. Thus, a realization for the $SL(2, \mathbf{R})$ generators is obtained containing three derivatives plus a free parameter t!. Choosing a bilinear combination of $\tilde{J}_a$ as $\mathcal{H}_{"G"}$ we get a differential operator of order higher than two, with the Lie-algebraic structure built in, which should be completely equivalent to the corresponding second order differential operator (the same bilinear combination of J_a). Both have the same algebraic sector, the eigenfunctions being related to each other by an obvious transformation. In other words one can say that this given class of higher-order QES operators is reducible to the second-order QES operators in an "appropriate space".

3. QES Spectral Problems and Conformal Field Theory

Deep parallels between QES systems and CFT have first been noted in [8]. Essentially, there are three aspects where similarity is quite obvious and immediately comes to one's mind. Let me list them in a natural order – more transparent relations come first: (i) calculation of the eigenvalues of QES Hamiltonians and the set of conformal weights in CFT; (ii) calculation of the eigenfunctions belonging to the algebraic sector and the correlation functions of certain primary fields in CFT; (iii) the operator product expansion (fusion rules) in CFT and corresponding relations between the wave functions in quantum mechanics. In the first two issues one starts from what is trivial in the quantal spectral problem and applies this knowledge to CFT. In the third issue the direction is reversed. The operator product expansion is a respectable and very well studied element of the CFT "establishment" – as a matter of fact, it is one of the basic tools in CFT – while its counterpart in QES problems seems to be rather obscure. Previously one could hardly even suspect that something of this kind exists in quantum mechanics. Needless to say that the question is *terra incognita*; only the first steps are being made now.

All three issues are rather technical and, hence, might seem uninteresting to many. I would like to emphasize that the whole subject is rather technical, and any new insight, from whatever point it might come, must be considered as potentially important. The observed correspondence between the QES problems and CFT can lead to a breakthrough either in one theory or in the other, to say nothing about the possibility of mutually applying methods specific to each particular theory beyond their original frameworks.

The three-year-old results of ref. [8] have been elaborated and essentially expanded in recent works [9, 10] which seem to lead to a qualitatively new level of understanding. These publications are very short, and I am not sure that I understand all details or, even, that all details are worked out. Still the topic seems very interesting, the results are exciting, and I hope that the discussion below will stimulate further efforts in this direction. The main assertions are formulated as follows. (i) In the quasiclassical limit $c \to \infty$ (where c is the Virasoro central charge) the decoupling equations of CFT with the null vector at the second level are identical to the "pole" equations in QES (see eq. (21)). The quantized parameter j figuring in QES plays the role of the genus of the Riemann surface on which CFT is formulated. (ii) The QES spectral problem is directly reducible to the calculation of the conformal weights *via* an extended Sugawara-type construction. (iii) The quantal analog of the operator product expansion is a set of fusion rules in the parameter space for the QES wave functions which relates the wave functions of a given system to that of a daughter system, with one parameter less. Below I will discuss all these issues, but at first let me remind a few facts from CFT crucial for understanding.

3.1. Elements of CFT.

Discussion in this subsection will be fragmentary and by no means should be considered as explanatory. Rather, it is just a description of notations standard for CFT. The reader interested in a systematic presentation is advised to turn to numerous reviews devoted to CFT, see e.g. [33].

Two-dimensional conformal field theories form a specific class of theories with the conserved and traceless energy-momentum tensor. These properties generate a powerful symmetry – the conformal group – which in two dimensions (unlike, say, four) contains an infinite number of parameters. The corresponding algebra goes under the name of the Virasoro algebra,

$$[L_n L_m] = (n-m)L_{n+m} + \frac{c}{12}(n^3 - n)\delta_{n+m,0} \tag{50}$$

where L_n are the coefficients of the Laurent expansion of the energy-momentum tensor $T(z)$,

$$T(z) = \sum_{n=-\infty}^{\infty} L_n z^{-n-2},$$

and the constant c is the central charge of the Virasoro algebra. Similar relations hold, of course, for the complex conjugate quantities, $\bar{T}(\bar{z})$ and $\bar{L}_n$. Eq. (49) summarizes the standard commutation relations for the energy-momentum tensor, including the so called Schwinger term [34]. The latter corresponds to the central charge in eq. (49).

Furthermore, apart from the conformal symmetry, one more invariance, chiral, is usually inherent to two-dimensional CFT. The currents generating this invariance, $J_a(z)$, satisfy the commutation relations known in physics as *"current algebra"* (for a review see [35]). The current algebra reflects the chiral group structure and also includes the Schwinger term [36]. Being expanded in the Laurent series,

$$J^a(z) = \sum_{n=-\infty}^{\infty} J_n^a z^{-n-1}$$

the current commutators give rise to the Kač-Moody algebra,

$$[J_m^a J_n^b] = i f^{abc} J_{m+n}^c + km\delta^{ab}\delta_{m+n,0}$$

where f^{abc} are the structure constants of the chiral group and k is the central charge, sometimes called the level of the Kač-Moody algebra. As in the case of the Virasoro algebra k is related to the Schwinger term [36]. The Kač-Moody and Virasoro algebras are obviously infinitely-dimensional.

In contradistinction with all conventional approaches to the field theory (where one would deal with Lagrangians, equations of motion, canonic quantization, etc.)

the ultimate dream of the CFT people is to find a *purely algebraic* solution of the theory which would be complete and close the issue. Such a revolutionary program has been formulated in [37] where an ideal goal has been put forward. Assume somebody comes and tells you the values of two constants, c and k. You switch on your algebraic machine – click, click – the output is all scaling dimensions and correlation functions of the theory.

So far, the program turns out to be successful (supplemented by the requirement of unitarity) essentially only at $c \leq 1$. At $c > 1$ (presumably) isolated points in the space of all two-dimensional CFT are tractable algebraically by virtue of the so called generalized Sugawara construction (see below).

The very fact that the modern approach to two-dimensional CFT is based on algebraic methods brings it together with the quantal QES problems, although this general relationship does not seem to be direct. What is direct, however, is the kinship of the Sugawara construction to eq. (38), the observation first made in [8]. As we will discuss later (see eq. (66)) the heart of the Sugawara-type approach is a relation connecting the energy-momentum tensor with the chiral currents.

All fields in CFT are grouped in the conformal families. Each family starts from a primary field, i.e. the one which transforms multiplicatively

$$\phi(z) \to \left(\frac{df}{dz}\right)^h \phi(z)$$

under the conformal transformation $z \to f(z)$; the real number h appearing in this equation is called the conformal weight. Apart from the primary field the conformal family includes an infinite number of descendants (say, the derivative of a primary field is a descendant) forming altogether the irreducible representation of the conformal group. The primary field is nothing else than the highest weight of the representation[6].

As has been noticed in [37] the complete information on the algebraic properties of the fields is encoded in the operator fusion rules. The conformal primary field ϕ can be alternatively defined through the fusion rule with the energy-momentum tensor,

$$T(z)\phi(w) = \frac{h}{(z-w)^2}\phi(w) + \frac{1}{z-w}\partial_z\phi(w) + \phi^{(-2)}(w) + (z-w)\phi^{(-3)}(w) + \ldots$$

(51)

$$\equiv \frac{1}{(z-w)^2}\hat{L}_0\phi(w) + \frac{1}{z-w}\hat{L}_{(-1)}\phi(w) + \hat{L}_{(-2)}\phi(w) + (z-w)\hat{L}_{(-3)}\phi(w) + \ldots .$$

Here h is the conformal weight, and the coefficients in front of the terms which are non-singular in $z - w$ define the descendent (secondary) fields $\hat{L}_{(-n)}\phi$, or, which is the same, $\phi^{(-2)}$, $\phi^{(-3)}$, etc. appearing at the level 2, 3, etc. The level 1 descendant $\phi^{(-1)} = \partial\phi$. (A side remark: $T(z)$ is a level 2 descendant of the unit operator). The conformal weight of the field $\hat{L}_{(-n)}\phi$ is $h + n$.

[6]Likewise, one can introduce the notion of the primary field with respect to the Kač-Moody algebra.

The correlation functions of the primary fields with the energy-momentum tensor satisfy the Ward identities stemming from the global conformal invariance,

$$< T(z)\phi_1(w_1)...\phi_n(w_n) >$$

$$= \sum_{j=1}^{n} \left(\frac{h_j}{(z-w_j)^2} + \frac{1}{z-w_j}\frac{\partial}{\partial w_j} \right) < \phi_1(w_1)...\phi_n(w_n) > . \tag{52}$$

These Ward identities combined with the fusion rules for $T(z)\phi(w)$, eq. (50), imply that the correlation functions involving a secondary field can be always expressed in terms of those including only primary fields. Indeed, substituting eq. (50) in (51), assuming that z is close to w_n and separating the terms of the appropriate power in $z - w_n$ – which can be done e.g. by the contour integration – we arrive at the following expression

$$< \phi_1(w_1)...\phi_n(w_n)\phi^{(-k)}(z) >= \mathcal{L}_{-k} < \phi_1(w_1)...\phi_{n-1}(w_{n-1})\phi(z) > \tag{53}$$

where

$$\mathcal{L}_{-k} = -\sum_{j=1}^{n-1} \left(\frac{(1-k)h_j}{(w_j - z)^k} + \frac{1}{(w_j-z)^{k-1}}\frac{\partial}{\partial w_j} \right). \tag{54}$$

Now, the primary field ϕ of conformal weight h acting on the vacuum produces the state $|h >= \phi(0)|0 >$ satisfying the conditions

$$L_0|h >= h|h >, \ L_n|h >= 0 \ for \ positive \ n.$$

The state $|h>$ is the highest weight state. Then $L_{-n_1}...L_{-n_k}|h>$ where all n_k are positive are descendant states. This set of states forms the Verma module. It may well happen that not all states from the Verma module are linearly independent. If there is a linear combination of states that vanishes it is called the null state.

The complete information specifying two-dimentional CFT is given by the set of the conformal weights of the Virasoro highest weight states and the set of coefficients in the fusion rules for the primary fields.

Some linear combinations of descendent fields at certain level vanish in many CFT. The null states exist, for instance, in the so called minimal models introduced in [37], the paper pioneering in many aspects of CFT and a constant source of inspiration for CFT practitioners. In the minimal models the Virasoro central charge $c < 1$, and they possess a closed operator algebra with a finite number of the primary fields. If we additionally require the representations of the Virasoro algebra to be unitary (this is what is called the discrete unitary series, see [38]) the Virasoro central charge and the conformal weights of the primary fields are expressible in terms of integer numbers,

$$c = 1 - \frac{6}{m(m+1)}, \ m = 3, 4, ...; \ h_{p,q} = \frac{[(m+1)p - mq]^2 - 1}{4m(m+1)}$$

where p and q are integers satisfying $1 \le p \le m-1, \ 1 \le q \le p$. In these models the primary field with the weight $h_{p,q}$ necessarily has a null descendent at level pq. Let us consider, for example, the Ising model – a simplest representative of the discrete unitary series. In this model $m = 3$ and, correspondingly, $c = \frac{1}{2}$. The identity operator ($h_{1,1} = 0$) has a null descendent at level 1 – which is trivial, of course. Two non-trivial operators with $h_{2,1} = \frac{1}{2}$ and $h_{1,2} = \frac{1}{16}$ have null descendents at level two – alternatively one can say that they are degenerate at level 2.

Since the correlation functions with the secondary fields are expressible through those with the primary fields by virtue of the differential procedure (52) it is clear that the existence of the null vectors implies that the correlators of the primary fields obey differential equations which are called decoupling equations.

It is not difficult to check that if a primary field is degenerate at level two and has the conformal weight h then

$$\left(\hat{L}_{-2} - \frac{3}{2(2h+1)}\hat{L}_{-1}^2\right)\phi = 0, \tag{55}$$

and

$$c = \frac{2h(5-8h)}{2h+1}. \tag{56}$$

The differential equation in this case takes the form

$$\frac{3}{2(h+1)}\frac{\partial^2}{\partial z^2} < \phi(z)\phi_1(w_1)...\phi_j(w_j)... >=$$

$$\sum_j \left(\frac{h_j}{(z-w_j)^2} + \frac{1}{z-w_j}\frac{\partial}{\partial w_j}\right) < \phi(z)\phi_1(w_1)...\phi_j(w_j)... > . \tag{57}$$

Thus, for any primary field degenerate at level 2 we get a differential equation of the second order for n-point functions including this field. We are, of course, interested in the case when the number n is sufficiently large. Indeed, all two- and three-point functions are fixed in the general case[7] by conformal invariance in terms of the conformal weights (up to multiplicative constants). The four- point functions depend only on one variable, the bi-harmonic ratio.

The solutions to eq. (56) generically reduce to the standard hypergeometric functions. The situation simplifies considerably if we pass to the classical limit of large c. From eq. (54) we see that infinitely large Virasoro central charge appears in the limit $h \to -1/2$, i.e. for non-unitary theories. The fact that CFT we are going to consider is non-unitary does not necessarily make the analysis uninteresting – it is well- known that in different applications non-unitary CFT play an important role.

The quasiclassical limit of the decoupling equation (56) has been studied in [39]. Analysis carried out in this work assumes that CFT is formulated on sphere. Omitting details which would lead me far astray I will just quote the equation for a correlation function with the primary field $\phi_{2,1}$, supplementing it by a few remarks explaining notations. The subscript "2, 1" marks the conformal weight of the field before the classical reduction. The reduced equation referring to $c = \infty$ has the form

$$\left(\frac{\partial^2}{\partial z^2} + \frac{1}{4}\sum_i \frac{1-m_i^2}{(z-w_i)^2} + \frac{1}{2}\sum_i \frac{\sigma_i}{z-w_i}\right) y_C(z) = 0. \tag{58}$$

[7] I mean, in the case when the fields are not necessarily degenerate at a certain level.

Here the function y_C is related to the original correlation function in a certain way which need not concern us here, integer parameters m determine the classical conformal weights of the participating vertex operators (conformal fields), and σ_i can be expressed in terms of m_i and w_i as follows:

$$\sigma_i = -\sum_{j\neq i} \frac{(1-m_i)(1-m_j)}{w_i - w_j}. \tag{59}$$

3.2. Decoupling equations and all that versus QES.

Let us compare now eqs.(21), (22) for QES wave functions with eq. (57) stemming, in the classical limit, from the fact of the existence of the null vector in the $\phi_{2,1}$ Verma module. It is clear, that under a certain identification of parameters both equations do not differ from each other.

First we observe that the coordinates of the force centers a_α play the same role as the coordinates w_i in the correlation function

$$< \phi_{2,1}(z)\phi_1(w_1)\phi_2(w_2)... > \tag{60}$$

In the quantal case we have at most four parameters a_α – two parameters appear when we deal with ES systems, three and four parameters generate QES systems. This means that the counterpart of ES wave function is three-point function in CFT, while four- and five-point functions correspond to QES systems.

Now, comparing the second term we easily see that

$$b_\alpha = \frac{1}{2}(m_\alpha - 1). \tag{61}$$

This identification implies, in turn, that

$$\sum_{\alpha=1}^{4}(m_\alpha - 1) = \sum_{\alpha=1}^{4}(2b_\alpha) = 2(2j-1). \tag{62}$$

This constraint on m_α coincides with the so called neutrality condition for the conformal weights in CFT! More exactly, the neutrality condition in the classical limit presented in ref. [39] reads $\sum_\alpha(m_\alpha - 1) = -2$, and the correspondence takes place only at $j = 0$. One should not forget however, that the theory considered in [39] is formulated on sphere. In general, this equation is modified,

$$\sum_{\alpha=1}^{4}(m_\alpha - 1) = 2(g-1), \tag{63}$$

where g is the genus of the surface considered. We see that $2j$ plays the role of the genus!

Finally, $4\tilde{c}_\alpha$ must coincide with $-\sigma_\alpha$. Unfortunately, I do not have explicit expressions for the coefficients σ_α for an arbitrary surface. For genus zero (sphere) these coefficients are given in eq. (58), and they are identical to $-4\tilde{c}_\alpha$, eq. (22), under the condition that $c_\alpha = 0$, which is indeed the case for $j = 0$ (see e.g. [4]). The conjecture [9] is that the coefficients σ_α continue to coincide with $-4\tilde{c}_\alpha$ for genus 1, 2, 3, etc.

Once the one-to-one correspondence between the QES wave functions and CFT correlators is established, it is not difficult to understand the meaning of the operator product expansion in quantum mechanics. OPE is absolutely transparent and familiar to everybody in the field theory, with no obvious counterpart in QES systems.

Following [9] let us consider the four-point function

$$< \phi_{2,1}(z)\phi_1(w_1)\phi_2(w_2)\phi_3(w_3) > \tag{64}$$

which, as we already know, is proportional to the wave function of a QES system with three parameters a_α, $\Psi^{QES}(z; w_1, w_2, w_3)$, where z is a coordinate variable while $w_\alpha \equiv a_\alpha$ are numerical parameters.

Now, we assume that two points in the correlator (63), say w_1 and w_2, are close to each other, and fuse the corresponding fields, according to the standard rules,

$$\phi_{\Delta_1}(w_1)\phi_{\Delta_2}(w_2) = \sum_{\Delta_3} \frac{C^{\Delta_3}_{\Delta_1\Delta_2}}{(z_1 - z_2)^{\Delta_1+\Delta_2-\Delta_3}} \phi_{\Delta_3}(w_1) \tag{65}$$

where Δ_i are the conformal weights and $C^{\Delta_3}_{\Delta_1\Delta_2}$ are the constants appearing in the operator product expansion. As a result, in eq. (63) we will have a sum of three-point functions times numerical coefficients. On the other hand, three-point correlators correspond to the wave functions of ES systems. Thus, the fusion rules in CFT are to be understood as fusion rules in the space of parameters of QES systems which relate the wave function of a QES system to those of exactly-solvable systems [9],

$$\lim_{w_1 \to w_2} \Psi^{QES}_{b_1,b_2,b_3}(z; w_1, w_2, w_3) = \sum_b \frac{C^b_{b_1 b_2}}{(w_1 - w_2)^\alpha} \Psi^{ES}_{b,b_3}(z; w_1, w_3). \tag{66}$$

Notice, that within the "twin" scenario $QES - CFT$ under discussion the parameters b_α are to be treated, according to eq. (60), as integers. The structure constants in the CFT operator product expansion can be read off if we know the coefficients of expansion of QES wave functions in ES wave functions! In the electrostatic analog picture (subsection 2.2) the fusion rules (64) correspond to two charges (or dyons) merging in the complex plane, with possible creation of multipoles.

3.3. Generalized Sugawara construction.

The last topic I would like to discuss is the spectral problem in CFT and its reduction to QES. A related question – which QES systems can be elevated to conformal field theories – will also find its solution [10].

If at $c < 1$ the algebraic analysis of CFT is quite successful and informative at $c > 1$ calculation of the conformal weights of the Virasoro primary fields is essentially an open problem. There exists an important case, however, when a complete solution can be found. The possibility of constructing the complete solution is based on a combination of the Virasoro and Kač- Moody algebras.

To make our discussion more concrete let us limit ourselves to $SL(2, \mathbf{R})$ currents although one could certainly choose another chiral group as well. A class of conformal field theories is generated if one postulates that the energy-momentum tensor is expressible in terms of the $SL(2,\mathbf{R})$ currents, [8]

$$T(z) = C^{ab} : J_a J_b : + C^a J_a + d^a \partial J_a \tag{67}$$

where C and d are numerical coefficients and $J_a(z)$ is the $SL(2,\mathbf{R})$ current. The dots in the product of the currents denote the normal-ordering prescription which in the case at hand is equivalent to subtracting a singular part of the product,

$$: J_a(z)J_b(z) := \lim_{\epsilon\to 0}(J_a(z+\epsilon)J_b(z) - \textit{singular terms}).$$

The form of the singular terms to be subtracted stems from the fusion rule

$$J_a(z)J_b(w) = \frac{k\delta_{ab}}{(z-w)^2} + \frac{if_{abc}\sqrt{2}J_c(w)}{z-w} + \textit{non} - \textit{singular terms}$$

equivalent to the Kač-Moody algebra.

The coefficients C and d are not arbitrary, they must satisfy the so called master equations – non-linear algebraic equations on $\{C, d\}$ ensuring the fact that the stress tensor (66) generates the Virasoro algebra while the chiral currents obey the fusion rule quoted above. The master equations have been first derived in [40, 8], see also [41, 10, 42, 43]. Let us note a straightforward analogy between the stress tensor (66) and the QES Hamiltonian (38), the starting point of ref. [8]. The whole approach based on eq. (66) and the master equations is known at present as the generalized Sugawara construction [44]. This class includes, as particular subclasses, the original Sugawara stress tensor [44] and the coset G/H construction [45, 46], the only solutions known for years. Now the old solutions are supplemented by a huge number of new, with rather unusual properties [8, 54, 40, 48, 49, 50, 51, 10]. A characteristic feature of the new solution of the master equation is the fact that, given the chiral group, these solutions exist only for exceptional (magic) values of k and depend on free parameters so that for these magic values we actually deal with a family of solutions (continuous deformations). For instance, the example found in [8] refers to $SL(2,\mathbf{R})$ and $k = 4$ and is a one-parameter family. The interested reader is referred for details to the original works and the review paper [43]. A general analysis of the magic values of k admitting continuous deformations for arbitrary chiral groups is carried out in [52].

The recent breakthrough in the generalized Sugawara construction resulted in several developments which deserve mentioning here. First, the master equations have been shown [41] to present an Einstein-like system of equations on the group manifold.

Second, the algebraic approach of ref. [8] turned out to be extremely helpful in analyzing [53] the geometry of the target space of the sigma models realizing the coset G/H stress tensors. As a matter of fact, literally the same expressions as those of sect. 2.3 (see eqs. (28), (29) and the preceding discussion) appear in the sigma models and immediately explain [53] peculiar properties of the background metric and the dilaton field in the sigma models (e.g. why the metric obtained is not G-invariant and why the combination $\sqrt{g}\exp(-2\phi)$ does not depend on the coefficients C^{ab}).

[8] It is conceivable that eq. (66) or its generalizations describe, for different choices of the coefficients C and d, all CFT. This conjecture is not proven, although it is discussed in the literature, see e.g. [10].

Third, considerable progress has been achieved in the issue of complete classification of the generalized Sugawara stress tensors (66) and the corresponding CFT's. The analysis of $SL(2,\mathbf{R})$-based CFT started in ref. [8] has been carried out in full [10], with the complete list of the allowed sets of coefficients C and d. Once these sets are established, computation of the conformal weights of all Kač-Moody primary fields – which are automatically Virasoro primaries, of course – becomes a simple problem *identical* to the spectral problem in the $SL(2,\mathbf{R})$ QES quantal systems. Indeed, the conformal weights are the eigenvalues of the operator L_0, the zero harmonics of the energy-momentum tensor. Projecting out the zero harmonics of the currents one arrives at the conclusion that the conformal weights are determined by the spectral equation with the operator

$$L_0 = C^{ab}(J_a)_0(J_b)_0 - d^a(J_a)_0 \tag{68}$$

where the subscript "0" marks the zero harmonics. Notice that the zero harmonics of the currents satisfy precisely the same $SL(2,\mathbf{R})$ algebra as the operators (5). Therefore, we have identical spectral problems and identical solutions. Most of the allowed sets of coefficients C and d found in [10] correspond, as a matter of fact, to ES systems. For instance, the energy-momentum tensor of the form

$$T = \frac{1}{k} : J_0^2 : +\mu : J_-^2 : + \frac{k-4}{2k}\partial J_0 \tag{69}$$

leads to the spectral problem with

$$L_0 = \frac{1}{k}(J_0)_0^2 + \mu(J_-)_0^2 + \frac{k-4}{2k}(J_0)_0 . \tag{70}$$

One easily recognizes in eq. (68) the Lie-algebraic form of the Pöschl-Teller potential (see, e.g. Table 1 in ref. [4]), with the well-known spectrum,

$$\Delta_n = \left(\frac{1+2(j-n)}{\sqrt{k}} - \sqrt{k}\right)^2 , \quad n < j + \frac{4-k}{2}, \tag{71}$$

where j is the $SL(2,\mathbf{R})$ spin.

Possible deformations for other chiral groups are studied in [52]. For instance, these authors show that for $SO(n)$ the magic value of k is $k = 2$. They also describe the space of deformations and give a general construction allowing one to obtain multi-parameter families of solutions of the master equation for the magic values of k.

As we know from the previous section not all Lie-algebraic systems are physically reasonable. In quantum mechanics there is an additional condition of normalizability of the wave functions. Is there any counterpart to this constraint in CFT, and if yes, what is its meaning? A hint that something of this kind may take place in CFT is provided by the following argument. Two-dimensional conformal field theories were shown [54] to be equivalent, under certain conditions, to three-dimensional Chern-Simons theories formulated on appropriate three-dimensional manifolds. Trying to solve the Chern-Simons theory we do geometrical quantization. It turns out then that the conformal blocks are the elements of the Hilbert space which should have finite norm. Of course, there is no obvious *direct* relation to two-dimensional CFT in this observation. The question of the direct analog of the normalizability condition in two-dimensional CFT remains open so far.

Concluding let me mention that a complete list of $SL(2,\mathbf{R})$ solutions of the master equation is given in ref.[10]. It can be interpreted as an exhaustive list of quantal QES (or ES) problems admitting a direct generalization to 2-dimensional CFT.

4. Acknowledgements

I am grateful to A. Gorsky, B.M. Levitan, P. Olver and A. Turbiner for discussions.

References

1. A. Turbiner, Comm. Math. Phys. **118** (1988) 467.
2. A. Ushveridze, Fiz. Elem. Chast. Atom. Yad. **20** (1989) 1185 [Sov. Journ. Part. Nucl. **20** (1989) 504].
3. N. Kamran and P. Olver, Journ. Math. Anal. Appl. **145** (1990) 342
4. M. Shifman, Int. Journ. Mod. Phys. **A4** (1989) 2897.
5. M. Shifman and A. Turbiner, Comm. Math. Phys. **126** (1989) 347.
6. A. González-López, N. Kamran and P. Olver, *Lie algebras of differential operators in two complex variables*, Amer. J. of Math., to appear.
7. A. González-López, N. Kamran and P. Olver, J. Phys., **A24** (1991) 3995.
8. A. Morozov, A. Perelomov, A. Rosly, M. Shifman and A. Turbiner, Int. Journ. Mod. Phys. **A5** (1990) 803.
9. A. Gorsky, Pisma ZhETF, **54** (1991) 296 [JETP Lett., **54** (1991) 289].
10. A. Gorsky and K. Selivanov, Mod. Phys. Lett. **A7** (1992) 2601.
11. P. Olver, *Quantization of Lie algebras, cohomology and quasi-exactly solvable Schrödinger operators*, Review Talk at the 873 Meeting of the American Mathematical Society, Springfield, March 20-22, 1992.
12. A. González-López, N. Kamran and P. Olver, *Normalizability of one-dimensional QES Schrödinger operators*, Preprint, School of Mathematics, University of Minnesota, 1992.
13. M. Gaudin, *La fonction d'onde de Bethe*, Masson, 1983.
14. A. Ushveridze, Mod. Phys. Lett. **A5** (1990) 1891;
T. Maglaperidze and A. Ushveridze, Mod. Phys. Lett. **A5** (1990) 1883.
15. A. Ushveridze, Fiz. Elem. Chast. Atom. Yad. **23** (1992) 58 [Sov. Journ. Part. Nucl. **23** (1992) 25].
16. A.A. Kirillov, *Elements of the theory of representations*, (in Russian), Nauka Publishers, Moscow 1972, Chapter 15 [Eng. transl. Springer-Verlag, Berlin-New York, 1976];
B. Kostant, *Quantization and representation theory of Lie groups*, Oxford, 1977; London Math. Lecture Notes Series, vol. 34, 1979;
N. Hurt, Geometrical quantization in action, D. Riedel Publishing Co., 1983.
17. A. González-López, N. Kamran and P. Olver, *Lie algebras of vector fields in the real plane*, Proc. of London Math. Society, to appear.
18. A. González-López, N. Kamran and P. Olver, *New quasi-exactly-solvable Hamiltonians in two dimensions*, University of Maryland preprint, 1993.
19. A. González-López, J.C. Hurtubise, N. Kamran and P. Olver, *Quantification de la cohomologie des algbres de Lie de champes de vecteures et fibrés en droites sur des surfaces complexes compactes*, Comptes Rendus, to appear.
20. O. Ogievetsky and A. Turbiner, *$SL(2,R)_q$ and quasi- exactly solvable problems*, Preprint CERN-TH-6212, 1991, unpublished.
21. M. Shifman, Int. Journ. Mod. Phys. **A4** (1989) 3311.
22. M. Shifman, Int. Journ. Mod. Phys. **A4** (1989) 3305.
23. E. Witten, Nucl. Phys. **B188** (1981) 513.
24. P. Jatkar, C. Kumar and A. Khare, Phys. Lett. **A142** (1989) 200.
25. P. Roy and Y. Varshni, Mod. Phys. Lett. **A6** (1991) 1257.
26. U. Paz, Mod. Phys. Lett. **A7** (1992) 2059.

27. A. Ushveridze, Mod. Phys. Lett. **A6** (1991) 739.
28. V. Ovsienko and A. Turbiner, Comptes Rendus Serie 1, **314** (1992) 13.
29. I.M. Gelfand and B.M. Levitan, Dokl. AN SSSR, **77** (1951) 557.
30. H. McKean and E. Trubowitz, Comm. Math. Phys. **82** (1981/82) 471.
31. B.M. Levitan, Matem. Sbornik, **132(174)** (1987) 73.
32. J. Pöschl and E. Trubowitz, Pure and Appl. Math. vol. 130, Academic Press, Boston, 1987.
33. P. Ginsparg, *Applied conformal field theory*, Lectures at 1988 Les Houch Summer School, Preprint HUTP-88/A054, 1988, in Proc. of the School, 1988.
34. J. Schwinger, Phys. Rev. **130** (1963) 406.
35. S. Treiman, R. Jackiw, B. Zumino and E. Witten, *Current Algebra and Anomalies*, Princeton University Press, 1985.
36. J. Schwinger, Phys. Rev. Lett. **3** (1959) 296.
37. A. Belavin, A. Polyakov and A. Zamolodchikov, Nucl. Phys. **B241** (1984) 333.
38. D. Friedan, Z. Qiu and S. Shenker, Phys. Rev. Lett. **52** (1984) 1575.
39. C. Gomez and G. Sierra, Phys. Lett. **255B** (1991) 51.
40. M.B. Halpern and E. Kiritsis, Mod. Phys. Lett. **A4** (1989) 1373; (E) 1797.
41. M.B. Halpern and J.P. Yamron, Nucl. Phys. **B332** (1990) 411.
42. A. Giveon, M.B. Halpern, E.B. Kiritsis and N.A. Obers, Int. Journ. Mod. Phys. **A7** (1992) 947.
43. M.B. Halpern, *Recent developments in the Virasoro master equation*, Talk at the Conference *Strings and Symmetries 1991*, Stony Brook, 1991, Preprint LBL-31192, 1991.
44. H. Sugawara, Phys. Rev. **170** (1968) 1659;
C. Sommerfield, Phys. Rev. **176** (1968) 2019.
45. M.B. Halpern, Phys. Rev. **D4** (1971) 2398.
46. P. Goddard, H. Kent and D. Olive, Phys. Lett. **B152** (1985) 88; Comm. Math. Phys. **103** (1986) 105.
47. E. Witten, in V. Knizhnik memorial volume, eds. L. Brink et al., World Sci., 1990.
48. M.B. Halpern, E. Kiritsis, N. Obers M. Porrati and J. Yamron, Int. Journ. Mod. Phys. **A5** (1990) 2275.
49. M.B. Halpern and N. Obers, Int. Journ. Mod. Phys. **A6** (1991) 1835.
50. S. Schrans and W. Troost, Nucl. Phys. **B345** (1990) 584.
51. M.B. Halpern and N. Obers, Nucl. Phys. **B345** (1990) 607; Comm. Math. Phys. **138** (1991) 63; Ann. Phys. (N.Y.) **212** (1991) 28; Journ. Math. Phys. **32** (1991) 3231.
52. A.A. Belov and Yu. E. Lozovik, Yad. Fiz. **53** (1991) 1464 [Sov. J. Nucl. Phys. **53** (1991) 905].
53. A.A. Tseytlin, *Conformal sigma models corresponding to gauged Wess- Zumino-Witten theories*, preprint CERN-TH-6804/93.
54. E. Witten, Comm. Math. Phys. **121** (1989) 351.

Theoretical Physics Institute, University of Minnesota, 116 Church Street SE, Minneapolis, Minnesota 55455

E-mail address: SHIFMAN@UMNACVX

Contemporary Mathematics
Volume **160**, 1994

LIE-ALGEBRAS AND LINEAR OPERATORS WITH INVARIANT SUBSPACES

ALEXANDER TURBINER

ABSTRACT. A general classification of linear differential and finite-difference operators possessing a finite-dimensional invariant subspace with a polynomial basis (the generalized Bochner problem) is given. The main result is that any operator with the above property must have a representation as a polynomial element of the universal enveloping algebra of some algebra of differential (difference) operators in finite-dimensional representation plus an operator annihilating the finite-dimensional invariant subspace. In low dimensions a classification is given by algebras $sl_2(\mathbf{R})$ (for differential operators in $\mathbf{R}$) and $sl_2(\mathbf{R})_q$ (for finite-difference operators in $\mathbf{R}$), $osp(2,2)$ (operators in one real and one Grassmann variable, or equivalently, 2×2 matrix operators in $\mathbf{R}$), $sl_3(\mathbf{R})$, $sl_2(\mathbf{R})\oplus sl_2(\mathbf{R})$ and $gl_2(\mathbf{R})\ltimes \mathbf{R}^{r+1}$, r a natural number (operators in $\mathbf{R}^2$). A classification of linear operators possessing infinitely many finite-dimensional invariant subspaces with a basis in polynomials is presented. A connection to the recently-discovered quasi-exactly-solvable spectral problems is discussed.

S. Bochner (1929) asked about a classification of differential equations

$$T\varphi = \epsilon\varphi \tag{0}$$

where T is a linear differential operator of kth order in one real variable $x \in \mathbf{R}$ and ϵ is the spectral parameter, having an infinite sequence of *orthogonal* polynomial solutions [1] (see also [2]).

Definition 0.1. Let us give the name of the *generalized Bochner problem* to the problem of classification of linear differential (difference) operators, for which the eigenvalue problem (0) has a certain number of eigenfunctions in the form of a finite-order polynomial in some variables.

Following this definition the original Bochner problem is simply a particular case. In [3] a general method has been formulated for generating eigenvalue problems for linear differential operators, linear matrix differential operators and linear finite-difference operators in one and several variables possessing polynomial solutions. The method was based on considering the eigenvalue problem for the representation of a polynomial element of the universal enveloping algebra of the Lie algebra in a

1991 *Mathematics Subject Classification.* 81C05, 81C40, 17B15.

Supported in Part by the Swiss National Foundation.
This paper is in final form and no version will be submitted for publication elsewhere.

finite-dimensional, 'projectivized' representation of this Lie algebra [3]. Below it is shown that this method provides both necessary and sufficient conditions for the existence of polynomial solutions of linear differential equations and a certain class of finite-difference equations.

The generalized Bochner problem can be subdivided into two parts: (i) a classification of linear operators possessing an invariant subspace (subspaces) with a basis in polynomials of finite degree and (ii) a description of the conditions under which such operators are symmetrical. This paper will be devoted to a solution of the first problem; as for the second one, only the first step has been done (see below). The plan of the paper is the following: Section 1 is devoted to the case of differential operators in one real variable; in Section 2 finite-difference operators in $\mathbf{R}$ are treated; operators in one real and one Grassmann variables are considered in Section 3, while the case of operators in $\mathbf{R}^2$ is given in Section 4. The general situation is described in the Conclusion.

1. Ordinary differential equations

Consider the space of all polynomials of order n

$$\mathcal{P}_n = \langle 1, x, x^2, \dots, x^n \rangle, \tag{1}$$

where n is a non-negative integer and $x \in \mathbf{R}$.

Definition 1.1. Let us name a linear differential operator of the kth order, T_k *quasi-exactly-solvable*, if it preserves the space $\mathcal{P}_n$. Correspondingly, the operator E_k, which preserves the infinite flag $\mathcal{P}_0 \subset \mathcal{P}_1 \subset \mathcal{P}_2 \subset \dots \subset \mathcal{P}_n \subset \dots$ of spaces of all polynomials, is named *exactly-solvable.*

Lemma 1.1. *(i) Suppose $n > (k-1)$. Any quasi-exactly-solvable operator T_k, can be represented by a k-th degree polynomial of the operators*

$$\begin{aligned} J^+ &= x^2\partial_x - nx, \\ J^0 &= x\partial_x - \frac{n}{2}\,, \\ J^- &= \partial_x\,, \end{aligned} \tag{2}$$

(the operators (2) obey the $sl_2(\mathbf{R})$ commutation relations [1]). If $n \le (k-1)$, the part of the quasi-exactly-solvable operator T_k containing derivatives up to order n can be represented by an nth degree polynomial in the generators (2).

(ii) Conversely, any polynomial in (2) is quasi-exactly solvable.

(iii) Among quasi-exactly-solvable operators there exist exactly-solvable operators $E_k \subset T_k$.

Comment 1. If we define the universal enveloping algebra U_g of a Lie algebra g as the algebra of all ordered polynomials in generators, then T_k at $k < n+1$ is simply an element of the universal enveloping algebra $U_{sl_2(\mathbf{R})}$ of the algebra $sl_2(\mathbf{R})$ taken in representation (2). If $k \ge n+1$, then T_k is represented as an element of

[1]The representation (2) is one of the 'projectivized' representations (see [3]). This realization of $sl_2(\mathbf{R})$ has been derived at the first time by Sophus Lie.

$U_{sl_2(\mathbf{R})}$ plus $B\frac{d^{n+1}}{dx^{n+1}}$, where B is any linear differential operator of order not higher than $(k-n-1)$.

Proof. The essential part of the proof is based on the Burnside theorem (see, e.g., [4][2]):

> Let $A_1, \ldots, A_k$ be linear operators in a linear space E over real numbers, $dimE < \infty$. Let us assume that there is no linear space L over real numbers, $0 < dimL < dimE$, such that $A_i : L \mapsto L$ for all $i = 1, 2, \ldots, k$. Then any linear operator acting in E can be represented as a polynomial in $A_1, \ldots, A_k$.

The operators $J^{\pm,0}$ act in $\mathcal{P}_n$ irreducibly and therefore the theorem can be applied. Also there exist operators B having $\mathcal{P}_n$ as a kernel, $B : \mathcal{P}_n \mapsto 0$. Clearly, those operators have a form $B(x, \partial_x)\frac{d^{n+1}}{dx^{n+1}}$, where $B(x, \partial_x)$ is any linear differential operator, and they make no contribution in T_k if $k < n+1$. It completes the proof of part (i). Parts (ii) and (iii) are easy to prove based on part (i).[3] □

Since $sl_2(\mathbf{R})$ is a graded algebra, let us introduce the grading of generators (2):

$$deg(J^+) = +1\ ,\ deg(J^0) = 0\ ,\ deg(J^-) = -1, \tag{3}$$

hence

$$deg[(J^+)^{n_+}(J^0)^{n_0}(J^-)^{n_-}]\ =\ n_+ - n_-. \tag{4}$$

The grading allows us to classify the operators T_k in the Lie-algebraic sense.

Lemma 1.2. *A quasi-exactly-solvable operator $T_k \subset U_{sl_2(\mathbf{R})}$ has no terms of positive grading, if and only if it is an exactly-solvable operator.*

Theorem 1.1. *Let n be a non-negative integer. Take the eigenvalue problem for a linear differential operator of the kth order in one variable*

$$T_k\varphi\ =\ \varepsilon\varphi\ , \tag{5}$$

where T_k is symmetric. The problem (5) has $(n+1)$ linearly independent eigenfunctions in the form of a polynomial in variable x of order not higher than n, if and only if T_k is quasi-exactly-solvable. The problem (5) has an infinite sequence of polynomial eigenfunctions, if and only if the operator is exactly-solvable.

Comment 2. The " if " part of the first statement is obvious. The " only if " part is a direct corollary of Lemma 1.1 .

This theorem gives a general classification of differential equations

$$\sum_{j=0}^{k} a_j(x)\varphi^{(j)}(x)\ =\ \varepsilon\varphi(x) \tag{6}$$

[2] Burnside theorem is a particular case of more general Jacobson theorem (see [5], Chapter XVII.3). I am grateful to V. Kac for this comment.

[3] A detailed, technical proof is presented in [13].

having at least one polynomial solution in x. The coefficient functions $a_j(x)$ must have the form

$$a_j(x) \;=\; \sum_{i=0}^{k+j} a_{j,i}x^i \tag{7}$$

The explicit expressions (7) for coefficient function in (6) are obtained by the substitution (2) into a general, kth degree polynomial element of the universal enveloping algebra $U_{sl_2(\mathbf{R})}$. Thus the coefficients $a_{j,i}$ can be expressed through the coefficients of the kth degree polynomial element of the universal enveloping algebra $U_{sl_2(\mathbf{R})}$. The number of free parameters of the polynomial solutions is defined by the number of parameters characterizing a general kth degree polynomial element of the universal enveloping algebra $U_{sl_2(\mathbf{R})}$. A rather straightforward calculation leads to the following formula

$$par(T_k) = (k+1)^2 \tag{8}$$

where we denote the number of free parameters of operator T_k by the symbol $par(T_k)$. For the case of an infinite sequence of polynomial solutions expression (7) simplifies to

$$a_j(x) \;=\; \sum_{i=0}^{j} a_{j,i}x^i \tag{9}$$

in agreement with the results of H.L. Krall's classification theorem [6](see also [2]). In this case the number of free parameters is equal to

$$par(E_k) = \frac{(k+1)(k+2)}{2} \tag{10}$$

In the present approach Krall's theorem is simply a description of differential operators of kth order in one variable preserving a finite flag $\mathcal{P}_0 \subset \mathcal{P}_1 \subset \mathcal{P}_2 \subset \cdots \subset \mathcal{P}_k$ of spaces of polynomials. One can easily show that the preservation of such a set of polynomial spaces implies the preservation of an infinite flag of such spaces.

One may ask a more general question: which non-degenerate linear differential operators have a finite-dimensional invariant sub-space of the form

$$\langle \alpha(z), \alpha(z)\beta(z), \ldots, \alpha(z)\beta(z)^n \rangle \ , \tag{11}$$

where $\alpha(z), \beta(z)$ are some functions. Such operators are easily obtained from the quasi-exactly-solvable operators (see Lemma 1.1) making the change of variable $x = \beta(z)$ and the "gauge" transformation $\tilde{T} = \alpha(z)T\alpha(z)^{-1}$. Since any one- or two-dimensional invariant sub-space can be presented in the form (11), the important general statement takes place: *there are no linear operators possessing one- or two-dimensional invariant sub-space with an explicit basis other than given by Lemma 1.1* . Therefore for any linear operator possessing one- or two-dimensional invariant sub-space with an explicit basis the eigenvalue problem (5) can be reduced to the form (16) or (15), respectively (see below).

Let us consider the second-order differential equation (6), which can possess polynomial solutions. From Theorem 1.1 it follows that the corresponding differential operator must be quasi-exactly-solvable and can be represented as

$$T_2 = c_{++}J^+J^+ + c_{+0}J^+J^0 + c_{+-}J^+J^- + c_{0-}J^0J^- + c_{--}J^-J^- +$$

$$c_+J^+ + c_0J^0 + c_-J^- + c, \tag{12}$$

where $c_{\alpha\beta}, c_\alpha, c \in R$. The number of free parameters is $par(T_2) = 9$. Under the condition $c_{++} = c_{+0} = c_+ = 0$, the operator T_2 becomes exactly-solvable (see Lemma 1.2) and the number of free parameters is reduced to $par(E_2) = 6$.

Lemma 1.3. *If the operator (12) is such that*

$$c_{++} = 0 \quad and \quad c_+ = (\frac{n}{2} - m)c_{+0} \ , \ at \ some \ m = 0, 1, 2, \ldots \tag{13}$$

then the operator T_2 preserves both $\mathcal{P}_n$ and $\mathcal{P}_m$. In this case the number of free parameters is $par(T_2) = 7$.

In fact, Lemma 1.3 means that $T_2(J^\alpha(n), c_{\alpha\beta}, c_\alpha)$ can be rewritten as $T_2(J^\alpha(m), c'_{\alpha\beta}, c'_\alpha)$. As a consequence of Lemma 1.3 and Theorem 1.1, in general, among polynomial solutions of (6) there are polynomials of order n and order m.

Remark. From the Lie-algebraic point of view Lemma 1.3 means the existence of representations of second-degree polynomials in the generators (2) possessing two invariant sub-spaces. In general, if n in (2) is a non-negative integer, then among representations of kth degree polynomials in the generators (2), lying in the universal enveloping algebra, there exist representations possessing $1, 2, ..., k$ invariant sub-spaces. Even starting from an infinite-dimensional representation of the original algebra (n in (2) is *not* a non-negative integer), one can construct the elements of the universal enveloping algebra having finite-dimensional representation (e.g., the parameter n in (13) is non-integer, however T_2 has the invariant sub-space of dimension $(m + 1)$). Also this property implies the existence of representations of the polynomial elements of the universal enveloping algebra, which can be obtained starting from different representations of the original algebra.

Substituting (2) into (12) and then into (6), we obtain

$$- P_4(x)\partial_x^2\varphi(x) \ + \ P_3(x)\partial_x\varphi(x) \ + \ P_2(x)\varphi(x) \ = \ \varepsilon\varphi(x), \tag{14}$$

where the $P_j(x)$ are polynomials of jth order with coefficients related to $c_{\alpha\beta}, c_\alpha$ and n. In general, problem (14) has $(n + 1)$ polynomial solutions. If $n = 1$, as a consequence of Lemma 1.1, a more general spectral problem than (14) arises

$$- F_3(x)\partial_x^2\varphi(x) \ + \ Q_2(x)\partial_x\varphi(x) \ + \ Q_1(x)\varphi(x) \ = \ \varepsilon\varphi(x), \tag{15}$$

where F_3 is an arbitrary real function of x and $Q_j(x), j = 1, 2$ are polynomials of order j, possessing only two polynomial solutions of the form $(ax+b)$. For the case $n = 0$ (one polynomial solution) the spectral problem (6) becomes

$$- F_2(x)\partial_x^2\varphi(x) \ + \ F_1(x)\partial_x\varphi(x) \ + \ Q_0\varphi(x) \ = \ \varepsilon\varphi(x), \tag{16}$$

where $F_{2,1}(x)$ are arbitrary real functions and Q_0 is a real constant. After the transformation

$$\Psi(z) = \varphi(x(z))e^{-A(z)}, \tag{17}$$

where $x = x(z)$ is a change of the variable and $A(z)$ is a certain real function, one can reduce (14)–(16) to the Sturm-Liouville-type problem

$$(-\partial_z^2 + V(z))\Psi(z) = \varepsilon\Psi(z), \tag{18}$$

with the potential, which is equal to

$$V(z) = (A')^2 - A'' + P_2(x(z)) ,$$

if

$$A = \int (\frac{P_3}{P_4})dx - log z' \ , \ z = \int \frac{dx}{\sqrt{P_4}} \ .$$

for the case of (14). If the functions (17), obtained after transformation, belong to the $\mathcal{L}_2(\mathcal{D})$-space[4] , we reproduce the recently discovered quasi-exactly-solvable problems [7], where a finite number of eigenstates was found algebraically. For example,

$$T_2 = -4J^0J^- + 4aJ^+ + 4bJ^0 - 2(n+1+2k)J^- + b(2n+1+2k) \tag{19}$$

leads to the spectral problem (18) ($x = z^2$) with the potential

$$V(z) = a^2z^6 + 2abz^4 + [b^2 - (4n+3+2k)a]z^2, \tag{20}$$

for which at $k = 0 (k = 1)$ the first $(n + 1)$ eigenfunctions, even (odd) in x, can be found algebraically. Of course, the number of those 'algebraized' eigenfunctions is nothing but the dimension of the irreducible representation (1).

Taking different exactly-solvable operators E_2 for the eigenvalue problem (6) one can reproduce the equations having the Hermite, Laguerre, Legendre and Jacobi polynomials as solutions [3].[5] It is worth noting that for general exactly-solvable operator E_2 the eigenvalues are quadratic in number of eigenstate and can be presented as follows

$$\epsilon_n = c_{00}n^2 + c_0 n + const$$

(for details see [3]).

Under special choices of the general element $E_4(E_6, E_8)$, one can reproduce all known fourth-(sixth-, eighth-)order differential equations giving rise to infinite sequences of orthogonal polynomials (see [2] and other papers in this volume).

Recently, A. González-Lopéz, N. Kamran and P. Olver [9] gave the complete description of second-order polynomial elements of $U_{sl_2(R)}$ leading to the square-integrable eigenfunctions of the Sturm-Liouville problem (18) after transformation (17). Consequently, for second-order ordinary differential equations (14) the combination of their result and Theorem 1.1 gives a general solution of the Bochner problem as well as the more general problem of classification of equations possessing a finite number of orthogonal polynomial solutions.

---

[4]Depending on the change of variable $x = x(z)$, the space $\mathcal{D}$ can be whole real line, semi-line and a finite interval.

[5]For instance, setting the parameter $a = 0$ in (19), the equation (15) converts to the Hermite equation (after some substitution).

2. Finite-difference equations in one variable

Let us introduce the finite-difference analogue of the differential operators (2) [10]

$$\tilde{J}^{+} = x^2 D - \{n\}x$$

$$\tilde{J}^{0} = xD - \hat{n} \tag{21}$$

$$\tilde{J}^{-} = D,$$

where $\hat{n} \equiv \frac{\{n\}\{n+1\}}{\{2n+2\}}$, $\{n\} = \frac{1-q^n}{1-q}$ is the quantum symbol (or q-number), $q \in \mathbf{R}$ is a number characterizing the deformation, $Df(x) = \frac{f(x)-f(qx)}{(1-q)x}$ is a shift or a finite-difference operator (or the so-called Jackson symbol (see [11])). The operators (21) after multiplication by some factors

$$\tilde{j}^{0} = \frac{q^{-n}}{q+1}\frac{\{2n+2\}}{\{n+1\}}\tilde{J}^{0}$$

$$\tilde{j}^{\pm} = q^{-n/2}\tilde{J}^{\pm}$$

(see [10]) form a quantum $sl_2(\mathbf{R})_q$ algebra with the following commutation relations

$$q\tilde{j}^{0}\tilde{j}^{-} - \tilde{j}^{-}\tilde{j}^{0} = -\tilde{j}^{-}$$

$$q^2\tilde{j}^{+}\tilde{j}^{-} - \tilde{j}^{-}\tilde{j}^{+} = -(q+1)\tilde{j}^{0} \tag{22}$$

$$\tilde{j}^{0}\tilde{j}^{+} - q\tilde{j}^{+}\tilde{j}^{0} = \tilde{j}^{+}$$

(this algebra corresponds to the second Witten quantum deformation of sl_2 in the classification of C. Zachos [12]). If $q \to 1$, the commutation relations (22) reduce to the standard $sl_2(\mathbf{R})$ ones. A remarkable property of generators (21) is that, if n is a non-negative integer, they form the finite-dimensional representation. For q others than root of unity this representation is irreducible.

Similarly as for differential operators one can introduce quasi-exactly-solvable $\tilde{T}_k$ and exactly-solvable operators $\tilde{E}_k$.

Lemma 2.1. *(i) Suppose $n > (k-1)$. Any quasi-exactly-solvable operator $\tilde{T}_k$, can be represented by a kth degree polynomial of the operators (21). If $n \leq (k-1)$, the part of the quasi-exactly-solvable operator $\tilde{T}_k$ containing derivatives up to order n can be represented by a nth degree polynomial in the generators (21).*

(ii) Conversely, any polynomial in (21) is quasi-exactly solvable.

(iii) Among quasi-exactly-solvable operators there exist exactly-solvable operators $\tilde{E}_k \subset \tilde{T}_k$.

Comment 3. If we define an analogue of the universal enveloping algebra U_g for the quantum algebra $\tilde{g}$ as an algebra of all ordered polynomials in generators, then a quasi-exactly-solvable operator $\tilde{T}_k$ at $k < n+1$ is simply an element of the 'universal enveloping algebra' $U_{sl_2(\mathbf{R})_q}$ of the algebra $sl_2(\mathbf{R})_q$ taken in representation (21). If $k \geq n+1$, then $\tilde{T}_k$ is represented as an element of $U_{sl_2(\mathbf{R})_q}$ plus BD^{n+1}, where B is any linear difference operator of order not higher than $(k-n-1)$.

Similar to $sl_2(\mathbf{R})$, one can introduce the grading of generators (21) of $sl_2(\mathbf{R})_q$ (see (3)) and, hence, of monomials of the universal enveloping $U_{sl_2(\mathbf{R})_q}$ (see (4)).

Lemma 2.2. *A quasi-exactly-solvable operator $\tilde{T}_k \subset U_{sl_2(\mathbf{R})_q}$ has no terms of positive grading, iff it is an exactly-solvable operator.*

Theorem 2.1. *Let n be a non-negative integer. Take the eigenvalue problem for a linear difference operator of the k-th order in one variable*

$$\tilde{T}_k \varphi(x) \; = \; \varepsilon \varphi(x), \tag{23}$$

where $\tilde{T}_k$ is symmetric. The problem (23) has $(n+1)$ linearly independent eigenfunctions in the form of a polynomial in variable x of order not higher than n, if and only if T_k is quasi-exactly-solvable. The problem (23) has an infinite sequence of polynomial eigenfunctions, if and only if the operator is exactly-solvable $\tilde{E}_k$.

Comment 4. Saying the operator $\tilde{T}_k$ is symmetric, we imply that, considering the action of this operator on a space of polynomials of degree not higher than n, one can introduce a positively-defined scalar product, and the operator $\tilde{T}_k$ is symmetric with respect to it.

This theorem gives a general classification of finite-difference equations

$$\sum_{j=0}^{k} \tilde{a}_j(x) D^j \varphi(x) \; = \; \varepsilon \varphi(x) \tag{24}$$

having polynomial solutions in x. The coefficient functions must have the form

$$\tilde{a}_j(x) \; = \; \sum_{i=0}^{k+j} \tilde{a}_{j,i} x^i. \tag{25}$$

In particular, this form occurs after substitution (21) into a general kth degree polynomial element of the universal enveloping algebra $U_{sl_2(\mathbf{R})_q}$. It guarantees the existence of at least a finite number of polynomial solutions. The coefficients $\tilde{a}_{j,i}$ are related to the coefficients of the kth degree polynomial element of the universal enveloping algebra $U_{sl_2(\mathbf{R})_q}$. The number of free parameters of the polynomial solutions is defined by the number of free parameters of a general k-th order polynomial element of the universal enveloping algebra $U_{sl_2(\mathbf{R})_q}$.[6] A rather straightforward calculation leads to the following formula

$$par(\tilde{T}_k) = (k+1)^2 + 1$$

(for the second-order finite-difference equation $par(\tilde{T}^2) = 10$). For the case of an infinite sequence of polynomial solutions the formula (25) simplifies to

$$\tilde{a}_j(x) \; = \; \sum_{i=0}^{j} \tilde{a}_{j,i} x^i \tag{26}$$

[6]For quantum $sl_2(\mathbf{R})_q$ algebra there are no polynomial Casimir operators (see, e.g., [12]). However, in the representation (21) the relationship between generators analogous to the quadratic Casimir operator

$$q\tilde{J}^+\tilde{J}^- - \tilde{J}^0\tilde{J}^0 + (\{n+1\} - 2\hat{n})\tilde{J}^0 = \hat{n}(\hat{n} - \{n+1\})$$

appears. It reduces the number of independent parameters of the second-order polynomial element of $U_{sl_2(\mathbf{R})_q}$. It becomes the standard Casimir operator at $q \to 1$.

and the number of free parameters is given by

$$par(\tilde{E}_k) = \frac{(k+1)(k+2)}{2} + 1$$

(for $k = 2$, $par(\tilde{E}^2) = 7$). The increase in the number of free parameters compared to ordinary differential equations is due to the presence of the deformation parameter q. In [3] one can find a description in the present approach of the q-deformed Hermite, Laguerre, Legendre and Jacobi polynomials (for definitions of these polynomials see [11]).

Lemma 2.3. *If the operator $\tilde{T}_2$ (see (12)) is such that*

$$\tilde{c}_{++} = 0 \quad \text{and} \quad \tilde{c}_{+} = (\hat{n} - \{m\})\tilde{c}_{+0}\ , \ \text{at some } m = 0, 1, 2, \ldots \tag{27}$$

then the operator $\tilde{T}_2$ preserves both $\mathcal{P}_n$ and $\mathcal{P}_m$, and polynomial solutions in x with 8 free parameters occur.

As usual in quantum algebras, a rather outstanding situation occurs if the deformation parameter q is equal to the root of unity. For instance, the following statement holds.

Lemma 2.4. *If a quasi-exactly-solvable operator $\tilde{T}_k$ preserves the space $\mathcal{P}_n$ and the parameter q satisfies to the equation*

$$q^n = 1\ , \tag{28}$$

then the operator $\tilde{T}_k$ preserves an infinite flag of polynomial spaces $\mathcal{P}_0 \subset \mathcal{P}_n \subset \mathcal{P}_{2n} \subset \cdots \subset \mathcal{P}_{kn} \subset \ldots$.

It is worth emphasizing that, in the limit as q tends to one, Lemmas 2.1,2,3 and Theorem 2.1 coincide with Lemmas 1.1,2,3 and Theorem 1.1, respectively. Thus the case of differential equations in one variable can be treated as particular case of finite-difference ones. Evidently, one can consider the finite-difference operators, which are a mixture of generators (21) with the same value of n and different q's.

3. Operators in one real and one Grassmann variable

Define the following space of polynomials in x, θ

$$\mathcal{P}_{N,M} = \langle x^0, x^1, \ldots, x^N, x^0\theta, x^1\theta, \ldots, x^M\theta \rangle \tag{29}$$

where N, M are non-negative integers, $x \in \mathbf{R}$ and θ is a Grassmann (anticommuting) variable.

The projectivized representation of the algebra $osp(2,2)$ is given as follows.

The algebra $osp(2,2)$ is characterized by four bosonic generators $T^{\pm,0}, J$ and four fermionic generators $Q_{1,2}, \overline{Q}_{1,2}$ and given by the commutation relations

$$[T^0, T^{\pm}] = \pm T^{\pm} \quad , \quad [T^+, T^-] = -2T^0 \quad , \quad [J, T^{\alpha}] = 0 \quad , \alpha = +, -, 0$$
$$\{Q_1, \overline{Q}_2\} = -T^- \quad , \quad \{Q_2, \overline{Q}_1\} = T^+ \quad ,$$
$$\frac{1}{2}(\{\overline{Q}_1, Q_1\} + \{\overline{Q}_2, Q_2\}) = +J \quad , \quad \frac{1}{2}(\{\overline{Q}_1, Q_1\} - \{\overline{Q}_2, Q_2\}) = T^0 \quad ,$$
$$\{Q_1, Q_1\} = \{Q_2, Q_2\} = \{Q_1, Q_2\} = 0 \quad ,$$

$$\{\overline{Q}_1,\overline{Q}_1\} = \{\overline{Q}_2,\overline{Q}_2\} = \{\overline{Q}_1,\overline{Q}_2\} = 0 \quad ,$$

$$[Q_1,T^+] = Q_2 \quad , \quad [Q_2,T^+] = 0 \quad , \quad [Q_1,T^-] = 0 \quad , \quad [Q_2,T^-] = -Q_1 \ ,$$

$$[\overline{Q}_1,T^+] = 0 \quad , \quad [\overline{Q}_2,T^+] = -\overline{Q}_1 \quad , \quad [\overline{Q}_1,T^-] = \overline{Q}_2 \quad , \quad [\overline{Q}_2,T^-] = 0 \ ,$$

$$[Q_{1,2},T^0] = \pm\frac{1}{2}Q_{1,2} \quad , \quad [\overline{Q}_{1,2},T^0] = \mp\frac{1}{2}\overline{Q}_{1,2}$$

$$[Q_{1,2},J] = -\frac{1}{2}Q_{1,2} \quad , \quad [\overline{Q}_{1,2},J] = \frac{1}{2}\overline{Q}_{1,2} \tag{30}$$

This algebra has the algebra $sl_2(\mathbf{R}) \oplus \mathbf{R}$ as sub-algebra.

The algebra (30) possesses the projectivized representation [14]

$$T^+ = x^2\partial_x - nx + x\theta\partial_\theta,$$

$$T^0 = x\partial_x - \frac{n}{2} + \frac{1}{2}\theta\partial_\theta \ , \tag{31}$$

$$T^- = \partial_x \ .$$

$$J = -\frac{n}{2} - \frac{1}{2}\theta\partial_\theta$$

for bosonic (even) generators and

$$Q = \begin{bmatrix} Q_1 \\ Q_2 \end{bmatrix} = \begin{bmatrix} \partial_\theta \\ x\partial_\theta \end{bmatrix} , \ \bar{Q} = \begin{bmatrix} \bar{Q}_1 \\ \bar{Q}_2 \end{bmatrix} = \begin{bmatrix} x\theta\partial_x - n\theta \\ -\theta\partial_x \end{bmatrix}, \tag{32}$$

for fermionic (odd) generators, where x is a real variable and θ is a Grassmann one. Inspection of the generators shows that if n is a non-negative integer, the representation (31), (32) is a finite-dimensional representation of dimension $(2n+1)$. The polynomial space $\mathcal{P}_{n,n-1}$ describes the corresponding invariant sub-space.

Definition 3.1. Let us name a linear differential operator of the k-th order, $T_k(x,\theta)$, *quasi-exactly-solvable* if it preserves the space $\mathcal{P}_{n,n-1}$. Correspondingly, the operator $E_k(x,\theta) \in T_k(x,\theta)$, which preserves the infinite flag $\mathcal{P}_{0,0} \subset \mathcal{P}_{1,0} \subset \mathcal{P}_{2,1} \subset \cdots \subset \mathcal{P}_{n,n-1} \subset \ldots$ of spaces of all polynomials, is named *exactly-solvable*.

Lemma 3.1. *Consider the space $\mathcal{P}_{n,n-1}$.*

(i) Suppose $n > (k-1)$. Any quasi-exactly-solvable operator $T_k(x,\theta)$, can be represented by a kth degree polynomial of the operators (31), (32). If $n \leq (k-1)$, the part of the quasi-exactly-solvable operator $T_k(x,\theta)$ containing derivatives in x up to order n can be represented by an nth degree polynomial in the generators (31), (32).

(ii) Conversely, any polynomial in (31), (32) is a quasi-exactly solvable operator.

(iii) Among quasi-exactly-solvable operators there exist exactly-solvable operators $E_k \subset T_k(x,\theta)$.

Let us introduce the grading of the bosonic generators (31)

$$deg(T^+) = +1 \ , \ deg(J, T^0) = 0 \ , \ deg(J^-) = -1 \tag{33}$$

and fermionic generators (32)

$$deg(Q_2, \overline{Q}_1) = +\frac{1}{2} \ , \ deg(Q_1, \overline{Q}_2) = -\frac{1}{2} \tag{34}$$

Hence the grading of monomials of the generators (31), (32) is equal to

$$deg[(T^+)^{n_+}(T^0)^{n_0}(J)^{\overline{n}}(T^-)^{n_-}{Q_1}^{m_1}{Q_2}^{m_2}{\overline{Q}_1}^{\overline{m}_1}{\overline{Q}_2}^{\overline{m}_2}] =$$

$$(n_+ - n_-) \ - \ (m_1 - m_2 - \overline{m}_1 + \overline{m}_2)/2 \tag{35}$$

The n's can be arbitrary non-negative integers, while the m's are equal to either 0 or 1. The notion of grading allows us to classify the operators $T_k(x, \theta)$ in the Lie-algebraic sense.

Lemma 3.2. *A quasi-exactly-solvable operator $T_k(x, \theta) \subset U_{osp(2,2)}$ has no terms of positive grading other than monomials of grading +1/2 containing the generator Q_1 or Q_2, iff it is an exactly-solvable operator.*

Theorem 3.1. *Let n be a non-negative integer. Take the eigenvalue problem for a linear differential operator in one real and one Grassmann variable*

$$T_k(x, \theta)\varphi \ = \ \varepsilon\varphi \tag{36}$$

where T_k is symmetric. In general, the problem (36) has $(2n+1)$ linearly independent eigenfunctions in the form of polynomials in variables x, θ of order not higher than n, if and only if T_k is quasi-exactly-solvable. The problem (36) has an infinite sequence of polynomial eigenfunctions, if and only if the operator is exactly-solvable.

This theorem gives a general classification of differential equations

$$\sum_{i,j=0}^{i=k,j=1} a_{i,j}(x, \theta)\varphi_{x,\theta}^{(i,j)}(x, \theta) \ = \ \varepsilon\varphi(x, \theta) \ , \tag{37}$$

where the notation $\varphi_{x,\theta}^{(i,j)}$ means the ith order derivative with respect to x and jth order derivative with respect to θ, having at least one polynomial solution in x, θ, thus resolving the generalized Bochner problem. Suppose that $k > 0$, then the coefficient functions $a_{i,j}(x, \theta)$ should have the form

$$a_{i,0}(x, \theta) \ = \ \sum_{p=0}^{k+i} a_{i,0,p} \ x^p \ + \ \theta \sum_{p=0}^{k+i-1} \overline{a}_{i,0,p} \ x^p$$

$$a_{i,1}(x, \theta) \ = \ \sum_{p=0}^{k+i-1} a_{i,1,p} \ x^p \ + \ \theta \sum_{p=0}^{k+i-1} \overline{a}_{i,1,p} \ x^p \tag{38}$$

The explicit expressions (38) are obtained by substituting (31), (32) into a general, the kth order, polynomial element of the universal enveloping algebra $U_{osp(2,2)}$. Thus the coefficients $a_{i,j,p}$ can be expressed through the coefficients of the k-th order polynomial element of universal enveloping algebra $U_{osp(2,2)}$. The number of free parameters of the polynomial solutions is defined by the number of parameters characterizing a general, kth order polynomial element of the universal enveloping algebra $U_{osp(2,2)}$. However, in counting parameters some relations between generators should be taken into account, specific for the given representation (31), (32), like

$$
\begin{aligned}
2T^+J - \overline{Q}_1 Q_2 &= nT^+ ,\\
T^+Q_1 - T^0 Q_2 &= -Q_2 ,\\
T^+\overline{Q}_2 + T^0\overline{Q}_1 &= \frac{(1-n)}{2}\overline{Q}_1 ,\\
JQ_2 &= \frac{n}{2} Q_2 ,\\
J\overline{Q}_1 &= \frac{(n+1)}{2}\overline{Q}_1 ,\\
T^+T^- - T^0T^0 - JJ + T^0 &= -\frac{n}{2}(n-1) ,\\
JJ &= (n+\frac{1}{2})J - \frac{n}{4}(n+1) ,\\
Q_1\overline{Q}_1 + Q_2\overline{Q}_2 - 2nJ &= -n(n+1) ,\\
2T^0J + Q_1\overline{Q}_1 - (n+1)T^0 - nJ &= -\frac{n}{2}(n+1) ,\\
T^-Q_2 - T^0Q_1 &= (\frac{n}{2}+1)Q_1 ,\\
T^-\overline{Q}_1 - T^0\overline{Q}_2 &= \frac{(n-1)}{2})\overline{Q}_2 ,\\
JQ_1 &= \frac{n}{2}Q_1 ,\\
J\overline{Q}_2 &= \frac{n+1}{2}\overline{Q}_2 ,
\end{aligned}
$$

$$2JT^- - Q_1\overline{Q}_2 = (n+1)T^- \tag{39}$$

between quadratic expressions in generators (and the ideals generated by them). Straightforward analysis leads to the following formula for the number of free parameters

$$par(T_k(x,\theta)) = 4k(k+1)+1 , \tag{40}$$

For the case of an exactly-solvable operator (an infinite sequence of polynomial solutions of Eq. (37)), the expressions (38) simplify and reduce to

$$
\begin{aligned}
a_{i,0}(x,\theta) &= \sum_{p=0}^{i} a_{i,0,p}x^p + \theta\sum_{p=0}^{i-1}\overline{a}_{i,0,p}x^p \\
a_{i,1}(x,\theta) &= \sum_{p=0}^{i} a_{i,1,p}x^p + \theta\sum_{p=0}^{i-1}\overline{a}_{i,1,p}x^p
\end{aligned} \tag{41}
$$

Correspondingly, the number of free parameters reduces to

$$par(E_k(x,\theta)) = 2k(k+2)+1 \tag{42}$$

Hence, Eq. (37) with the coefficient functions (41) gives a general form of eigenvalue problem for the operator T_k, which can lead to an infinite family of orthogonal polynomials as eigenfunctions. If in (41) we put formally all coefficients, $\overline{a}_{i,0,p}$ and $a_{i,1}(x,\theta)$ equal to zero, we reproduce the eigenvalue problem for the differential operators in one real variable, which gives rise to all known families of orthogonal polynomials in one real variable (see [13]).

3.1. Second-order differential equations in x,θ. Now let us consider in more detail the second-order differential equation Eq. (37), which can possess polynomial solutions. As follows from Theorem 3.1, the corresponding differential operator $T_2(x,\theta)$ should be quasi-exactly-solvable. Hence, this operator can be expressed in terms of $osp(2,2)$ generators taking into account the relations (39)

$$\begin{gathered}T_2 = c_{++}T^+T^+ + c_{+0}T^+T^0 + c_{+-}T^+T^- + c_{0-}T^0T^- + c_{--}T^-T^- + \\ c_{+J}T^+J + c_{0J}T^0J + c_{-J}T^-J + \\ c_{+\bar{1}}T^+\overline{Q}_1 + c_{+2}T^+Q_2 + c_{+1}T^+Q_1 + c_{+\bar{2}}T^+\overline{Q}_2 + c_{01}T^0Q_1 + \\ c_{0\bar{2}}T^0\overline{Q}_2 + c_{-1}T^-Q_1 + c_{-\bar{2}}T^-\overline{Q}_2 + \\ c_+T^+ + c_0T^0 + c_-T^- + c_JJ + c_1Q^1 + c_2Q^2 + c_{\bar{1}}\overline{Q}_1 + c_{\bar{2}}\overline{Q}_2 + c\end{gathered} \tag{43}$$

where $c_{\alpha\beta}, c_\alpha, c$ are parameters. Following Lemma 3.2, under the conditions

$$c_{++} = c_{+0} = c_{+\bar{1}} = c_{+\bar{2}} = c_{\bar{1}} = c_{+2} = c_{+J} = c_+ = 0\ , \tag{44}$$

the operator $T_2(x,\theta)$ becomes exactly-solvable (see (41)).

Now we proceed to the detailed analysis of the quasi-exactly-solvable operator $T_2(x,\theta)$. Set

$$c_{++} = 0 \tag{45}$$

in Eq. (43). The remainder possesses an exceptionally rich structure. The whole situation can be subdivided into three cases

$$c_{+2} \neq 0\ ,\ c_{+\bar{1}} \ = \ 0\ (case\ I) \tag{46}$$

$$c_{+2} \ = \ 0\ ,\ c_{+\bar{1}} \neq 0\ (case\ II) \tag{47}$$

$$c_{+2} \ = \ 0\ ,\ c_{+\bar{1}} \ = \ 0\ (case\ III) \tag{48}$$

We emphasize that we keep the parameter n in the representation (31), (32) as a fixed, non-negative integer.

Case I. The conditions (45) and (46) are fulfilled (see Fig. 3.I).

Case I.1.1. If

$$(n+2)c_{+0} + nc_{+J} + 2c_{+} = 0 \ ,$$
$$c_{+\bar{2}} = c_{\bar{1}} = 0 \ ,$$
$$(n+1)c_{0\bar{2}} + 2c_{\bar{2}} = 0 \ , \tag{49}$$

then $T_2(x,\theta)$ preserves $\mathcal{P}_{n,n-1}$ and $\mathcal{P}_{n+1,n-1}$.

Case I.1.2. If

$$(n+4+2m)c_{+0} + nc_{+J} + 2c_{+} = 0 \ ,$$
$$c_{+\bar{2}} = c_{\bar{1}} = 0 \ ,$$
$$c_{0\bar{2}} = c_{\bar{2}} = c_{-\bar{2}} = 0 \ , \tag{50}$$

at a certain integer $m \geq 0$, then $T_2(x,\theta)$ preserves $\mathcal{P}_{n,n-1}$ and $\mathcal{P}_{n+2+m,n-1}$. If m is non-integer, then $T_2(x,\theta)$ preserves $\mathcal{P}_{n,n-1}$ and $\mathcal{P}_{\infty,n-1}$.

Case I.1.3. If

$$(n+1)c_{+J} + 2c_{+} = 0 \ ,$$
$$c_{+0} = 0 \ ,$$
$$c_{+\bar{2}} = c_{\bar{1}} = 0 \ ,$$
$$c_{0\bar{2}} = c_{\bar{2}} = c_{-\bar{2}} = 0 \ , \tag{51}$$

then $T_2(x,\theta)$ preserves the infinite flag of polynomial spaces the $\mathcal{P}_{n+m,n-1}$, $m = 0,1,2,\ldots$.

Case I.2.1. If

$$(n-3)c_{+0} + (n+1)c_{+J} + 2c_{+} = 0 \ ,$$
$$(n-1)c_{+\bar{2}} = c_{\bar{1}} \ ,$$
$$(n-1)c_{0\bar{2}} + 2c_{\bar{2}} = 0 \ , \tag{52}$$

then $T_2(x,\theta)$ preserves $\mathcal{P}_{n,n-1}$ and $\mathcal{P}_{n,n-2}$.

Case I.2.2. If

$$3c_{+0} - c_{+J} = 0 \ ,$$
$$(2k+2n+4)c_{+0} + 2c_{+} = 0 \ ,$$
$$c_{+\bar{2}} = c_{\bar{1}} = 0 \ ,$$
$$(2k-n+3)c_{0\bar{2}} + 2c_{\bar{2}} = 0 \ , \tag{53}$$

at a certain integer $k \geq 0$, then $T_2(x,\theta)$ preserves $\mathcal{P}_{n,n-1}$ and $\mathcal{P}_{k+2,k}$.

Case I.2.3. If

$$c_{+0} = c_{+J} = c_{+} = 0 \ ,$$
$$c_{+\bar{2}} = c_{\bar{1}} = 0 \ ,$$
$$c_{0\bar{2}} = c_{\bar{2}} = 0 \ , \tag{54}$$

then $T_2(x,\theta)$ preserves $\mathcal{P}_{n,n-1}$ and the infinite flag of the polynomial spaces $\mathcal{P}_{k+2,k}$, $k = 0,1,2,\ldots$. Note in general for this case $c_{-\bar{2}} \neq 0$.

Case I.3.1. If

$$(n-5-2m)c_{+0}+(n+1)c_{+J}+2c_{+}=0\ ,$$
$$c_{+\bar{2}}=c_{\bar{1}}=0\ ,$$
$$c_{0\bar{2}}=c_{\bar{2}}=c_{-\bar{2}}=0\ , \tag{55}$$

at a certain integer $0 \le m \le (n-3)$, then $T_2(x,\theta)$ preserves $\mathcal{P}_{n,n-1}$ and $\mathcal{P}_{n,n-3-m}$.

Case I.3.2. If

$$c_{+0}=0\ ,$$
$$(n+1)c_{+J}+2c_{+}=0\ ,$$
$$c_{+\bar{2}}=c_{\bar{1}}=0\ ,$$
$$c_{0\bar{2}}=c_{\bar{2}}=c_{-\bar{2}}=0\ , \tag{56}$$

then $T_2(x,\theta)$ preserves $\mathcal{P}_{n,n-1}$ and the sequence of the polynomial spaces $\mathcal{P}_{n,n-3-m}$, $m=0,1,2,\dots,(n-3)$.

Case I.3.3. If

$$(2k+1-n)c_{+0}+(n+1)c_{+J}+2c_{+}=0\ ,$$
$$(2m+5)c_{+0}-c_{+J}=0\ ,$$
$$c_{+\bar{2}}=c_{\bar{1}}=0\ ,$$
$$c_{0\bar{2}}=c_{\bar{2}}=c_{-\bar{2}}=0\ , \tag{57}$$

at certain integers $k,m\ge 0$, then $T_2(x,\theta)$ preserves $\mathcal{P}_{n,n-1}$ and $\mathcal{P}_{k+3+m,k}$.

Case I.3.4. If

$$c_{+0}=c_{+J}=c_{+}=0\ ,$$
$$c_{+\bar{2}}=c_{\bar{1}}=0\ ,$$
$$c_{0\bar{2}}=c_{\bar{2}}=c_{-\bar{2}}=0\ , \tag{58}$$

then $T_2(x,\theta)$ preserves $\mathcal{P}_{n,n-1}$ and the infinite flag of polynomial spaces $\mathcal{P}_{k+3+m,k}$, $k,m=0,1,2,\dots$ (cf. *Cases I.1.3 and I.2.3*).

Case II. The conditions (45) and (47) are fulfilled (see Fig. 3.II).

Case II.1.1. If

$$(n+1)c_{+0}+(n+1)c_{+J}+2c_{+}=0\ ,$$
$$c_2=0, \tag{59}$$

then $T_2(x,\theta)$ preserves $\mathcal{P}_{n,n-1}$ and $\mathcal{P}_{n,n}$.

Case II.1.2. If

$$(n+3)c_{+0}+(n+1)c_{+J}+2c_{+}=0\ ,$$
$$(n+2)c_{01}+2c_1=0\ ,$$
$$c_{+1}=c_2=0\ , \tag{60}$$

then $T_2(x,\theta)$ preserves $\mathcal{P}_{n,n-1}$ and $\mathcal{P}_{n,n+1}$.

Case II.1.3. If

$$(2k+5+n)c_{+0}+(n+1)c_{+J}+2c_{+}=0\ ,$$
$$c_{01}=c_1=0\ ,$$
$$c_{+1}=c_2=0\ ,$$
$$c_{-1}=0\ , \tag{61}$$

at a certain integer $k\geq 0$, then $T_2(x,\theta)$ preserves $\mathcal{P}_{n,n-1}$ and $\mathcal{P}_{n,n+2+k}$.

Case II.1.4. If

$$c_{+0}=0,$$
$$(n+1)c_{+J}+2c_{+}=0\ ,$$
$$c_{01}=c_1=0\ ,$$
$$c_{+1}=c_2=0\ ,$$
$$c_{-1}=0, \tag{62}$$

then $T_2(x,\theta)$ preserves $\mathcal{P}_{n,n-1}$ and the infinite flag of polynomial spaces $\mathcal{P}_{n,n+k}$, $k=0,1,2,\dots$ (cf. *Cases I.1.3, I.2.3 and I.3.4*).

Case II.2.1. If

$$(n-2)c_{+0}+nc_{+J}+2c_{+}=0\ ,$$
$$c_{+1}=c_2\ , \tag{63}$$

then $T_2(x,\theta)$ preserves $\mathcal{P}_{n,n-1}$ and $\mathcal{P}_{n-1,n-1}$.

Case II.2.2. If

$$(n+1)c_{+0}-c_{+}=0\ ,$$
$$3c_{+0}+c_{+J}=0\ ,$$
$$nc_{01}+2c_1=0\ ,$$
$$c_{+1}=c_2=0\ , \tag{64}$$

then $T_2(x,\theta)$ preserves $\mathcal{P}_{n,n-1}$ and $\mathcal{P}_{n-1,n}$.

Case II.2.3. If

$$(n-2)c_{+0}+nc_{+J}+2c_{+}=0\ ,$$
$$(2k+1)c_{+0}+c_{+J}=0\ ,$$
$$c_{+1}=c_2=0\ ,$$
$$c_{01}=c_1=c_{-1}=0\ , \tag{65}$$

at a certain integer $k\geq 0$, then $T_2(x,\theta)$ preserves $\mathcal{P}_{n,n-1}$ and $\mathcal{P}_{n-1,n+k+1}$.

Case II.2.4. If

$$c_{+0}=c_{+J}=c_{+}=0\ ,$$
$$c_{+1}=c_2=0\ ,$$
$$c_{01}=c_1=c_{-1}=0\ , \tag{66}$$

then $T_2(x,\theta)$ preserves $\mathcal{P}_{n,n-1}$ and the infinite flag of the polynomial spaces $\mathcal{P}_{n-1,n+k}$, $k=0,1,2,\ldots$ (cf. *Cases I.1.3, I.2.3, I.3.4 and II.1.4*).

Case II.3.1. If

$$(n-4)c_{+0}+nc_{+J}+2c_{+}=0\ ,$$
$$(n-2)c_{01}+2c_1=0\ ,$$
$$c_{+1}=c_2=0\ , \tag{67}$$

then $T_2(x,\theta)$ preserves $\mathcal{P}_{n,n-1}$ and $\mathcal{P}_{n-2,n-1}$.

Case II.4.1. If

$$(m-2n)c_{+0}+c_{+}=0\ ,$$
$$3c_{+0}+c_{+J}=0\ ,$$
$$(2m+2-n)c_{01}+2c_1=0\ ,$$
$$c_{+1}=c_2=0\ , \tag{68}$$

at a certain integer $m\geq 0$, then $T_2(x,\theta)$ preserves $\mathcal{P}_{n,n-1}$ and $\mathcal{P}_{m,m+1}$.

Case II.4.2. If

$$(2m-n)c_{+0}+nc_{+J}+2c_{+}=0\ ,$$
$$(2k+5)c_{+0}+2c_{+J}=0\ ,$$
$$c_{+1}=c_2=0,$$
$$c_{01}=c_1=c_{-1}=0 \tag{69}$$

at certain integers $k\geq 0, m\geq 0$, then $T_2(x,\theta)$ preserves $\mathcal{P}_{n,n-1}$ and $\mathcal{P}_{m,m+2+k}$.

Case II.4.3. If

$$c_{+0}=c_{+J}=c_{+}=0\ ,$$
$$c_{+1}=c_2=0,$$
$$c_{01}=c_1=c_{-1}=0 \tag{70}$$

then $T_2(x,\theta)$ preserves $\mathcal{P}_{n,n-1}$ and the infinite flag of the polynomial spaces $\mathcal{P}_{m,m+1+k}$, $m,k=0,1,2,\ldots$ (cf. *Cases I.1.3, I.2.3, I.3.4, II.1.4 and II.2.4*).

Case III. The conditions (45) and (48) are fulfilled (see Fig. 3.III).

Case III.1.1. If

$$(2m-n)c_{+0}+nc_{+J}+2c_{+}=0\ ,$$
$$c_{+0}+c_{+J}=0\ ,$$
$$(m-n)c_{+1}+c_2=0, \tag{71}$$

at a certain integer $m\geq 0$, then $T_2(x,\theta)$ preserves $\mathcal{P}_{n,n-1}$ and $\mathcal{P}_{m,m}$.

Case III.1.2. If

$$c_{+0} = c_{+J} = c_{+} = 0 \ ,$$

$$c_{+1} = c_2 = 0, \tag{72}$$

then $T_2(x,\theta)$ preserves $\mathcal{P}_{n,n-1}$ and the infinite flag of the polynomial spaces $\mathcal{P}_{m,m}$, $m = 0,1,2,\ldots$ (cf. *Cases I.1.3, I.2.3, I.3.4, II.1.4, II.2.4 and II.4.3*).

Case III.2.1. If

$$(2m-n)c_{+0} + nc_{+J} + 2c_{+} = 0 \ ,$$

$$c_{+0} - c_{+J} = 0 \ ,$$

$$mc_{+\bar{2}} - c_{\bar{1}} = 0, \tag{73}$$

at a certain integer $m \geq 0$, then $T_2(x,\theta)$ preserves $\mathcal{P}_{n,n-1}$ and $\mathcal{P}_{m,m-1}$.

Case III.2.2. If

$$c_{+0} = c_{+J} = c_{+} = 0 \ ,$$

$$c_{+\bar{2}} = c_{\bar{1}} = 0 \tag{74}$$

then $\mathbf{T}_2(x,\theta)$ preserves $\mathcal{P}_{n,n-1}$ and the infinite flag of polynomial spaces $\mathcal{P}_{m,m-1}$, $m = 0,1,2,\ldots$ (cf. *Cases I.1.3, I.2.3, I.3.4, II.1.4, II.2.4, II.4.3 and III.1.2*). This case corresponds to exactly-solvable operators E_k.

In [13] it has been shown that under a certain condition some quasi-exactly-solvable operators $T_2(x)$ in one real variable can preserve two polynomial spaces of different dimensions n and m (see Lemma 1.3). It has been shown that those quasi-exactly-solvable operators $T_2(x)$ can be represented through the generators of $sl_2(\mathbf{R})$ in a projectivized representation characterized either by the mark n or by the mark m. The above analysis shows that the quasi-exactly-solvable operators $T_2(x,\theta)$ in two variables (one real and one Grassmann) possess an extremely rich variety of internal properties. They are characterized by different structures of invariant sub-spaces. However, generically the quasi-exactly-solvable operators $T_2(x,\theta)$ can preserve either one, or two, or infinitely many polynomial spaces. For the latter, those operators become 'exactly-solvable' (see *Cases I.1.3, I.2.3, I.3.4, II.1.4, II.2.4, II.4.3 and III.1.2*) giving rise to eigenvalue problems (36) possessing infinite sequences of polynomial eigenfunctions. In general, for the two latter cases the interpretation of $T_2(x,\theta)$ in terms of $osp(2,2)$ generators characterized by different marks does not exist, unlike the case of quasi-exactly-solvable operators in one real variable. The only exceptions are given by *Case III.2.1* and *Case III.2.2.*

FIGURES. 3.0–3.III

Newton diagrams describing invariant subspaces $\mathcal{P}_{N,M}$ of the second-order polynomials in the generators of $osp(2,2)$. The lower line corresponds to the part of the space of zero degree in θ and the upper line of first degree in θ. The letters without brackets indicate the maximal degree of the polynomial in x. The letters in brackets indicate the maximal (or minimal) possible degree, if the degree can be varied. The *thin* line displays schematically the length of polynomial in x (the number of monomials). The *thick* line shows that the length of polynomial can not be more (or less) than that size. The *dashed* line means that the length of polynomial can take any size on this line. If the dashed line is unbounded, it means that the degree of the polynomial can be arbitrary up to infinity. The numbering of the figures I-III corresponds to the cases, which satisfy the conditions (45), (46) (*Case I*); (45), (47) (*Case II*) and (45), (48) (*Case III*).

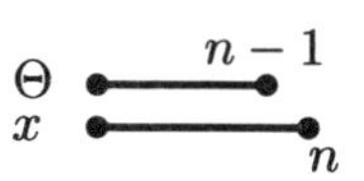

Fig. 3.0. Basic subspace

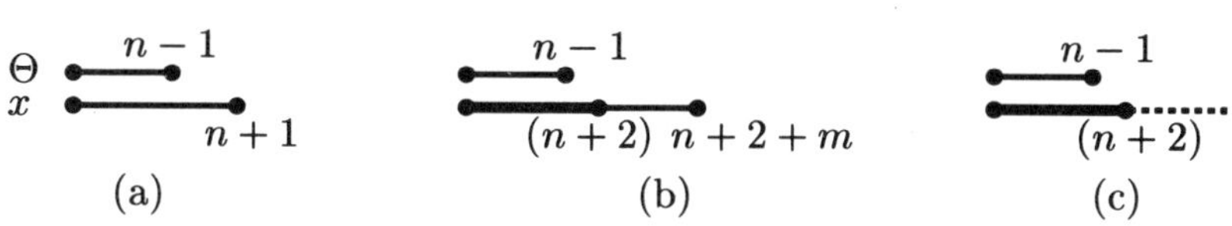

Fig. 3.I.1 . Subspaces for the *Case I.1*

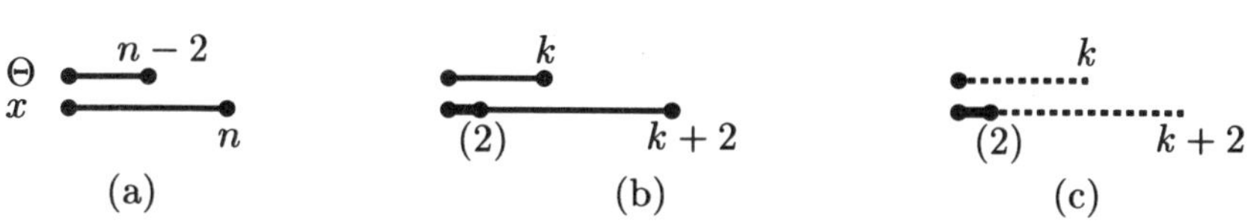

Fig. 3.I.2 . Subspaces for the *Case I.2*

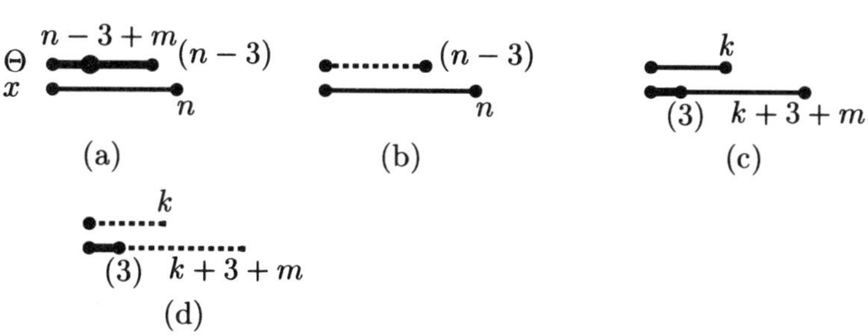

Fig. 3.I.3 . Subspaces for the *Case I.3*

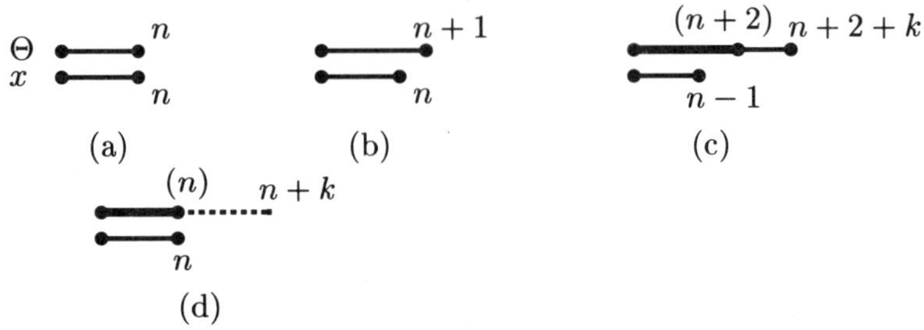

Fig. 3.II.1 . Subspaces for the *Case II.1*

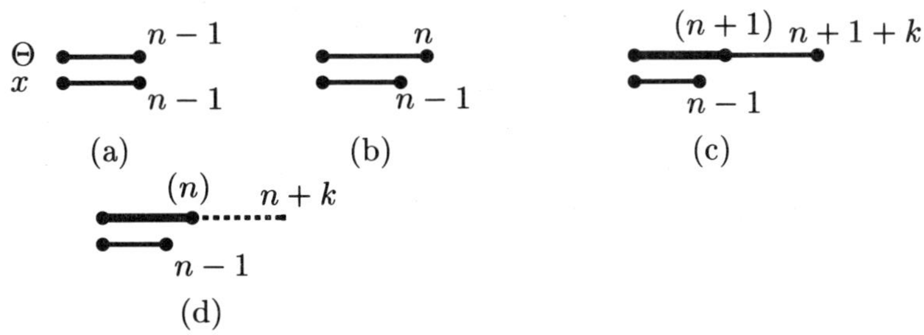

Fig. 3.II.2 . Subspaces for the *Case II.2*

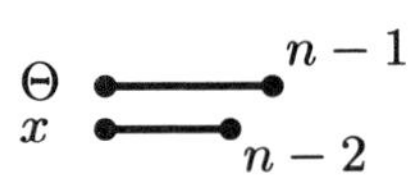

Fig. 3.II.3 . Subspace for the *Case II.3*

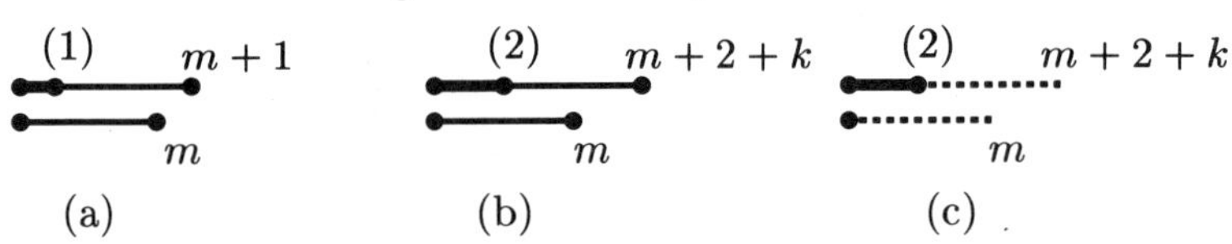

Fig. 3.II.4 . Subspaces for the *Case II.4*

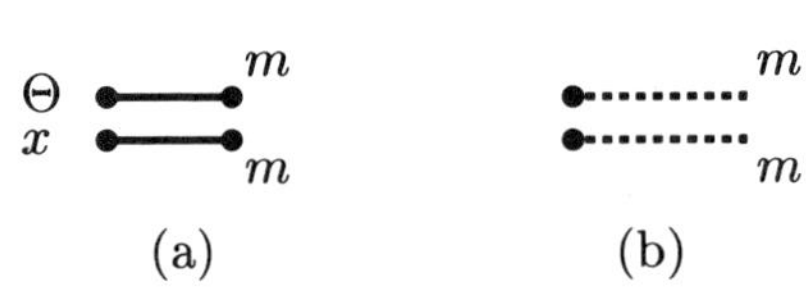

Fig. 3.III.1 . Subspaces for the *Case III.1*

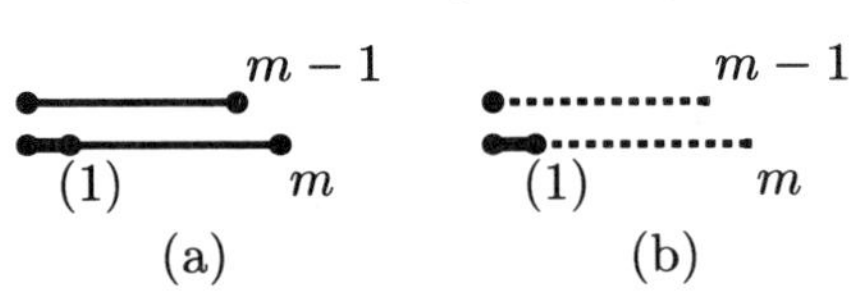

Fig. 3.III.2 . Subspaces for the *Case III.2*

3.2. 2 x 2 matrix differential equations in x. It is well known that anti-commuting variables can be represented by matrices. In our case the matrix representation is as follows: substitute θ and ∂_θ in the generators (31), (32) by the Pauli matrices σ^+ and σ^-, respectively, acting on two-component spinors. In fact, all main notations are preserved such as quasi-exactly-solvable and exactly-solvable operator, grading etc.

In the explicit form the fermionic generators (32) in matrix representation are written as follows:

$$Q = \begin{bmatrix} \sigma^- \\ x\sigma^- \end{bmatrix}, \ \bar{Q} = \begin{bmatrix} x\sigma^+\partial_x - n\sigma^+ \\ -\sigma^-\partial_x \end{bmatrix}. \tag{75}$$

The representation (75) assumes that in the spectral problem (36) an eigenfunction $\varphi(x)$ is treated as a two-component spinor

$$\varphi(x) = \begin{bmatrix} \varphi_1(x) \\ \varphi_2(x) \end{bmatrix}, \tag{76}$$

In the matrix formalism, the polynomial space (29) has the form:

$$\mathcal{P}_{N,M} = \left\langle \begin{array}{c} x^0, x^1, \ldots, x^M \\ x^0, x^1, \ldots, x^N \end{array} \right\rangle \tag{77}$$

where the terms of zero degree in θ come in as the lower component and the terms of first degree in θ come in as the upper component. The operator $T_k(x, \theta)$ becomes a 2×2 matrix differential operator $\mathbf{T}_k(x)$ having derivatives in x up to k-th order. In order to distinguish the matrix operator in x from the operator in x, θ, we will denote the former as $\mathbf{T}_k(x)$. Finally, as a consequence of Theorem 3.1, we arrive at the eigenvalue problem for a 2×2 matrix quasi-exactly-solvable differential operator $\mathbf{T}_k(x)$, possessing in general $(2n+1)$ polynomial solutions of the form $\mathcal{P}_{n,n-1}$. This eigenvalue problem can be written in the form (cf. Eq.(37))

$$\sum_{i=0}^{i=k} \mathbf{a}_{k,i}(x)\varphi_x^{(i)}(x) = \varepsilon\varphi(x) , \tag{78}$$

where the notation $\varphi_x^{(i)}$ means the ith order derivative with respect to x of each component of the spinor $\varphi(x)$ (see Eq. (76)). The coefficient functions $\mathbf{a}_{k,i}(x)$ are given by 2×2 matrices and generically for the kth order quasi-exactly-solvable operator their matrix elements are polynomials. Suppose that $k > 0$. Then the matrix elements are given by the following expressions

$$\mathbf{a}_{k,i}(x) = \begin{pmatrix} A_{k,i}^{[k+i]} & B_{k,i}^{[k+i-1]} \\ C_{k,i}^{[k+i+1]} & D_{k,i}^{[k+i]} \end{pmatrix} \tag{79}$$

at $k > 0$, where the superscript in square brackets displays the order of the corresponding polynomial.

It is easy to calculate the number of free parameters of a quasi-exactly-solvable operator $\mathbf{T}_k(x)$

$$par(\mathbf{T}_k(x)) = 4(k+1)^2 \tag{80}$$

(cf. Eq.(40)).

For the case of exactly-solvable problems, the matrix elements (79) of the coefficient functions are modified

$$\mathbf{a}_{k,i}(x) = \begin{pmatrix} A_{k,i}^{[i]} & B_{k,i}^{[i-1]} \\ C_{k,i}^{[i+1]} & D_{k,i}^{[i]} \end{pmatrix} \tag{81}$$

where $k > 0$. An infinite family of orthogonal polynomials as eigenfunctions of Eq. (78), if they exist, will occur, if and only if the coefficient functions have the form (81). The number of free parameters of an exactly-solvable operator $\mathbf{E}_k(x)$ and, correspondingly, the maximal number of free parameters of the 2×2 matrix orthogonal polynomials in one real variable, is equal to

$$par(\mathbf{E}_k(x)) = 2k(k+3) + 3 \tag{82}$$

(cf. Eq.(42)).

The increase in the number of free parameters for the 2×2 matrix operators with respect to the case of the operators in x, θ is connected to the occurrence of extra monomials of degree $(k+1)$ in generators of $osp(2,2)$ (see Eqs.(31), (32), (75)), leading to the kth order differential operators in x.

Thus, the above formulas describe the coefficient functions of matrix differential equations (78), which can possess polynomials in x as solutions, resolving the analogue of the generalized Bochner problem Eq. (0) for the case of 2×2 matrix differential equations in one real variable.

Now let us take the quasi-exactly-solvable matrix operator $\mathbf{T}_2(x)$ and try to reduce Eq. (36) to the Schroedinger equation

$$[-\frac{1}{2}\frac{d^2}{dy^2} + \mathbf{V}(y)]\Psi(y) = E\Psi(y) \tag{83}$$

where $\mathbf{V}(y)$ is a two-by-two *hermitian* matrix, by making a change of variable $x \mapsto y$ and "gauge" transformation

$$\Psi = \mathbf{U}\varphi \tag{84}$$

where $\mathbf{U}$ is an arbitrary 2×2 matrix depending on the variable y. In order to get some "reasonable" Schroedinger equation one should fulfill two requirements: (i) the potential $\mathbf{V}(y)$ must be hermitian and (ii) the eigenfunctions $\Psi(y)$ must belong to a certain Hilbert space.

Unlike the case of quasi-exactly-solvable differential operators in one real variable (see [9]), this problem has no complete solution so far. Therefore it seems instructive to display a particular example [14].

Consider the quasi-exactly-solvable operator

$$\mathbf{T}_2 = -2T^0T^- + 2T^-J - i\beta T^0 Q_1 +$$

$$\alpha T^0 - (2n+1)T^- - \frac{i\beta}{2}(3n+1)Q^1 + \frac{i}{2}\alpha\beta Q^2 - i\beta\overline{Q}_1 \, , \tag{85}$$

where α and β are parameters. Upon introducing a new variable $y = x^2$ and after straightforward calculations one finds the following expression for the matrix U in

Eq. (84)

$$\mathbf{U} = \exp(-\frac{\alpha y^2}{4} + \frac{i\beta y^2}{4}\sigma_1) \tag{86}$$

and for the potential $\hat{V}$ in Eq. (83)

$$\mathbf{V}(y) = \frac{1}{8}(\alpha^2 - \beta^2)y^2 + \sigma_2[-(n+\frac{1}{4})\beta + \frac{\alpha\beta}{4}y^2 - \frac{\alpha}{4}\tan\frac{\beta y^2}{2}]\cos\frac{\beta y^2}{2} +$$

$$\sigma_3[-(n+\frac{1}{4})\beta + \frac{\alpha\beta}{4}y^2 - \frac{\alpha}{4}\cot\frac{\beta y^2}{2}]\sin\frac{\beta y^2}{2} \tag{87}$$

It is easy to see that the potential $\mathbf{V}$ is hermitian; $(2n+1)$ eigenfunctions have the form of polynomials multiplied by the exponential factor U and they are obviously normalizable.

4. Polynomials in two real variables

For further consideration it is convenient to illustrate the space of polynomials of finite degree in several variables through Newton diagrams. In order to do this, let us introduce a d-dimensional integer lattice in $\mathbf{R}^d$ and put into correspondence to each node with the coordinates $(k_1, k_2, \dots k_d)$ the monomial ${x_1}^{k_1}{x_2}^{k_2}\dots{x_d}^{k_d}$.

4.1. Polynomials of the first type. Now let us describe the projectivized representation of the algebra $sl_3(\mathbf{R})$ in the differential operators of the first order acting on functions of two real variables. It is easy to show that the generators have the form [14]

$$J_3^1 = y^2\partial_y + xy\partial_x - ny\ , \ J_2^1 = x^2\partial_x + xy\partial_y - nx\ ,$$

$$J_3^2 = -y\partial_x\ , \ J_1^2 = -\partial_x\ , \ J_1^3 = -\partial_y\ , \ J_2^3 = -x\partial_y\ ,$$

$$J_d = y\partial_y + 2x\partial_x - n\ , \ \tilde{J}_d = 2y\partial_y + x\partial_x - n\ , \tag{88}$$

where x, y are the real variables and n is a real number. If n is a non-negative integer, the representation becomes finite-dimensional of dimension $(1+n)(1+n/2)$. The invariant sub-space has a polynomial basis and is presented as a space of all polynomials of the following type

$$\mathcal{P}_n^{(I)} = \langle 1; x, y; x^2, xy, y^2; \dots; x^n, x^{n-1}y, \dots, xy^{n-1}, y^n\rangle \tag{89}$$

or, graphically, the space (89) is given by the Newton diagram of Figure 4.1.

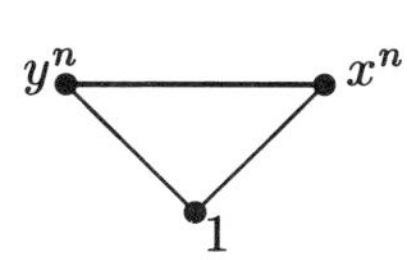

Fig. 4.1 . Graphical representation (Newton diagram) of the space $\mathcal{P}_n^{(I)}$ (see (89)).

Definition 4.1. Let us name a linear differential operator of the k-th order a *quasi-exactly-solvable of the first type*,$T_k^{(I)}(x,y)$, if it preserves the space $\mathcal{P}_n^{(I)}$. Correspondingly, the operator $E_k^{(I)}(x,y) \in T_k^{(I)}(x,y)$, which preserves the infinite flag $\mathcal{P}_0^{(I)} \subset \mathcal{P}_1^{(I)} \subset \mathcal{P}_2^{(I)} \subset \cdots \subset \mathcal{P}_n^{(I)} \subset \ldots$ of spaces of all polynomials of the type (89) , is named an *exactly-solvable of the first type.*

Lemma 4.1. *Consider the space $\mathcal{P}_n^{(I)}$.*

(i) Suppose $n > (k-1)$. Any quasi-exactly-solvable operator of the first type $T_k^{(I)}(x,y)$, can be represented by a k-th degree polynomial of the generators (88). If $n \le (k-1)$, the part of the quasi-exactly-solvable operator $T_k^{(I)}(x,y)$ of the first type containing derivatives in x, y up to the order n can be represented by a n-th degree polynomial in the generators (88).

(ii) Conversely, any polynomial in (88) is a quasi-exactly solvable operator of the first type.

(iii) Among quasi-exactly-solvable operators of the first type there exist exactly-solvable operators of the first type $E_k^{(I)}(x,y) \subset T_k^{(I)}(x,y)$.

Comment 5. The meaning of the lemma is the following: $T_k^{(I)}(x,y)$ at $k < n+1$ is simply an element of the universal enveloping algebra $U_{sl_3(\mathbf{R})}$ of the algebra $sl_3(\mathbf{R})$ in realization (88). If $k \ge n+1$, then $T_k(x,y)$ is represented as a polynomial of nth degree in (88) plus $B\frac{\partial^{n+1}}{\partial x^{n-m+1}\partial y^m}$, where $m = 0, 1, \ldots (n+1)$ and B is any linear differential operator of order not higher than $(k-n-1)$.

Let us introduce the grading of the generators (88) in the following way. The generators are characterized by the two-dimensional grading vectors

$$deg(J_3^1) = (0,+1)\ ,\ deg(J_2^1) = (+1,0)\ ,$$
$$deg(J_3^2) = (-1,+1)\ ,\ deg(J_2^3) = (+1,-1)\ ,$$
$$deg(J_d) = (0,0)\ ,\ deg(\tilde{J}_d) = (0,0)\ ,$$
$$deg(J_1^3) = (0,-1)\ ,\ deg(J_1^2) = (-1,0). \tag{90}$$

It is apparent that, the grading vector of a monomial in the generators (88) can be defined by the grading vectors of the generators by the rule

$$\tilde{deg}(T) \equiv deg[(J_3^1)^{n_{13}}(J_2^1)^{n_{12}}(J_3^2)^{n_{23}}(J_2^3)^{n_{32}}(J_d)^{n_d}(\tilde{J}_d)^{n_{\tilde{d}}}(J_1^3)^{n_{31}}(J_1^2)^{n_{21}}] =$$

$$(n_{13}+n_{32}-n_{23}-n_{31}\ ,\ n_{12}+n_{23}-n_{32}-n_{21}) \equiv (deg_x(T)\ ,\ deg_y(T)) \tag{91}$$

Here the n's can be arbitrary non-negative integers.

Definition 4.2. Let us name the grading of a monomial T in generators (75) the number

$$deg(T) = deg_x(T) + deg_y(T)\ .$$

We will say that a monomial T possesses positive grading if this number is positive. If this number is zero, then a monomial has zero grading. The notion of grading allows one to classify the operators $T_k^{(I)}(x,y)$ in a Lie-algebraic sense.

Lemma 4.2. *A quasi-exactly-solvable operator $T_k^{(I)} \subset U_{sl_3(\mathbf{R})}$ has no terms of positive grading if and only if it is an exactly-solvable operator of the first type.*

It is worth noting that among exactly-solvable operators there exists a certain important class of degenerate operators, which preserve an infinite flag of spaces of all homogeneous polynomials

$$\tilde{\mathcal{P}}_m^{(I)} = \langle x^m, x^{m-1}y, \ldots, xy^{m-1}, y^m\rangle \tag{92}$$

(represented by a horizontal line in Fig. 4.1).

Lemma 4.3. *A linear differential operator $T_k(x,y)$ preserves the infinite flag $\tilde{\mathcal{P}}_0^{(I)} \subset \tilde{\mathcal{P}}_1^{(I)} \subset \tilde{\mathcal{P}}_2^{(I)} \subset \cdots \subset \tilde{\mathcal{P}}_n^{(I)} \subset \ldots$ of spaces of all polynomials of the type (89), if and only if it is an exactly-solvable operator having terms of zero grading only. Any operators of such a type can be represented as a polynomial in the generators $J_3^2, J_2^3, J_d, \tilde{J}_d$ (see (88)), which form the algebra $so_3 \oplus R$. If such an operator contains only terms with zero grading vectors, this operator preserves any space of polynomials.*

Theorem 4.1. *Let n be a non-negative integer. In general, the eigenvalue problem for a linear symmetric differential operator in two real variables $T_k(x,y)$:*

$$T_k(x,y)\varphi(x,y) \;=\; \varepsilon\varphi(x,y) \tag{93}$$

has $(n+1)(n/2+1)$ eigenfunctions in the form of a polynomial in variables x,y belonging to the space (89), if and only if $T_k(x,y)$ is a symmetric quasi-exactly-solvable operator of the first type. The problem (93) has an infinite sequence of eigenfunctions in the form of polynomials of the form (89), if and only if the operator is a symmetric exactly-solvable operator.

This theorem gives a general classification of differential equations

$$\sum_{m=0}^{k}\sum_{i=0}^{m} a_{i,m-i}^{(m)}(x,y)\frac{\partial^m\varphi(x,y)}{\partial x^i\partial y^{m-i}} \;=\; \varepsilon\varphi(x,y) \tag{94}$$

having at least one eigenfunction in the form of a polynomial in x,y of the type (89). In general, the coefficient functions $a_{i,m-i}^{(m)}(x,y)$ have quite cumbersome functional structure and we do not display them here (below we will give their explicit form for $T_2^{(I)}(x,y)$). They are polynomials in x,y of order $(k+m)$ and always contain a general inhomogeneous polynomial of order m as a part. The explicit expressions for those polynomials are obtained by substituting (88) into a general, kth order polynomial element of the universal enveloping algebra $U_{sl_3(\mathbf{R})}$ of the algebra $sl_3(\mathbf{R})$. Thus, the coefficients in the polynomials $a_{i,m-i}^{(m)}(x,y)$ can be expressed through the coefficients of the k-th order polynomial element of the universal enveloping algebra $U_{sl_3(\mathbf{R})}$. The number of free parameters of the polynomial solutions is defined by the number of parameters characterizing a general, kth order polynomial element of the universal enveloping algebra $U_{sl_3(\mathbf{R})}$. In counting free parameters some relations between generators should be taken into account, specifically for the given representation (88)

$$\begin{aligned}
J_2^1 J_d \;-\; 2J_2^1\tilde{J}_d \;-\; 3J_3^1J_2^3 \;&=\; nJ_2^1\\
J_3^1\tilde{J}_d \;-\; 2J_3^1 J_d \;-\; 3J_2^1J_3^2 \;&=\; nJ_3^1\\
J_2^3 J_d \;+\; J_2^3\tilde{J}_d \;-\; 3J_2^1J_1^3 \;&=\; (n+3)J_2^3
\end{aligned}$$

$$J_3^2 J_d \;+\; J_3^2 \tilde{J}_d \;-\; 3J_3^1 J_1^2 \;=\; (n+3)J_3^2$$

$$3(J_2^1 J_1^2 \;+\; J_3^1 J_1^3 \;+\; J_2^3 J_3^2) \;+\; J_d J_d \;+ \tilde{J}_d \tilde{J}_d \;-\; J_d \tilde{J}_d \;=\; 3J_d \;+\; 3n \;+\; n^2$$

$$2J_d J_d \;+\; 2\tilde{J}_d \tilde{J}_d \;-\; 5J_d \tilde{J}_d \;+\; 9J_2^3 J_3^2 \;=\; (n+6)J_d \;+\; (n-3)\tilde{J}_d \;+\; n^2 \;+\; 3n$$

$$J_d J_d \;-\; \tilde{J}_d \tilde{J}_d \;+\; 3(J_2^1 J_1^2 \;-\; J_3^1 J_1^3) \;=\; (n+3)(J_d \;-\; \tilde{J}_d)$$

$$J_d J_1^2 \;-\; 2\tilde{J}_d J_1^2 \;-\; 3J_3^2 J_1^3 \;=\; nJ_1^2$$

$$\tilde{J}_d J_1^3 \;-\; 2J_d J_1^3 \;-\; 3J_2^3 J_1^2 \;=\; nJ_1^3 \tag{95}$$

between quadratic expressions in generators (and the ideals generated by them).[7] For the case of exactly-solvable problems, the coefficient functions $a^{(m)}_{i,m-i}(x,y)$ take the form

$$a^{(m)}_{i,m-i}(x,y) \;=\; \sum_{p,q=0}^{p+q\leq m} a_{i,m-i,p,q}x^p y^q \tag{96}$$

with arbitrary coefficients.

Now let us proceed to the case of the second-order differential equations. The second-order polynomial in the generators (88) can be represented as

$$T_2 \;=\; c_{\alpha\beta,\gamma\delta} J^\alpha_\beta J^\gamma_\delta \;+\; c_{\alpha\beta} J^\alpha_\beta \;+\; c \quad ,$$

where we imply summation over all repeating indices; $\alpha, \beta, \gamma, \delta$ correspond to the indices of operators in (88) and for the Cartan generators we suppose both indices simulate d or $\tilde{d}$, all c's are set to be real numbers. After substitution the expressions (88) into T_2 the explicit form of the quasi-exactly-solvable operator is given by

$$T_2^{(I)}(x,y) \;=$$

$$[x^2 P^{xx}_{2,2}(x,y) \;+\; xP^{xx}_{2,1}(x,y) \;+\; \tilde{P}^{xx}_{2,0}(x,y)]\frac{\partial^2}{\partial x^2} \;+$$

$$[xyP^{xy}_{2,2}(x,y) \;+\; P^{xy}_{3,1}(x,y) \;+\; \tilde{P}^{xy}_{2,0}(x,y)]\frac{\partial^2}{\partial x \partial y} \;+$$

$$[y^2 P^{yy}_{2,2}(x,y) \;+\; yP^{yy}_{2,1}(x,y) \;+\; \tilde{P}^{yy}_{2,0}(x,y)]\frac{\partial^2}{\partial y^2} \;+$$

$$[xP^{x}_{2,2}(x,y) \;+\; P^{x}_{2,1}(x,y) \;+\; \tilde{P}^{x}_{1,0}(x,y)]\frac{\partial}{\partial x} \;+$$

$$[yP^{y}_{2,2}(x,y) \;+\; P^{y}_{2,1}(x,y) \;+\; \tilde{P}^{y}_{1,0}(x,y)]\frac{\partial}{\partial y} +$$

$$[P^{0}_{2,2}(x,y) \;+\; P^{0}_{1,1}(x,y) \;+\; \tilde{P}^{0}_{0,0}(x,y)] \tag{97}$$

where $P^c_{k,m}(x,y)$ and $\tilde{P}^c_{k,m}(x,y)$ are homogeneous and inhomogeneous polynomials of order k, respectively, the index m numerates them, the superscript $\underline{c}$ characterizes the order of derivative, which this coefficient function corresponds to. For the case

[7] It was shown that there exist 56 cubic relations between the generators (88). However, it turned out that all of them (including standard cubic Casimir operator in given representation (88)) are functionally-dependent on quadratic relations (95) (Personal communication by G. Post).

of the second-order-exactly-solvable operator $E_2^{(I)}(x,y)$, the structure of the coefficient function is similar to (97), except for the fact that all tilde-less polynomials disappear. The number of free parameters is equal to

$$par(T_2^{(I)}(x,y)) = 36 \ .$$

while for the case of an exactly-solvable operator (an infinite sequence of polynomial eigenfunctions in the problem (93))

$$par(E_2^{(I)}(x,y)) = 25 \ .$$

An important particular case is when the quasi-exactly-solvable operator $T_2^{(I)}(x,y)$ possesses two invariant sub-spaces of the type (89). This situation is described by the following lemma:

Lemma 4.4. *Suppose $T_2^{(I)}(x,y)$ has no terms of grading 2*

$$c_{12,12} = c_{13,13} = c_{12,13} = 0 \ . \tag{98}$$

If there exist some coefficient c's and a non-negative integer N such that the conditions

$$c_{12} = (n-N-m)c_{12,d} + (n-2N+m)c_{12,\tilde{d}} + (N-m)c_{12,32} \ ,$$

$$c_{13} = (n-N-m)c_{13,d} + (n-2N+m)c_{13,\tilde{d}} + mc_{13,23} \ , \tag{99}$$

are fulfilled at all $m = 0,1,2,\dots,N$, then the operator $T_2^{(I)}(x,y)$ preserves both $\mathcal{P}_n$ and $\mathcal{P}_N$, and $par(T_2^{(I)}(x,y)) = 31$.

Now let us proceed to the important item: under what conditions on the coefficients in (97) the second-order-quasi-exactly-solvable operators can be reduced to a form of the Schroedinger operator after some gauge transformation (conjugation)

$$f(x,y)e^{t(x,y)}T_2(x,y)^{(I)}e^{-t(x,y)} \ = \ -\Delta_g \ + \ V(x,y,n) \tag{100}$$

where f,t,V are some functions in $\mathbf{R}^2$, and Δ_g is the Laplace-Beltrami operator with some metric tensor $g_{\mu\nu}$; $\mu,\nu = 1,2$ [8]. This is a difficult problem which has no complete solution yet. In [14, 16] a few multi-parametric examples have been constructed.

In general, if one can perform the transformation (100), we arrive at potentials $V(x,y,n)$ containing explicitly a dependence on the parameter n and hence the dimension of the initial invariant subspace (89). Finally, we end up at quasi-exactly-solvable Schroedinger equations with a certain number of eigenfunctions known algebraically. Taking an exactly-solvable operator $E_2^{(I)}(x,y)$ instead of the quasi-exactly-solvable one and performing the transformation (100), a certain exactly-solvable Schroedinger operator emerges. Evidently, they will have no dependence on the parameter n and correspondently an infinite sequence of eigenfunctions of the Schroedinger equation has a form of polynomials.

Also it is worth noting that there exists an important particular class of quasi-exactly-solvable operators, lying in a certain sense in between quasi-exactly-solvable

[8] Our further consideration will be restricted the case $f = 1$ only.

and exactly-solvable operators. The algebra $sl_3(\mathbf{R})$ in realization (88) contains the algebra $so_3(\mathbf{R})$ as a sub-algebra in such a way that

$$J^1 = (1+y^2)\partial_y + xy\partial_x - ny, \quad J^2 = (1+x^2)\partial_x + xy\partial_y - nx,$$

$$J^3 = x\partial_y - y\partial_x \tag{101}$$

which is not graded anymore. If the parameter n is a non-negative integer (and coincides with that in (88)), the same finite-dimensional invariant sub-space $\mathcal{P}_n^{(I)}$ (see (89) and Fig. 4.1) as in the original $sl_3(\mathbf{R})$ occurs. For this case a corresponding finite-dimensional representation is reducible and unitary. It has been proven [14, 17], that any symmetric bilinear combination of generators (101), $T_2^{(e)} = c_{\alpha\beta}J^\alpha J^\beta$, where $c_{\alpha\beta} = c_{\beta\alpha}$ are real numbers, can be reduced to a form of the Laplace-Beltrami operator plus a scalar function making a certain gauge transformation (conjugation). [9] In general, the metric tensor $g_{\mu\nu}$ is not degenerate [10] and the obtained potential $V(x,y)$ is given by a rational function of the coordinates. Moreover, the potential $V(x,y)$ has no dependence on the label n! All n-dependent terms contain no x,y- dependence and can be included into redefinition of the reference point for eigenvalues. At first sight, those quasi-exactly-solvable operators look as if they are exactly-solvable, but is not so. As a consequence of the fact that the quadratic Casimir operator for $so_3(\mathbf{R})$ in the realization (101) is non-trivial and commutes with $T_2^{(e)}$, the functional space of $T_2^{(e)}$ is subdivided into the finite-dimensional blocks of increasing sizes, corresponding to the irreducible representations of $so_3(\mathbf{R})$ (unlike truly exactly-solvable operators, for which the sizes of blocks are fixed and equal to one). We name such operators *exactly-solvable of the second type.*

Another interesting property appears, if the matrix $c_{\alpha\beta}$ has one vanishing eigenvalue: a certain mysterious relation holds [14] (see also [17, 18])

$$V(x,y,\{c\}) = \frac{3}{16}R(x,y,\{c\}), \tag{102}$$

where $R(x,y,\{c\})$ is the scalar curvature calculated through the metric tensor attached in the Laplace-Beltrami operator. This can imply that the corresponding Schroedinger operator has the form of a purely geometrical object! The real meaning of this fact is not understood so far.

4.2. Polynomials of the second type. In [13] we studied the quasi-exactly-solvable operators in one real variable. It turned out that the solution to this problem was found using a connection with the projectivized representation of the algebra $sl_2(\mathbf{R})$. As a natural step in developing the original idea, let us consider the projectivized representation of the direct sum of two algebras $sl_2(\mathbf{R})$.

[9] In fact, in [17] has been proven a more general theorem, that any symmetric bilinear combination of generators of unitary representation of the semisimple Lie algebra, realized in the first-order differential operators, can be reduced to a form of the Laplace-Beltrami operator with some metric tensor plus a scalar function making a certain gauge transformation (conjugation).

[10] Degeneracy of $g_{\mu\nu}$ occurs, for example, if $c_{\alpha\beta}$ has two vanishing eigenvalues.

The algebra $sl_2(\mathbf{R}) \oplus sl_2(\mathbf{R})$ taken in projectivized representation acts on functions of two real variables. The generators have the form (see (2))

$$J_x^+ = x^2\partial_x - nx \quad , \quad J_y^+ = y^2\partial_y - my \ ,$$

$$J_x^0 = x\partial_x - \frac{n}{2} \quad , \quad J_y^0 = y\partial_y - \frac{m}{2} \ ,$$

$$J_x^- = \partial_x \quad , \quad J_y^- = \partial_y \ , \tag{103}$$

where x, y are the real variables and n, m are numbers. If n, m are non-negative integers, there exists the finite-dimensional representation of dimension $(n+1)(m+1)$. Evidently, the invariant sub-space has a polynomial basis and is presented as a space of all polynomials described by the Newton diagram of Fig. 4.2. We denote this space as $\mathcal{P}^{(II)}_{n,m}$.

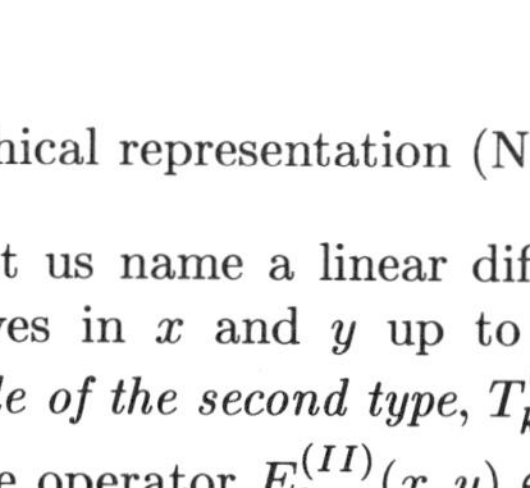

Fig. 4.2. Graphical representation (Newton diagram) of the space $\mathcal{P}^{(II)}_{n,m}$.

Definition 4.3. Let us name a linear differential operator of the (k,p)th order, containing derivatives in x and y up to kth and pth orders, respectively[11], a *quasi-exactly-solvable of the second type*, $T^{(II)}_{k,p}(x,y)$, if it preserves the space $\mathcal{P}^{(II)}_{n,m}$. Correspondingly, the operator $E^{(II)}_{k,p}(x,y) \in T^{(II)}_{k,p}(x,y)$, which preserves either the infinite flag $\mathcal{P}^{(II)}_{0,m} \subset \mathcal{P}^{(II)}_{1,m} \subset \mathcal{P}^{(II)}_{2,m} \subset \cdots \subset \mathcal{P}^{(I)}_{n,m} \subset \ldots$, or the infinite flag $\mathcal{P}^{(II)}_{n,0} \subset \mathcal{P}^{(II)}_{n,1} \subset \mathcal{P}^{(II)}_{n,2} \subset \cdots \subset \mathcal{P}^{(I)}_{n,m} \subset \ldots$ of spaces of all polynomials, is named an *exactly-solvable of the type* 2_x or 2_y, respectively.

[11] So a leading derivative has a form $\frac{\partial^{(k+p)}}{\partial_x^k \partial_y^p}$. Also we will use a notation through this section $T_N(x,y)$ implying that in general all derivatives of the order N are presented.

Lemma 4.5. *Consider the space* $\mathcal{P}^{(II)}_{n,m}$.

(i) Suppose $n > (k-1)$ *and* $m > (p-1)$. *Any quasi-exactly-solvable operator of the second type* $T^{(II)}_{k,p}(x,y)$, *can be represented by a* (k,p)*th degree polynomial of the generators (103). If* $n \leq (k-1)$ *and/or* $m \leq (p-1)$, *the part of the quasi-exactly-solvable operator* $T^{(II)}_{k,p}(x,y)$ *of the second type containing derivatives in* x, y *up to order* n, m, *respectively, can be represented by an* (n,m)*th degree polynomial in the generators (103).*

(ii) Conversely, any polynomial in (103) is a quasi-exactly solvable operator of the second type.

(iii) Among quasi-exactly-solvable operators of the second type there exist exactly-solvable operators of the second type $E^{(II)}_{k,p}(x,y) \subset T^{(II)}_{k,p}(x,y)$.

Similarly, as it has been done for the algebra $sl_3(\mathbf{R})$, one can introduce the notion of grading:

$$deg(J^+_x) = (+1,0)\ ,\ deg(J^+_y) = (0,+1)\ ,$$
$$deg(J^0_x) = (0,0)\ ,\ deg(J^0_y) = (0,0)\ ,$$
$$deg(J^-_x) = (-1,0)\ ,\ deg(J^-_y) = (0,-1). \tag{104}$$

It is apparent that, the grading vector of a monomial in the generators (90) can be defined by the grading vectors (104) of the generators by the rule

$$\tilde{deg}(T) \equiv deg[(J^+_x)^{n_{x+}}(J^0_x)^{n_{x0}}(J^-_x)^{n_{x-}}(J^+_y)^{n_{y+}}(J^0_y)^{n_{y0}}(J^-_y)^{n_{y-}}] =$$
$$(n_{x+} - n_{x-}\ ,\ n_{y+} - n_{y-}) \equiv (deg_x(T)\ ,\ deg_y(T)) \tag{105}$$

Here the n's can be arbitrary non-negative integers.

Definition 4.4. Let us name the x-grading, y-grading and grading of a monomial T in generators (103) the numbers $deg_x(T)$, $deg_y(T)$ and $deg(T) = deg_x(T) + deg_y(T)$, respectively. We say that a monomial T possesses positive x-grading (y-grading, grading), if the number $deg_x(T)$ ($deg_y(T)$, $deg(T)$) is positive. If $deg_x(T)(deg_y(T), deg(T)) = 0$, then a monomial has zero x-grading (y-grading, grading).

The notion of grading allows one to classify the operators $T^{(II)}_{k,p}(x,y)$ in the Lie-algebraic sense.

Lemma 4.6. *The quasi-exactly-solvable operator* $T^{(II)}_{k,p}(x,y)$ *preserves the infinite flag* $\mathcal{P}^{(II)}_{0,m} \subset \mathcal{P}^{(II)}_{1,m} \subset \mathcal{P}^{(II)}_{2,m} \subset \cdots \subset \mathcal{P}^{(II)}_{n,m} \subset \ldots$ *of spaces of all the polynomials, if and only if it is an exactly-solvable operator of type* 2_x *having no terms of positive* x*-grading,* $deg_x(T) > 0$.

The quasi-exactly-solvable operator $T^{(II)}_{k,p}(x,y)$ *preserves the infinite flag* $\mathcal{P}^{(II)}_{n,0} \subset \mathcal{P}^{(II)}_{n,1} \subset \mathcal{P}^{(II)}_{n,2} \subset \cdots \subset \mathcal{P}^{(II)}_{n,m} \subset \ldots$ *of all spaces of the polynomials, if and only if it is an exactly-solvable operator of type* 2_y *having no terms of positive* y*-grading,* $deg_y(T) > 0$.

If a quasi-exactly-solvable operator of the second type contains no terms of positive grading, this operator preserves the infinite flag $\mathcal{P}_0^{(I)} \subset \mathcal{P}_1^{(I)} \subset \mathcal{P}_2^{(I)} \subset \dots \subset \mathcal{P}_n^{(I)} \subset \dots$ *of spaces of all the polynomials of the type (76) and is attached to the exactly-solvable operator of the first type.*

Theorem 4.2. *Let* n, m *be non-negative integers. In general, the eigenvalue problem (93) for a linear symmetric differential operator in two real variables* $T_{k,p}(x,y)$ *has* $(n+1)(m+1)$ *eigenfunctions in the form of a polynomial in variables* x, y *belonging to the space* $\mathcal{P}_{n,m}^{(II)}$ *, if and only if* $T_{k,p}(x,y)$ *is a quasi-exactly-solvable symmetric operator of the second type. The problem (93) has an infinite sequence of eigenfunctions in the form of polynomials belonging the space* $\mathcal{P}_{n,m}^{(II)}$ *at fixed* m *(*n*), if and only if the operator is an exactly-solvable symmetric operator of type* 2_x *(*2_y*).*

This theorem gives a general classification of differential equations (94), having at least one eigenfunction in the form of a polynomial in x, y of type $\mathcal{P}_{n,m}^{(II)}$. In general, the coefficient functions $a_{i,m-i}^{(m)}(x,y)$ in (94) have a quite cumbersome functional structure and we do not display them here (below we will give their explicit form for $T_2^{(II)}(x,y)$). They are polynomials in x, y of the order $(k+m)$. The explicit expressions for those polynomials are obtained by substituting (103) into a general, kth order polynomial element of the universal enveloping algebra $U_{sl_2(\mathbf{R})\oplus sl_2(\mathbf{R})}$ of the algebra $sl_2(\mathbf{R}) \oplus sl_2(\mathbf{R})$. Thus, the coefficients in the polynomials $a_{i,m-i}^{(m)}(x,y)$ can be expressed through the coefficients of the k-th order polynomial element of the universal enveloping algebra $U_{sl_2(\mathbf{R})\oplus sl_2(\mathbf{R})}$. The number of free parameters of the polynomial solutions is defined by the number of parameters characterizing a general, kth order polynomial element of the universal enveloping algebra $U_{sl_2(\mathbf{R})\oplus sl_2(\mathbf{R})}$. In counting free parameters some relations between generators should be taken into account, specifically for the given representation (103)

$$\begin{aligned} J_x^+ J_x^- \; - \; J_x^0 J_x^0 \; + \; J_x^0 \; &= \; -\frac{n}{2}(\frac{n}{2}+1) \\ J_y^+ J_y^- \; - \; J_y^0 J_y^0 \; + \; J_y^0 \; &= \; -\frac{m}{2}(\frac{m}{2}+1) \end{aligned} \tag{106}$$

between quadratic expressions in generators (and the ideals generated by them). [12]

Now let us proceed to the case of the second-order differential equations. The second-order polynomial in the generators (103) can be represented as

$$T_2 \; = \; c_{\alpha\beta}^{xx} J_x^\alpha J_x^\beta \; + \; c_{\alpha\beta}^{xy} J_x^\alpha J_y^\beta \; + \; c_{\alpha\beta}^{yy} J_y^\alpha J_y^\beta \; + \; c_\alpha^x J_x^\alpha \; + \; c_\alpha^y J_y^\alpha \; + \; c \quad , \tag{107}$$

where we imply summation over all repeating indices and $\alpha, \beta = \pm, 0$; all c's are set to be real numbers. Taking (106) in account, it is easy to show that T_2 is characterized by 26 free parameters. Substituting (103) into (107), we obtain the explicit form of the second-order-quasi-exactly-solvable operator

$$T_2^{(II)}(x,y) \; =$$

$$\tilde{P}_{4,0}^{xx}(x)\frac{\partial^2}{\partial x^2} \; + \; [x^2y^2 P_{0,2}^{xy} \; + \; xy\tilde{P}_{1,2}^{xy}(x,y) \; + \; \tilde{P}_{2,0}^{xy}(x,y)]\frac{\partial^2}{\partial x \partial y} \; + \; \tilde{P}_{4,0}^{yy}(y)\frac{\partial^2}{\partial y^2} \; +$$

[12] For this case they correspond to the quadratic Casimir operators.

$$[\tilde{P}^x_{3,1}(x) + y\tilde{P}^x_{2,0}(x)]\frac{\partial}{\partial x} + [\tilde{P}^y_{3,1}(y) + x\tilde{P}^y_{2,0}(y)]\frac{\partial}{\partial y} + \tilde{P}^0_{2,0}(x,y) \tag{108}$$

where $P^c_{k,m}(x,y)$ and $\tilde{P}^c_{k,m}(x,y)$ are homogeneous and inhomogeneous polynomials of order k, respectively, the index m marks them, and the superscript $\underline{c}$ characterizes the order of the derivative corresponding to this coefficient function. For the case of the second-order-exactly-solvable operator of type 2_x

$$E_2^{(II)}(x,y) =$$

$$\tilde{Q}^{xx}_{2,0}(x)\frac{\partial^2}{\partial x^2} + [x\tilde{Q}^{xy}_{2,1}(y) + \tilde{Q}^{xy}_{2,0}(y)]\frac{\partial^2}{\partial x \partial y} + \tilde{Q}^{yy}_{4,0}(y)\frac{\partial^2}{\partial y^2} +$$

$$\tilde{Q}^x_{1,1}(x) + y\tilde{Q}^x_{1,0}(x)]\frac{\partial}{\partial x} + \tilde{Q}^y_{1,1}(y) + x\tilde{Q}^y_{1,0}(y)]\frac{\partial}{\partial y} + \tilde{Q}^0_{2,0}(y) \tag{109}$$

For the case of the second-order-exactly-solvable operator of type 2_y the functional form is similar to (109) with the interchange $x \leftrightarrow y$.

It is easy to show that the numbers of free parameters are

$$par(T_2^{(II)}(x,y)) = 26 \ .$$

$$par(E_2^{(II)}(x,y)) = 20 \ .$$

As in the previous cases, there is an important particular case of quasi-exactly-solvable operators of the second order, where they possess two invariant sub-spaces.

Lemma 4.7. *Suppose in (107) there are no terms of x-grading 2*

$$c^{xx}_{++} = 0 \ . \tag{110}$$

If there exist some coefficient c's and a non-negative integer N such that the conditions

$$c^{xy}_{++} = 0 \ ,$$

$$c^{xy}_{+-} = 0 \ ,$$

$$c^x_+ = (n/2 - N)c^{xx}_{+0} + (m/2 - k)c^{xy}_{+0} \ , \tag{111}$$

are fulfilled at all $k = 0, 1, 2, \ldots, m$, then the operator $T_2^{(II)}(x,y)$ preserves both $\mathcal{P}^{II}_{n,m}$ and $\mathcal{P}^{II}_{N,m}$, and $par(T_2^{(II)}(x,y)) = 22$.

Generically, the question of the reduction of quasi-exactly-solvable operators of the second type $T_2^{(II)}(x,y)$ to the form of the Schroedinger operator is still open. Recently, several multi-parametrical families of those Schroedinger operators were constructed in [14, 16]. As well as the case of the first type of quasi-exactly-solvable operators, corresponding Schroedinger operators contain in general the Laplace-Beltrami operator characterizing a non-flat space metric tensor.

The above analysis of linear differential operators preserving the space $\mathcal{P}^{(II)}_{n,m}$ can be naturally extended to the case of linear finite-difference operators defined through the Jackson symbol D (see Section 2) with the action:

$$Df(x) = \frac{f(x) - f(qx)}{(1-q)x} + f(qx)D$$

instead of a continuous derivative, where $f(x)$ is a real function and q is a number. All above-described results hold [13] with replacement of the algebra $sl_2(\mathbf{R}) \oplus sl_2(\mathbf{R})$ by the quantum algebra $sl_2(\mathbf{R})_q \oplus sl_2(\mathbf{R})_q$ in 'projectivized' representation (21) (see [3, 13]). Obviously, in the limit $q \to 1$ all results coincide to the results of the present Section.

4.3. Polynomials of the third type. The third case, which we shall discuss here, corresponds to the family of Lie algebras $gl_2(\mathbf{R}) \ltimes \mathbf{R}^{r+1}$ (semidirect sum of $gl_2(\mathbf{R})$ with a $(r+1)$-dimensional abelian ideal; Case 24 in the classification given in [19]). This family of Lie algebras, depending on an integer $r > 0$, can be realized in terms of the first-order differential operators

$$J^1 = \partial_x ,$$

$$J^2 = x\partial_x - \frac{n}{3} , \quad J^3 = y\partial_y - \frac{n}{3r} ,$$

$$J^4 = x^2\partial_x + rxy\partial_y - nx ,$$

$$J^{5+i} = x^i\partial_y , \quad i = 0, 1, \ldots, r , \tag{112}$$

where x, y are real variables and n is a real number. [14] If n is a non-negative integer, the representation becomes finite-dimensional. The invariant sub-space has a polynomial basis and is presented as a space of all polynomials of the form

$$\mathcal{P}^{(III)}_{r,n} = \sum_{i,j\geq 0}^{i+rj\leq n} a_{ij}x^i y^j \tag{113}$$

or, graphically, (113) is given by the Newton diagram of Figure 4.3. The general formula for the dimension of the corresponding finite-dimensional representation (113) is given by

$$dim\mathcal{P}^{(III)}_{r,n} = \frac{[n^2 + (r+2)n + \alpha_{r,n}]}{2r} \tag{114}$$

where for small r

$$\alpha_{1,n} = 2 ,$$

$$\alpha_{2,n} = \begin{cases} 3 & \text{at odd n} \\ 4 & \text{at even n} \end{cases}$$

$$\alpha_{3,n} = \begin{cases} 4 & \text{at (n+1)=0 (mod 3)} \\ 6 & \text{at other n} \end{cases}$$

$$\alpha_{4,n} = \begin{cases} 5 & \text{at (n+1)=0 (mod 4)} \\ 8 & \text{at other n} \\ 9 & \text{at (n+3)=0 (mod 4)} \end{cases}$$

[13] with minor modifications

[14] It is worth noting that at $r = 1$ the algebra $\{gl_2(\mathbf{R}) \ltimes \mathbf{R}^2\} \subset sl_3(\mathbf{R})$. Thus, this case is reduced to one about the first type polynomials (see Section 4.1). Hereafter, we include $r = 1$ into consideration just for the sake of completeness.

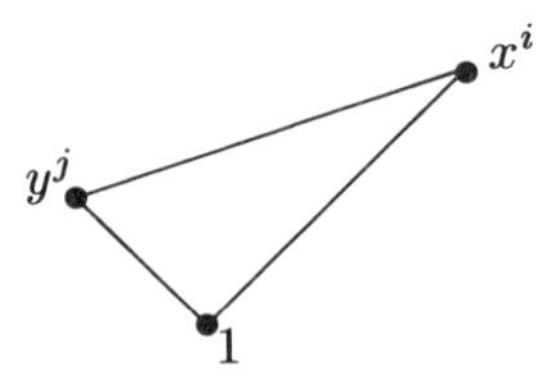

Fig. 4.3 . Graphical representation (Newton diagram) of the space $\mathcal{P}^{(III)}_{r,n}$ (see (113)).

Definition 4.5. Let us name a linear differential operator of the kth order a *quasi-exactly-solvable of the r-third type*, $T_k^{(r,III)}(x,y)$, if it preserves the space $\mathcal{P}^{(III)}_{r,n}$. Correspondingly, the operator $E_k^{(r,III)}(x,y) \in T_k^{(r,III)}(x,y)$, which preserves the infinite flag $\mathcal{P}^{(III)}_{r,0} \subset \mathcal{P}^{(III)}_{r,1} \subset \mathcal{P}^{(III)}_{r,2} \subset \cdots \subset \mathcal{P}^{(III)}_{r,n} \subset \ldots$ of spaces of all polynomials of the type (113) , is named an *exactly-solvable of the r-third type.*

Lemma 4.8. *Take the space* $\mathcal{P}^{(III)}_{r,n}$.

(i) Suppose $n > (k-1)$. Any quasi-exactly-solvable operator $T_k^{(r,III)}(x,y)$ can be represented by a kth degree polynomial of the generators (112). If $n \leq (k-1)$, the part of the quasi-exactly-solvable operator $T_k^{(r,III)}(x,y)$ of the r-third type containing derivatives in x, y up to order n can be represented by a n-th degree polynomial in the generators (112).

(ii) Conversely, any polynomial in (112) is a quasi-exactly solvable operator of the r-third type.

(iii) Among quasi-exactly-solvable operators of the r-third type there exist exactly-solvable operators of the r-third type $E_k^{(r,III)}(x,y) \subset T_k^{(r,III)}(x,y)$.

One can introduce the grading of the generators (112) in an analogous way as has been done before for the cases of the algebra $sl_3(\mathbf{R})$ (see (90)) and $sl_2(\mathbf{R}) \oplus sl_2(\mathbf{R})$ (see (104)). All generators are characterized by the two-dimensional grading vectors

$$deg(J^1) = (-1,0) \ ,$$
$$deg(J^2) = (0,0) \ , \ deg(J^3) = (0,0) \ ,$$
$$deg(J^4) = (1,0) \ ,$$

$$deg(J^5) = (0,-1) \ , \ deg(J^6) = (1,-1) \ , \ldots , \ deg(J^{5+r}) = (r,-1) \tag{115}$$

Similarly as before, the grading vector of a monomial in the generators can be defined through the grading vectors of the generators (112) (cf. (90), (105)).

Definition 4.6. Let us name the *grading* of a monomial T in generators (112) the number

$$deg(T) = deg_x(T) + r \ deg_y(T)$$

(cf. the case of $sl_3(\mathbf{R})$). We say that a monomial T possesses positive grading if $deg(T)$ is positive. If it is zero, then a monomial has zero grading.

The notion of grading allows us to classify the operators $T_k(x,y)$ in a Lie-algebraic sense.

Lemma 4.9. *A quasi-exactly-solvable operator* $T_k^{(r,III)} \subset U_{gl_2(R)\ltimes R^{r+1}}$ *has no terms of positive grading iff it is an exactly-solvable operator.*

Theorem 4.3. *Let n and $(r-1)$ be non-negative integers. In general, the eigenvalue problem (93) for a linear symmetric differential operator in two real variables $T_k(x,y)$ has a certain number of eigenfunctions in the form of a polynomial in variables x,y belonging to the space (113), if and only if $T_k(x,y)$ is a quasi-exactly-solvable, symmetric operator of the r-third type. The problem (93) has an infinite sequence of eigenfunctions in the form of polynomials belonging to (113), if and only if the operator is an exactly-solvable, symmetric operator of the r-third type.*

This theorem gives a general classification of differential equations (94), having at least one eigenfunction in the form of a polynomial in x,y of the type $\mathcal{P}_{r,n}^{(III)}$. In general, the coefficient functions $a_{i,m-i}^{(m)}(x,y)$ in (94) are polynomials in x,y and have a quite cumbersome functional structure and we do not display them here (below we will give their explicit form for $T_2^{(r,III)}(x,y)$). The explicit expressions for those polynomials are obtained by substituting (112) into a general, kth order polynomial element of the universal enveloping algebra $U_{gl_2(\mathbf{R})\ltimes\mathbf{R}^{r+1}}$ of the algebra $gl_2(\mathbf{R}) \ltimes \mathbf{R}^{r+1}$. Thus, the coefficients in the polynomials $a_{i,m-i}^{(m)}(x,y)$ can be expressed through the coefficients of the kth order polynomial element of the universal enveloping algebra $U_{gl_2(\mathbf{R})\ltimes\mathbf{R}^{r+1}}$. The number of free parameters of the polynomial solutions is defined by the number of parameters characterizing a general, kth order polynomial element of the universal enveloping algebra. In counting free parameters some relations between generators should be taken into account, specifically for a given representation (112)

$$J^2J^5 - J^1J^6 + \frac{n}{3}J^5 = 0\ ,$$

$$J^1J^4 - J^2J^2 - rJ^2J^3 - J^2 - r(\frac{n}{3}+1)J^3 = -\frac{n}{3}(\frac{n}{3}+1)\ , \tag{116}$$

$$J^2J^{6+i} + rJ^3J^{6+i} - J^4J^{5+i} - (\frac{n}{3}+1)J^{6+i} = 0\ ,$$

$$\textit{at } i = 0,1,2,\ldots,(r-1)\ , \tag{117}$$

$$J^1J^{7+i} - J^2J^{6+i} - (\frac{n}{3}+1)J^{6+i} = 0\ ,$$

$$\textit{at } i = 0,1,2,\ldots,(r-2)\ , \tag{118}$$

$$J^5J^{7+i} = \cdots = J^{5+k}J^{7+i-k}\ ,$$

$$\textit{at } k = 0,1,2,\ldots \textit{ and } 2k \le (2+i)\ , \textit{and } i = 0,1,2,\ldots,(2r-2) \tag{119}$$

between quadratic expressions in generators (and the ideals generated by them).

Now let us proceed to the case of the second-order differential equations. Taking the relations (116)-(119) in account, one can find the number of free parameters

$$par(T_2^{(r,III)}(x,y)) \;=\; 5(r+4) \tag{120}$$

and obtain the explicit form of the second-order-quasi-exactly-solvable operator

$$T_2^{(r,III)}(x,y) \;=$$

$$\tilde{P}^{xx}_{4,0}(x)\frac{\partial^2}{\partial x^2} \;+\; [\tilde{P}^{xy}_{r+2,1}(x) \;+\; y\tilde{P}^{xy}_{3,0}(x)]\frac{\partial^2}{\partial x\partial y} \;+\; [\tilde{P}^{yy}_{2r,1}(x) \;+\; y\tilde{P}^{yy}_{r+1,0}(x)]]\frac{\partial^2}{\partial y^2} \;+$$

$$[\tilde{P}^{x}_{3,0}(x)\,]\frac{\partial}{\partial x} \;+\; [\tilde{P}^{y}_{r+1,1}(x) \;+\; y\tilde{P}^{x}_{1,0}(x)]\frac{\partial}{\partial y} \;+\; \tilde{P}^{0}_{2,0}(x) \tag{121}$$

where $\tilde{P}^{c}_{k,m}(x,y)$ are inhomogeneous polynomials of order k, the index m numerates them, and the superscript $\underline{c}$ characterizes the order of the derivative corresponding to this coefficient function. For the case of the second-order-exactly-solvable operator of r-third type

$$E_2^{(r,III)}(x,y) \;=$$

$$\tilde{P}^{xx}_{2,0}(x)\frac{\partial^2}{\partial x^2} \;+\; [\tilde{P}^{xy}_{r+1,1}(x) \;+\; y\tilde{P}^{xy}_{1,0}(x)]\frac{\partial^2}{\partial x\partial y} \;+\; [\tilde{P}^{yy}_{2r,1}(x) \;+\; y\tilde{P}^{yy}_{r,0}(x)]]\frac{\partial^2}{\partial y^2} \;+$$

$$[\tilde{P}^{x}_{1,0}(x)\,]\frac{\partial}{\partial x} \;+\; [\tilde{P}^{y}_{r,1}(x) \;+\; y\tilde{P}^{x}_{0,0}(x)]\frac{\partial}{\partial y} \;+\; \tilde{P}^{0}_{0,0}(x) \tag{122}$$

In this case the number of free parameters is equal to

$$par(E_2^{(r,III)}(x,y)) \;=\; 5(r+3) \tag{123}$$

As in the previous cases, there is an important particular case of r-third type quasi-exactly-solvable operators, where they possess two invariant sub-spaces.

Lemma 4.10. *Suppose $E_2^{(r,III)}(x,y)$ has no terms of grading 2*

$$c_{44} = 0\ . \tag{124}$$

If there exist some coefficient c's and a non-negative integer N such that the conditions

$$c_{4,5+r} = 0\ ,$$

$$c_4 \;=\; (\frac{N}{3} - m)c_{24} \;+\; (\frac{N}{3} - k)c_{34}\ , \tag{125}$$

are fulfilled at all $m,k \;=\; 0,1,2,\ldots$ such that $m + rk \;=\; N$, then the operator $T_2^{(r,III)}(x,y)$ preserves both $\mathcal{P}^{III}_{r,n}$ and $\mathcal{P}^{III}_{r,N}$, and $par(T_2^{(r,III)}(x,y)) = 5r+17$.

Generically, the question of the reduction of the quasi-exactly-solvable operator $T_2^{(r,III)}(x,y)$ to the form of the Schroedinger operator is still open. Initially in [16] a few multi-parametrical families of those Schroedinger operators have been constructed. Similar to the case of quasi-exactly-solvable operators of the first and the second types, corresponding Schroedinger operators contain in general the Laplace-Beltrami operator with non-flat space metric tensor.

4.4. Discussion. Through all the above analysis, one very crucial requirement played a role: we considered the problem in general position, assuming the spaces of all polynomials contain polynomials with all possible real (complex) coefficients. If this requirement is not fulfilled, then the above connection of the finite-dimensional space of all polynomials and the finite-dimensional representation of the Lie algebra is lost and degenerate cases occur. This demands a separate investigation.

So, we described three types of finite-dimensional spaces (see Fig. 4.1-3) with a basis in polynomials of two real variables, which can be preserved by linear differential operators. The natural question emerges: is it possible to find linear differential operators having no representation through generators of a Lie algebra, which preserve the finite-dimensional space of polynomials, other than those shown in Fig. 4.1,2,3? The answer is given by the following conjecture [20].

Conjecture 1. *If a linear differential operator $T_k(x,y)$ does preserve a finite-dimensional space of all polynomials (presented by a convex Newton diagram) other than $\mathcal{P}_n^{(I)}, \mathcal{P}_{n,m}^{(II)}, \mathcal{P}_{r,n}^{(III)}$, then this operator preserves either one of those spaces as well, or a certain infinite flag of finite-dimensional spaces of all polynomials.*

The general question about reducibility of a second-order polynomial in generators of the algebras $sl_3(\mathbf{R})$, $sl_2(\mathbf{R}) \oplus sl_2(\mathbf{R})$ and $\{gl_2(\mathbf{R}) \ltimes \mathbf{R}^{r+1}\}, r = 2,3,4,\ldots$ to the form of the Laplace-Beltrami operator plus a potential is still open. For all cases where this reducibility has been performed, a metric tensor appeared which corresponded to non-flat space. All attempts to find the quasi-exactly-solvable or exactly-solvable problems in $\mathbf{R}^2$ corresponding to the flat space did lead to a problem with separability of variables. We believe that the following conjecture holds.

Conjecture 2. *In $\mathbf{R}^2$ there exist no quasi-exactly-solvable or exactly-solvable problems containing the Laplace-Beltrami operator with flat-space metric tensor, which are characterized by non-separable variables.*

Also the next question, under what conditions the eigenfunctions obtained belong to some L_2-space, is lacking a solution. The analysis analogous to that which has been performed in [16] for one-dimensional quasi-exactly-solvable problems should be done.

5. General considerations

As the main conclusion one can formulate as the following theorem.

(Main) Theorem. *Consider a Lie algebra g of the differential operators of the first order, which possesses a finite-dimensional irreducible representation $\mathcal{P}$. Any linear differential operator acting in $\mathcal{P}$ (having $\mathcal{P}$ as the invariant subspace) can be represented by a polynomial in generators of the algebra g plus an operator annihilating $\mathcal{P}$.*

As a probable extension of this theorem, we assume that the following conjecture holds [20].

Conjecture 3. *If a linear differential operator T acting on functions in $\mathbf{R}^k$ does possess a single finite-dimensional invariant subspace with polynomial basis (presented by a convex Newton diagram), this finite-dimensional space coincides with a certain finite-dimensional representation of some Lie algebra, realized by differential operators of the first order.*

The interesting question is how to describe finite modules of smooth functions in R^k which can serve as invariant sub-spaces of linear operators. Evidently, if we could find some realization of a Lie algebra g of differential operators, possessing an irreducible finite module of smooth functions, we can immediately consider the direct sum of several species of g acting on the functions given at the corresponding direct products of spaces, where the original algebra g acts in each of them. As an example, this procedure is presented in Section 4.2: the direct sum of two $sl_2(\mathbf{R})$ generates the linear differential operators acting on the functions at $\mathbf{R}^2$ having a rectangular $\mathcal{P}^{(II)}_{n,m}$ as the invariant sub-space. Taking the direct sum of k species of $sl_2(\mathbf{R})$ acting at $\mathbf{R}^k$, we arrive to the operators possessing a k-dimensional parallelopiped as the invariant sub-space. In this case the algebra $sl_2(\mathbf{R})$ plays the role of a "primary" algebra giving rise to a multidimensional convex geometrical figure (Newton diagram) as the invariant sub-space of some linear differential operators. In the space $\mathbf{R}^1$ there is only one such "irreducible" convex Newton diagram - a finite interval, in other words, a space of all polynomials in x of degree not higher than a certain positive integer. Thus, in $\mathbf{R}^2$ the rectangular Newton diagram is "reducible" stemming from the direct product of two intervals as one-dimensional Newton diagrams. In turn, the irreducible convex Newton diagrams in $\mathbf{R}^2$ are likely exhausted by different triangles, connected to the algebras $sl_3(\mathbf{R})$ and $gl_2(\mathbf{R}) \ltimes \mathbf{R}^{r+1}, r = 2, 3, 4, \ldots$ (see Figures 4.1, 4.3). Those algebras play a role of primary algebras. At least, we could not find any other convex Newton diagrams in $\mathbf{R}^2$ (see conjecture 1).

Before description of our present knowledge about convex Newton diagrams for the $\mathbf{R}^k$ case, let us recall the regular representation of the Lie algebra $sl_N(\mathbf{R})$ given on the flag manifold which acts on the smooth functions of $N(N-1)/2$ real variables $z_{i,i+q}$, $i = 1, 2, \ldots, (N-1)$, $q = 1, 2, \ldots, (N-i)$. The explicit formulas for generators in the Chevalley basis are given by (see [21])

$$D(e_i) = -\frac{\partial}{\partial z_{i,i+1}} - \sum_{q=i+2}^{N} z_{i+1,q}\frac{\partial}{\partial z_{i,q}}$$

$$D(f_i) = \sum_{q=1}^{i} z_{q,i+1}z_{i,i+1}\frac{\partial}{\partial z_{q,i+1}} + \sum_{q=1}^{i-1}(z_{q,i+1} - z_{q,i}z_{i,i+1})\frac{\partial}{\partial z_{q,i}}$$

$$- \sum_{q=i+2}^{N} z_{i,q}\frac{\partial}{\partial z_{i+1,q}} - n_i z_{i,i+1}$$

$$D(h_i) = - \sum_{q=i+1}^{N} z_{i,q}\frac{\partial}{\partial z_{i,q}} + \sum_{q=1}^{i-1} z_{q,i}\frac{\partial}{\partial z_{q,i}}+$$

$$\sum_{q=i+2}^{N} z_{i+1,q}\frac{\partial}{\partial z_{i+1,q}} - \sum_{q=1}^{i} z_{q,i+1}\frac{\partial}{\partial z_{q,i+1}} + n_i \tag{126}$$

where we use a notation $e_i, f_i, h_i,\ i = 1, 2, \dots, (N-1)$ for generators of positive and negative roots, and the Cartan generators, respectively. The algebra $sl_N(\mathbf{R})$ is generated by generators e_i, f_i, h_i. If n_i are non-negative integers, the finite-dimensional irreducible representation of $sl_N(\mathbf{R})$ will occur in the form of inhomogeneous polynomials in variables $z_{i,i+q}$. The highest weight vector is characterized by the integer numbers $n_i,\ i = 1, 2, \dots, (N-1)$. If some integers n_i become zero, the representation (126) is degenerated and acts on a space of dimension lower than $N(N-1)/2$ (see, e.g., (88) as an example of a degenerate representation of $sl_3(\mathbf{R})$).

In connection with the space $\mathbf{R}^3$, at least two irreducible convex Newton diagrams appear: a tetrahedron, related to a degenerate finite-dimensional representation of $sl_4(\mathbf{R})$ (the highest weight vector is characterized by two zeroing and one non-zeroing components), and certain geometrical bodies, corresponding to the regular representation of $sl_3(\mathbf{R})$ on the flag manifold (see (126) at $N = 3$) and a representation of $sp_2(\mathbf{R})$, respectively. We do not know whether those exhaust all irreducible Newton diagrams or not.

For the general case $\mathbf{R}^k$, firstly, we wish to indicate, that there also exist Newton polyhedra related to the almost degenerate representation of $sl_{k+1}(\mathbf{R})$ (in formulas (126) all integers except the one characterizing the highest weight vector become zero). Also there exist some convex geometrical figures related to some finite-dimensional degenerate and, sometimes, regular representations of $sl_{k+1-i}(\mathbf{R})$, $i > 0$.[15] It is easy to show that the algebra $so_{k+1}(\mathbf{R})$ acts on $\mathbf{R}^k$ in a degenerate representation

$$H_{ij} = x_i\frac{\partial}{\partial x_j} - x_j\frac{\partial}{\partial x_i}\ ,\ i > j\ ,\ i, j = 1, 2, \dots, k$$

$$G_i = \sum_{j=1, i\neq j}^{k} x_i x_j \frac{\partial}{\partial x_j} + (1 + x_i^2)\frac{\partial}{\partial x_i} - n x_i\ ,\ i = 1, 2, \dots, k \tag{127}$$

where $x_i, i = 1, 2, \dots, k$ are real variables and n is a number. If n is a non-negative integer, the algebra (127) possesses the finite-dimensional invariant sub-space, [16] which has the form of a polyhedron in $\mathbf{R}^k$ and coincides with a finite-dimensional invariant sub-space of the algebra $sl_{k+1}(\mathbf{R})$ in the almost degenerate representation, acting on $\mathbf{R}^k$. Making a gauge transformation, any symmetric bilinear combination in the generators (127) of the algebra $so_{k+1}(R)$ with real coefficients can be reduced to the form of a Laplace-Beltrami operator with some metric tensor plus a scalar function giving rise to exactly-solvable operators of the second type (for the proof see [17] and the discussion in the end of Section 4.1).

The above procedure can be extended to the case of the Lie super-algebras and also the quantum algebras. For the former case, our experience is very limited. The only statement one can make is that for linear operators acting on functions of one

[15]In order to decide what algebras act on $\mathbf{R}^k$, it is necessary to perform some analysis of the Young tableau. I am grateful to W. Rühl for an enlighting discussion of this question.

[16]The corresponding finite-dimensional representation is unitary and reducible (see, e.g., the discussion in the end of Section 4.1 and also in [17]).

real and one Grassmann variable (or, equivalently, on two-component spinors in on real variable) there can appear finite-dimensional invariant sub-spaces other than finite-dimensional representations of the algebra $osp(2,2)$ (see Section 3.1 and the discussion in [15]). Nevertheless, those sub-spaces are connected with the representations of polynomial elements of the universal enveloping algebra of the algebra $osp(2,2)$ and, correspondingly, they can be constructed through finite-dimensional representations of the original super-algebra $osp(2,2)$. Probably, there are no more irreducible convex Newton diagrams in this space. For the latter (quantum algebras), there exists an irreducible Newton diagram in $\mathbf{R}$ – an interval (see Section 2) – but we could not find any irreducible (non-trivial) Newton diagrams in $\mathbf{R}^k$ for $k > 1$; however, at least, in $\mathbf{R}^2$ they occur, if a quantum plane is introduced, $xy = qyx$, instead of an ordinary one, $xy = yx$, or a quantum hyperplane [30] for the case of more than two variables [17].

It is worth noting that one can construct quasi-exactly-solvable and exactly-solvable operators considering "mixed" algebras: a direct sum of the Lie algebras, Lie superalgebras and the quantum algebras, realized in the first-order differential and/or difference operators, possessing finite-dimensional invariant subspaces. In this case the Newton diagrams are reducible.

As for the cases considered in Sections 1-4, the problem of reducibility of general quasi-exactly-solvable and exactly-solvable operators to the Schroedinger-type operators is open.

Above we have described the linear operators which possess a finite-dimensional invariant sub-space with a polynomial basis. Certainly, making changes of variables and gauge transformations one can obtain the operators with an invariant sub-space with a non-polynomial basis (but emerging from a polynomial one, see e.g. (11)). Those 'induced' linear spaces we consider as the equivalent to the original, polynomial ones. A general problem can be posed: is it possible to describe all linear differential operators possessing a finite-dimensional invariant sub-space with a basis in a certain explicit form[18], which *cannot* be reduced to a basis of polynomials using a change of variables and a gauge transformation? Equivalently, this problem can be re-formulated as a problem of a description of finite-dimensional linear spaces of functions, where some linear differential operators can act. Since the original aim of the present investigation was mainly the construction of the quasi-exactly-solvable Schroedinger operators [7, 14, 17, 9, 16, 18], we did not touch on the general problem (except for a short remark on page 4). A solution of this problem plays a crucial role for a complete classification of the Schroedinger operators possessing a finite-dimensional invariant sub-space with an explicit basis in functions. Although it is rather surprising, it is very likely that this general and abstract problem has a constructive solution, at least for the one-dimensional Schroedinger operators. This work is in progress. [19]

---

[17] I am grateful to O.V. Ogievetsky for an interesting discussion of this subject and valuable comment.

[18] It is quite difficult to define precisely the meaning of the words "space with a basis in a certain explicit form." One of the examples is the linear space $< 1, x, f_1(x), f_2(x), \dots, f_n(x) >$, where $f_i, i = 1, 2, \dots n$ are any functions on R^1. Other examples of such spaces are the spaces (1), (11), (29), (77), (89), (113).

[19] Recently, a classification of linear differential operators possessing a finite-dimensional invariant sub-space with a basis in monomials was completed [23].

Concluding remarks.

(I). A crucial approach to the problem of classification of linear operators possessing a finite-dimensional invariant sub-space(s) with a basis in functions in an explicit form[20] is connected to a theory of Riemann surfaces.

Take a symmetric operator depending on a real parameter polynomially and possessing the infinite discrete set of eigenvalues. Searching of eigenvalues is equivalent to solving the transcendental secular equation. It is rather natural that the roots of this equation (eigenvalues) form an infinite-sheet Riemann surface as a parameter manifold. Generically, this surface is characterized by square-root branch points (see the excellent paper by Bender and Wu [25] and also [26, 27, 28]). Existance of N-dimensional invariant sub-space with an explicit basis implies, that the secular equation is reduced to two equations: an algebraic one and a transcendental one. It reflects the fact, that the original determinant factors into the product of two determinants. Equivalently, the infinite-sheet Riemann surface is split off into two disjoint Riemann surfaces: N-sheet one and an infinite-sheet one. If there exist k finite-dimensional invariant sub-spaces (for instance, which are embedded consequently one into the other, or, another case, any two have no intersection), the original, infinite-sheet Riemann surface is split off into k finite-sheet Riemann surfaces and an infinite-sheet one. Namely, such a situation occurs for a general quasi-exactly-solvable operator. As for exactly-solvable operators, the corresponding infinite-sheet Riemann surface is split off completely for separate sheets[21]. Following this idea, in [7, 29] a complete classification was done of second-order differential operators in $\mathbf{R}$ related with $sl_2(\mathbf{R})$-hidden algebra. This insight allowed to separate the quasi-exactly-solvable and exactly-solvable operators from generic operators, possessing abstract finite-dimansional invariant sub-spaces. I hope that such an approach can lead to an abstract classification of linear operators.

(II). The above-discussed questions about the existence of convex Newton diagrams (equivalently, the finite-dimensional linear spaces with polynomial basis) as an invariant sub-space of some linear differential operator are related to a classification of finite-dimensional Lie algebras of first-order differential operators possessing an invariant sub-space. Originally, this problem was formulated by Sophus Lie, who solved this problem for the algebras on $\mathbf{C}^1$ and gave a complete classification of the algebras of vector fields on $\mathbf{C}^2$ (see [22] and references therein). A complete classification of Lie algebras of first-order differential operators on $\mathbf{C}^2$ has been presented just recently [19], while the same problem for the algebras on $\mathbf{R}^2$ has not been solved so far, although the vector fields on $\mathbf{R}^2$ have been described [22]. The same problem for algebras on $\mathbf{C}^3$ has also been discussed by S. Lie, who presented 35 algebras of vector fields, emphasizing that this list is not complete. As far as we know at present, similar questions about a classification of the Lie super-algebras, both vector fields and first-order differential operators in real and Grassmann variables, are never discussed in literature. The same situation takes place for the algebras of first-order finite-difference operators on the real line, or a quantum hyperplane [30] .

---

[20]see footnote 18

[21]Of course, if the eigenfunctions related with the infinite flag of invariant sub-spaces form a complete set of the eigenfunctions of an original operator. In contrary, an additional Riemann surface is left as well.

ACKNOWLEDGEMENT

In closing, I am very indebted to V. Arnold, J. Froehlich, L. Michel, G. Post, W. Rühl and M. Shubin for their interest in the subject and numerous valuable discussions. Also I am very grateful to the Institute of Theoretical Physics, ETH-Honggerberg, Zurich, where this work was mainly done, and the Research Institute for Theoretical Physics, University of Helsinki and the Institut des Hautes Études Scientifiques, Bures-sur-Yvette, where this work was completed, for their kind hospitality extended to me. I am deeply appreciate to N. Kamran and P. Olver for their kind invitation to make a contribution to present volume.

REFERENCES

1. S. Bochner, "Über Sturm-Liouvillesche Polynomsysteme", *Math.Z.* **29** (1929) 730-736.
2. L.L. Littlejohn, "Orthogonal polynomial solutions to ordinary and partial differential equations", Proceedings of an International Symposium on Orthogonal Polynomials and their Applications, Segovia, Spain, Sept.22-27, 1986, Lecture Notes in Mathematics No.1329, M.Alfaro et al. (Eds.), Springer-Verlag (1988), pp. 98-124.
3. A.V. Turbiner, "On polynomial solutions of differential equations", Preprint CPT-91/P.2628 (1991); *J.Math.Phys.*, **33** (1992) 3989-3994.
4. H. Weyl, "The theory of groups and Quantum mechanics", Dover Publications, New-York, 1931.
5. S. Lang, "Algebra", Addison–Wesley Series in Mathematics, Addison–Wesley Publishing Company, Reading, Mass., 1965.
6. H.L. Krall, "Certain differential equations for Chebyshev polynomials", *Duke Math.J.* **4** (1938) 705-718.
7. A.V. Turbiner, "Spectral Riemannian surfaces of the Sturm-Liouville operators and Quasi-exactly-solvable problems", *Sov.Math.–Funk.Analysis i ego Prilogenia,* **22** (1988) 92-94;
   "Quantum Mechanics: the Problems Lying in between Exactly-Solvable and Non-Solvable", *Sov.Physics–ZhETF,* **94** (1988) 33-44;
   "Quasi-exactly-solvable problems and $sl(2,\mathbf{R})$ algebra", *Comm. Math. Phys.* **118** (1988) 467-474 (Preprint ITEP-197 (1987)).
8. D.P. Jatkar, C. Nagaraja Kumar, A. Khare, "A quasi-exactly-solvable problem without sl(2) symmetry", *Phys.Lett.***A142** (1989) 200-202.
9. A. González-Lopéz, N. Kamran and P.J. Olver, "Normalizability properties of one-dimensional quasi-exactly-solvable Schroedinger operators", Preprint of the School of Math., Univ. Minnesota, 1992
   (Invited talk at the Session of AMS No.873, March 20-21, 1992, Springfield, USA)
   "Normalizability of One-dimensional Quasi-exactly-solvable Schroedinger Operators", *Comm.Math.Phys.*, **153** (1993) 117-146
10. O. Ogievetsky and A. Turbiner, "$sl(2,\mathbf{R})_q$ and quasi-exactly-solvable problems", Preprint CERN-TH: 6212/91 (1991) (unpublished).
11. H. Exton, "q-Hypergeometrical functions and applications", Horwood Publishers, Chichester, 1983.
    G. Gasper, M. Rahman, "Basic Hypergeometric Series", Cambridge University Press, Cambridge, 1990.
12. C. Zachos, "Elementary paradigms of quantum algebras", Proceedings of the Conference on Deformation Theory of Algebras and Quantization with Applications to Physics, Contemporary Mathematics, J. Stasheff and M.Gerstenhaber (eds.), AMS, 1991.
13. A.V. Turbiner, "Lie-algebras and polynomials in one variable", *Journ.Phys.* **A25**, L1087-L1093 (1992); "Lie-algebraic approach to the theory of polynomial solutions. I. Ordinary differential equations and finite-difference equations in one variable", Preprint CPT-91/P.2679, 1991
    (Invited talk at the Session of AMS No.873, March 20-21, 1992, Springfield, USA).

14. M.A. Shifman and A.V. Turbiner, "Quantal problems with partial algebraization of the spectrum", *Comm. Math. Phys.* **126** 347-365 (1989).
15. A.V. Turbiner, "Lie-algebraic approach to the theory of polynomial solutions. II. Differential equations in one real and one Grassmann variables and 2×2 matrix differential equations", Preprint CPT-92/P.2708 and ETH-TH/92-21, 1992.
16. A. González-Lopéz, N. Kamran and P.J. Olver, "New Quasi-Exactly-Solvable Hamiltonians in Two Dimensions ", Preprint of the School of Math., Univ. Minnesota, 1992 (submitted to Ann.Phys.).
17. A.Yu. Morozov, A.M. Perelomov, A.A. Rosly, M.A. Shifman and A.V. Turbiner, "Quasi-Exactly-Solvable Problems: One-dimensional Analogue of Rational Conformal Field Theories", *Int.Journ.Mod. Phys.* **A5** (1990) 803-843.
18. A.S. Losev and A.V. Turbiner, "Multi-Dimensional-Exactly-Solvable Problems in Quantum Mechanics and Pullbacks of Affine Coordinates on Grassmannian", *Int.Journ.Mod. Phys.* **A7** (1992) 1449-1466.
19. A. González-Lopéz, N. Kamran and P.J. Olver, "Quasi-Exactly-Solvable Lie Algebras of the first order differential operators in Two Complex Variables ", *J.Phys.***A24** (1991) 3995-4008; "Lie algebras of differential operators in two complex variables ", IMA preprint series # 617 (February 1990), *American J. Math.* **114** (1992) 1163-1185
20. A.V. Turbiner, "Lie-algebraic approach to the theory of polynomial solutions. III. Differential equations in two real variables and general outlook", Preprint ETH-TH/92-34, 1992.
21. P. Bouwknegt, J. McCarthy, K. Pilch, "Quantum group structure in the Fock space resolutions of $\hat{sl}(n)$ representation", *Comm. Math. Phys.* **131** (1990) 125-155.
W. Rühl, "$SL(N)$ Kac-Moody Algebras and Wess-Zumino-Witten Models", *Ann. Phys.* **206** (1991) 368-408.
22. A. González-Lopéz, N. Kamran and P.J. Olver, "Lie Algebras of Vector Fields in the Real Plane", *Proc. Lond. Math. Soc.* **64** (1992) 339–368.
23. G. Post and A.V. Turbiner, "Classification of linear differential operators with an invariant subspace in monomials", Memorandum No. 1143, University of Twente (June, 1993) (submitted to Proc. Edinburg Math. Soc.)
24. A.V. Turbiner and G. Post, "Identités d'opérateurs et représentations d'algèbras", Preprint IHES/P/93/18 (1993), submitted to Comptes Rendus Acad. Sci. Paris; "The operator identities, representations of algebras and the problem of ordering", Preprint CRN 93-53 (1993), *Journ. Phys.***A** (in print)
25. C.M. Bender and T.T. Wu, "Anharmonic oscillator", *Phys. Rev.* **184** 1231-1260 (1969)
26. B. Simon, "Analiticity and the anharmonic oscillator", *Ann. Phys. (N.Y.)* **58** 76-136 (1970)
27. P.E. Shanley, "Spectral singularities of quartic anharmonic oscillator", *Phys. Lett.* **A117** 161-165 (1986)
28. A.V. Turbiner and A.G. Ushveridze, "Spectral singularities and the quasi-exactly-solvable problem", *Phys. Lett.* **A126** 181-183 (1987)
29. A.V. Turbiner, "Lame equation, sl_2 and isospectral deformation", *Journ. Phys.* **A22** L1-L3 (1989)
30. J. Wess and B. Zumino, "Covariant differential calculus on the quantum hyperplane", *Nucl. Phys.* **B18** 302 (1990) (Proc. Suppl.);
B. Zumino, "Deformation of the quantum mechanical phase space with bosonic or fermionic coordinates" *Mod. Phys. Lett.* **A6** 1225 (1991).

APPENDIX

In connection to the finite-dimensional representations of the algebras of first-order differential (difference) operators some operator identities occur [24].

1. The following *operator identity* holds:

$$(A.1)\quad (J_n^+)^{n+1} \equiv (x^2\partial_x - nx)^{n+1} = x^{2n+2}\partial_x^{n+1}\ ,\ \partial_x \equiv \frac{d}{dx},\ n = 0, 1, 2, \ldots$$

Proof. The proof is straightforward:

(i) the operator $(J_n^+)^{n+1}$ annihilates the space of all polynomials of power not higher than n, $\mathcal{P}_n(x) = Span\{x^i : 0 \le i \le n\}$;

(ii) in general, an $(n+1)$-th order linear differential operator annihilating $\mathcal{P}_n(x)$ must have the form $B(x)\partial_x^{n+1}$, where $B(x)$ is an arbitrary function and

(iii) since $(J_n^+)^{n+1}$ is a graded operator, $\deg(J_n^+) = +1$, [22] $\deg((J_n^+)^{n+1}) = n+1$, hence $B(x) = bx^{2n+2}$, while clearly the constant $b = 1$.

□

It is worth noting that taking the power in (A.1) different from $(n+1)$, the l.h.s. in (A.1), generally, will contain all derivative terms from zero up to $(n+1)$-th order with non-vanishing coefficients.

The identity (A.1) has a Lie-algebraic interpretation. The operator (J_n^+) is the positive-root generator of the algebra sl_2 of first-order differential operators (the other sl_2-generators are $J_n^0 = x\partial_x - n/2$, $J_n^- = \partial_x$, see (2)). Correspondingly, the space $\mathcal{P}_n(x)$ is nothing but the $(n+1)$-dimensional irreducible representation of sl_2. The identity (A.1) is a consequence of the fact that $(J_n^+)^{n+1} = 0$ in matrix representation [23].

Another Lie-algebraic interpretation of (A.1) is connected with occurence of some relations between elements of the universal enveloping algebra of three-dimensional Heisenberg algebra, H_1: $\{P, Q, 1\}$. Let $[P, Q] = 1$, then

$$(A.2)\qquad (Q^2P - nQ)^{n+1} = Q^{2n+2}P^{n+1},\ n = 0, 1, 2, \ldots$$

and certainly this identity is more general than (A.1). Taking different representations for P and Q in terms of differential operators (others than the standard one, $P = \partial_x$ and $Q = x$), one can get various families of operator identities, other than (A.1). Also the formula (A.2) has a meaning of a formula of an ordering: Q-operators are placed on the left, P-operators on the right.

One can easily check, that once two operators obey $[P, Q] = 1$, then

$$(A.3)\qquad \begin{aligned} J^+ &= Q^2P - nQ\ , \\ J^0 &= QP - n/2\ , \\ J^- &= P\ , \end{aligned}$$

---

[22] so J_n^+ maps x^k to a multiple of x^{k+1}

[23] It is a particular case of more general statement: *in the algebra gl_k a positive (negative) -root generator $J_{\pm\alpha}$ in a finite-dimensional representation of the dimension d, taken in power of of the dimension d is equal to zero, $(J_{\pm\alpha})^d = 0$.*

obey sl_2-algebra commutation relations. The representation (2) is a particular case of this representation. In fact, (A.3) is one of possible embeddings of the algebra sl_2 into the universal enveloping algebra of three-dimensional Heisenberg algebra.

There exist other algebras of differential or finite-difference operators (in more than one variable), which admit a finite-dimensional representation. This leads to more general and remarkable operator identities. In fact, (A.1) is one representative of an infinite family of identities for differential and finite-difference operators. Also there occur generalizations of the relation (A.2) for a certain polynomial elements of the universal enveloping algebra of $(2n+1)$-dimensional Heisenberg algebra.

2. The Lie-algebraic interpretation presented above allows us to generalize (A.1) for the case of differential operators of several variables, taking appropriate powers of the highest-positive-root generators of (super) Lie algebras of first-order differential operators, possessing a finite-dimensional invariant sub-space. First we consider the case of sl_3. There exists a representation of $sl_3(\mathbf{R})$ as differential operators on $\mathbf{R}^2$ (see Section 4.1, eq.(88)). One of the generators is

$$J_2^1(n) = x^2\partial_x \ + \ xy\partial_y - nx \ .$$

The space (89), $\mathcal{P}_n(x,y) = Span\{x^iy^j : 0 \le i+j \le n\}$ is a finite-dimensional representation for sl_3, and due to the fact that $(J_2^1(n))^{n+1} = 0$ on the space $\mathcal{P}_n(x,y)$, we have

$$(J_2^1(n))^{n+1} \equiv (x^2\partial_x \ + \ xy\partial_y - nx)^{n+1} = \tag{A.4}$$
$$\sum_{k=0}^{n+1} \binom{n+1}{k} x^{2n+2-k} y^k \partial_x^{n+1-k} \partial_y^k \ .$$

This identity is valid for $y \in \mathbf{R}$ (as described above), but also if y is a Grassmann variable, i.e. , $y^2 = 0$ [24]. In the last case, $J_2^1(n)$ is a generator T^+ of the algebra $osp(2,2)$ (see (31)).

More generally (using algebra sl_{k+1} instead of sl_3), the following operator identity holds:

$$(J_k^{k-1}(n))^{n+1} \equiv (x_1(\sum_{m=1}^{k} x_m\partial_{x_m} \ - n))^{n+1} = \tag{A.5}$$
$$x_1^{n+1} \sum_{j_1+j_2+\ldots+j_k=n+1} C^{n+1}_{j_1,j_2,\cdots,j_k} x_1^{j_1} x_2^{j_2} \ldots x_k^{j_k} \partial_{x_1}^{j_1} \partial_{x_2}^{j_2} \ldots \partial_{x_k}^{j_k} \ ,$$

where $C^{n+1}_{j_1,j_2,\cdots,j_k}$ are the generalized binomial (multinomial) coefficients. If $x \in \mathbf{R}^k$, then $J_k^{k-1}(n)$ is a generator of the algebra $sl_{k+1}(\mathbf{R})$, generated by $D(f_i)$ (see (126)). While some of the variables x's are Grassmann ones, the operator $J_k^{k-1}(n)$ is a generator of a certain super Lie algebra of first-order differential operators. The operator in the l.h.s. of (A.5) annihilates the linear space of polynomials $\mathcal{P}_n(x_1,x_2,\ldots x_k) = Span\{x_1^{j_1}x_2^{j_2}\ldots x_k^{j_k} : 0 \le j_1+j_2+\ldots+j_k \le n\}$, which is represented by tetrahedron Newton diagram in $\mathbf{R}^k$.

Considering $(2k+1)$-dimensional Heisenberg algebra:

$$[P_i, Q_j] = \delta_{ij}, \ [P_i, P_j] = 0, \ [Q_i, Q_j] = 0, \ i,j = 1,2,\ldots k$$

[24]In this case just two terms in the l.h.s. of (A.4) survive.

one can rewrite (A.5) in more general form

$$(Q_1(\sum_{m=1}^{k} Q_m P_m \ - \ n))^{n+1} =$$

$$\text{(A.6)} \quad Q_1^{n+1} \sum_{j_1+j_2+\ldots+j_k=n+1} C^{n+1}_{j_1,j_2,\cdots,j_k} Q_1^{j_1} Q_2^{j_2} \ldots Q_k^{j_k} P_1^{j_1} P_2^{j_2} \ldots P_k^{j_k} \ ,$$

(cf.(A.2)). It is clear, the formulas for sl_{k+1}-generators at $k > 1$ (see (126)) in terms of P_i, Q_i, analogous to (A.3), can be easily derived.

3. For the case of the algebra $gl_2(\mathbf{R}) \ltimes \mathbf{R}^{r+1}, r = 1, 2, \ldots$ (see (112)) the situation becomes rather sophisticated

$$(J_n^4)^{n+1} \equiv (x^2\partial_x \ + \ rxy\partial_y \ - \ nx)^{n+1} =$$

$$\text{(A.7)} \qquad \sum_{k=0}^{n+1} r^k \binom{n+1}{k} x^{2n+2-k} y^{n+1-k} \partial_x^k \partial_y^{n+1-k} + Q_{r,n} \ ,$$

where $Q_{1,n} = 0$ and also

$$Q_{r,n} \ = \ \begin{cases} 0 & \text{at n=0} \\ r(r-1)x^2y\partial_y & \text{at n=1} \\ r(r-1)x^3y[3ry\partial_y^2 + 3x\partial_{xy}^2 + (r-2)\partial_y] & \text{at n=2} \end{cases}$$

(cf. (A.4)). For arbitrary n and r

$$Q_{r,n} = \frac{r(r-1)n(n+1)}{2}$$

$$[x^{2n}y\partial_x^{n-1}\partial_y + r(n-1)x^{2n-1}y^2\partial_x^{n-2}\partial_y^2 + \frac{r^2}{4}(n-1)(n-2)x^{2n-2}y^3\partial_x^{n-3}\partial_y^3 + \ldots]$$

$$+ \ [\textit{lower order derivatives}].$$

The r.h.s. of (A.7) still has quite regular structure, which can be analyzed using the following trick: variables y's in the l.h.s. of (A.4), (A.7) can be related to each other through the very simple connection

$$y_{(A.7)} \ = \ y^r_{(A.4)} \ ,$$

and then J_2^1 (see (A.4)) coincides with J_n^4.

4. The above-described family of operator identities can be generalized for the case of finite-difference operators with the Jackson symbol, D_x (see, e.g., [11]), obeyed the following Leibnitz rule

$$D_x f(x) = \frac{f(x) - f(q^2x)}{(1-q^2)x} + f(q^2x)D_x \ ,$$

instead of the ordinary, continuous derivative.[25] Here, q is an arbitrary complex number. The following operator identity holds:

$$\text{(A.8)} \quad (\tilde{J}_n^+)^{n+1} \equiv (x^2 D_x - \{n\}x)^{n+1} = q^{2n(n+1)} x^{2n+2} D_x^{n+1}, n = 0, 1, 2, \ldots$$

---

[25]The definition of the Jackson symbol used here is slightly different than that presented in Section 2.

(cf. (A.1)), where $\{n\} = \frac{1-q^{2n}}{1-q^2}$ is a so-called q-number. The operator in the l.h.s. annihilates the space (1), $\mathcal{P}_n(x)$. The proof is similar to the proof of identity (A.1).

From the algebraic point of view the operator $\tilde{J}_n^+$ is the generator of a q-deformed algebra $sl_2(\mathbf{R})_q$ of first-order finite-difference operators on the line: $\tilde{J}_n^0 = xD - \hat{n}$, $\tilde{J}_n^- = D$, where $\hat{n} \equiv \frac{\{n\}\{n+1\}}{\{2n+2\}}$ (see [10] and also [3]), obeying the commutation relations (22).

One can show, that once two operators obey $\tilde{P}\tilde{Q} - q^2\tilde{Q}\tilde{P} = 1$, then

$$(\tilde{Q}^2\tilde{P} - \{n\}\tilde{Q})^{n+1} = q^{2n(n+1)}\tilde{Q}^{2n+2}\tilde{P}^{n+1}, \quad n = 0, 1, 2, \ldots \tag{A.9}$$

holds and

$$\begin{aligned} \tilde{J}^+ &= \tilde{Q}^2\tilde{P} - \{n\}\tilde{Q} , \\ \tilde{J}^0 &= \tilde{Q}\tilde{P} - \hat{n} , \\ \tilde{J}^- &= \tilde{P} , \end{aligned} \tag{A.10}$$

obey $sl_2(\mathbf{R})_q$-algebra commutation relations (22). It is evident, that this representation is more general than the representation (21).

An attempt to generalize (A.4) replacing continuous derivatives by Jackson symbols immediately leads to requirement to introduce the quantum plane and q-differential calculus [30]

$$\begin{gathered} xy = qyx , \\ D_x x = 1 + q^2 x D_x + (q^2 - 1) y D_y \quad , \quad D_x y = q y D_x , \\ D_y x = q x D_y \quad , \quad D_y y = 1 + q^2 y D_y , \\ D_x D_y = q^{-1} D_y D_x . \end{gathered} \tag{A.11}$$

The formulae analogous to (A.4) have the form

$$\begin{gathered} (\tilde{J}_2^1(n))^{n+1} \equiv (x^2 D_x + xyD_y - \{n\}x)^{n+1} = \\ \sum_{k=0}^{n+1} q^{2n^2 - n(k-2) + k(k-1)} \binom{n+1}{k}_q x^{2n+2-k} y^k D_x^{n+1-k} D_y^k , \end{gathered} \tag{A.12}$$

where

$$\binom{n}{k}_q \equiv \frac{\{n\}!}{\{k\}!\{n-k\}!} , \quad \{n\}! = \{1\}\{2\}\ldots\{n\}$$

are q-binomial coefficient and q-factorial, respectively. As in all previous cases, if $y \in \mathbf{R}$, the operator $\tilde{J}_2^1(n)$ is one of generators of the q-deformed algebra $sl_3(\mathbf{R})_q$ of finite-difference operators, acting on the quantum plane and having the linear space $\mathcal{P}_n(x,y) = Span\{x^i y^j : 0 \le i + j \le n\}$ as a finite-dimensional representation; the l.h.s. of (A.12) annihilates $\mathcal{P}_n(x,y)$. If y is a Grassmann variable, $\tilde{J}_2^1(n)$ is a generator of the q-deformed superalgebra $osp(2,2)_q$ possessing finite-dimensional representation.

Introducing two couples of the operators $\tilde{P}_{1,2}, \tilde{Q}_{1,2}$ obeying the following relations:

$$\begin{gathered} Q_1 Q_2 = q Q_2 Q_1 , \\ P_1 Q_1 = 1 + q^2 Q_1 P_1 + (q^2 - 1) Q_2 P_2 \quad , \quad P_1 Q_2 = q Q_2 P_1 , \end{gathered}$$

$$P_2Q_1 = qQ_1P_2 \quad , \quad P_2Q_2 = 1 + q^2Q_2P_2 \ ,$$

$$P_1P_2 = q^{-1}P_2P_1 \ . \tag{A.13}$$

It is easy to derive a family of abstract identities similar to (A.6) at $k = 2$

$$(Q_1^2P_1 \ + \ Q_1Q_2P_2 - \{n\}Q_1)^{n+1} =$$

$$\sum_{k=0}^{n+1} q^{2n^2-n(k-2)+k(k-1)} \binom{n+1}{k}_q Q_1^{2n+2-k} Q_2^k P_1^{n+1-k} P_2^k \ , \tag{A.14}$$

while (A.12) is a particular case corresponding a special choice of a representation (A.13) in the form of (A.11).

Introducing a quantum hyperplane [30], one can generalize the whole family of operator identities (A.5) replacing first of all continuous derivatives by finite-difference operators and then introducing the abstract operators like it was done above for the case of (A.11)-(A.12).

The Lie-algebraic interpretation of above operator identities allows to make a general conclusion that *an existence of finite-dimensional representations of the Lie algebra* gl_k *of differential operators leads to an appearence of some specific operator identities analogous to those families described above.*

INSTITUTE FOR THEORETICAL AND EXPERIMENTAL PHYSICS, MOSCOW 117259, RUSSIA
Current address: I.H.E.S., Bures-sur-Yvette, F-91440, France
E-mail address: TURBINER@VXCERN.CERN.CH or TURBINER@CERNVM

Recent Titles in This Series

(*Continued from the front of this publication*)

129 **John R. Graef and Jack K. Hale, Editors,** Oscillation and dynamics in delay equations, 1992

128 **Ridgley Lange and Shengwang Wang,** New approaches in spectral decomposition, 1992

127 **Vladimir Oliker and Andrejs Treibergs, Editors,** Geometry and nonlinear partial differential equations, 1992

126 **R. Keith Dennis, Claudio Pedrini, and Michael R. Stein, Editors,** Algebraic K-theory, commutative algebra, and algebraic geometry, 1992

125 **F. Thomas Bruss, Thomas S. Ferguson, and Stephen M. Samuels, Editors,** Strategies for sequential search and selection in real time, 1992

124 **Darrell Haile and James Osterburg, Editors,** Azumaya algebras, actions, and modules, 1992

123 **Steven L. Kleiman and Anders Thorup, Editors,** Enumerative algebraic geometry, 1991

122 **D. H. Sattinger, C. A. Tracy, and S. Venakides, Editors,** Inverse scattering and applications, 1991

121 **Alex J. Feingold, Igor B. Frenkel, and John F. X. Ries,** Spinor construction of vertex operator algebras, triality, and $E_8^{(1)}$, 1991

120 **Robert S. Doran, Editor,** Selfadjoint and nonselfadjoint operator algebras and operator theory, 1991

119 **Robert A. Melter, Azriel Rosenfeld, and Prabir Bhattacharya, Editors,** Vision geometry, 1991

118 **Yan Shi-Jian, Wang Jiagang, and Yang Chung-chun, Editors,** Probability theory and its applications in China, 1991

117 **Morton Brown, Editor,** Continuum theory and dynamical systems, 1991

116 **Brian Harbourne and Robert Speiser, Editors,** Algebraic geometry: Sundance 1988, 1991

115 **Nancy Flournoy and Robert K. Tsutakawa, Editors,** Statistical multiple integration, 1991

114 **Jeffrey C. Lagarias and Michael J. Todd, Editors,** Mathematical developments arising from linear programming, 1990

113 **Eric Grinberg and Eric Todd Quinto, Editors,** Integral geometry and tomography, 1990

112 **Philip J. Brown and Wayne A. Fuller, Editors,** Statistical analysis of measurement error models and applications, 1990

111 **Earl S. Kramer and Spyros S. Magliveras, Editors,** Finite geometries and combinatorial designs, 1990

110 **Georgia Benkart and J. Marshall Osborn, Editors,** Lie algebras and related topics, 1990

109 **Benjamin Fine, Anthony Gaglione, and Francis C. Y. Tang, Editors,** Combinatorial group theory, 1990

108 **Melvyn S. Berger, Editor,** Mathematics of nonlinear science, 1990

107 **Mario Milman and Tomas Schonbek, Editors,** Harmonic analysis and partial differential equations, 1990

106 **Wilfried Sieg, Editor,** Logic and computation, 1990

105 **Jerome Kaminker, Editor,** Geometric and topological invariants of elliptic operators, 1990

104 **Michael Makkai and Robert Paré,** Accessible categories: The foundations of categorical model theory, 1989

103 **Steve Fisk,** Coloring theories, 1989

102 **Stephen McAdam,** Primes associated to an ideal, 1989

101 **S.-Y. Cheng, H. Choi, and Robert E. Greene, Editors,** Recent developments in geometry, 1989

(See the AMS catalog for earlier titles)